Actions of Prolactin on Molecular Processes

Editor

James A. Rillema, Ph.D.
Professor
Department of Physiology
Wayne State University
Detroit, Michigan

CRC Press
Taylor & Francis Group
Boca Raton London New York

CRC Press is an imprint of the
Taylor & Francis Group, an **informa** business

First published 1987 by CRC Press
Taylor & Francis Group
6000 Broken Sound Parkway NW, Suite 300
Boca Raton, FL 33487-2742

Reissued 2018 by CRC Press

Library of Congress Cataloging-in-Publication Data

Actions of prolactin on molecular processes.

Includes index.
1. Prolactin—Physiological effect. I. Rillema, James A.
QP572.P74A26 1987 599'.019'27 86-17138
ISBN 0-8493-5376-9

A Library of Congress record exists under LC control number: 86017138

ISBN 13: 978-1-315-89035-7 (hbk)
ISBN 13: 978-1-351-06945-8 (ebk)

Visit the Taylor & Francis Web site at http://www.taylorandfrancis.com and the
CRC Press Web site at http://www.crcpress.com

PREFACE

Prolactin is an adenohypophyseal hormone which has multiple functions ranging from osmoregulation in the teleosts to the regulation of lactation in mammals. In addition, as research work on this hormone continues, further functions of prolactin on biological processes will undoubtedly be discovered. In the mammary gland as well as in other tissues, prolactin is known to regulate both differentiative and proliferative events. Prolactin also has been implicated in the induction and promotion of neoplastic processes in mammary tissues.

It is well-established that the initial interaction of prolactin with its target cells is with specific receptor molecules located on the external surface of the plasma membrane. Subsequent to the prolactin-receptor interaction, however, the molecular events that culminate in the regulation of biological processes are not totally understood. This book was assembled to review and analyze the currently available information relative to the molecular events involved in the actions of prolactin on cells. Hopefully, the assimilation of this information will provide the basis for ultimately determining the sequence of molecular reactions by which prolactin expresses its biological responses.

James A. Rillema, Ph.D.
Detroit, Michigan
January, 1986

THE EDITOR

James A. Rillema, Ph.D., is a Professor of Physiology at the Wayne State University School of Medicine, Detroit, Michigan.

Dr. Rillema was awarded the B.S. degree in Chemistry from Calvin College, Grand Rapids, Michigan, in 1964. His graduate studies (1964 to 1968) were in the area of cell physiology with Dr. W. L. Frantz at Michigan State University; the M.S. degree was awarded in 1966, and the Ph.D. in 1968. Research work in Dr. Frantz's laboratory was concerned with the prolactin regulation of protein metabolism in the pigeon crop sac. Dr. Rillema then spent 3 post-doctoral years (1968 to 1971) with Dr. J. L. Kostyo at Emory University; research in Dr. Kostyo's laboratory focused on the mechanism of action of growth hormone on skeletel muscle cells. Since 1971, Dr. Rillema has been a member of the Physiology Department faculty at Wayne State University; he was promoted to full professor in 1979. His research work has focused on the mechanism of action of prolactin on lactational processes in the mammary gland. From 1975 to 1980, Dr. Rillema was funded by a Career Development Award from the American Cancer Society. The 1980 to 1981 academic year was spent on a sabbatical leave with Dr. Peter Emmelot and Dr. M. Sluyser at The Netherlands Cancer Institute; on this sabbatical leave, Dr. Rillema was a Fogarty Senior International Fellow sponsored by the National Institutes of Health.

CONTRIBUTORS

Thomas R. Anderson, Ph.D.
Research Associate
Department of Physiology-Anatomy
and Cancer Research Laboratory
University of California, Berkeley
Berkeley, California

Mihir R. Banerjee, Ph.D.
Professor and Head
Tumor Biology Laboratory
School of Biological Sciences
University of Nebraska-Lincoln
Lincoln, Nebraska

Charles L. Brooks, Ph.D.
Assistant Professor
Department of Veterinary Pathobiology
The Ohio State University
Columbus, Ohio

Mark E. Costlow, Ph.D.
Associate Member
Department of Biochemistry
St. Jude Children's Research Hospital
Memphis, Tennessee

Jitendra R. Dave, Ph.D.
Visiting Associate
Laboratory of Clinical Studies
National Institute on Alcohol Abuse
and Alcoholism
Division of Intramural Clinical and
Biological Research
National Institutes of Health
Bethesda, Maryland

Nora J. Hebert, A.B.
Research Associate
Department of Physiology-Anatomy
and Cancer Research Laboratory
University of California, Berkeley
Berkeley, California

Nelson D. Horseman, Ph.D.
Associate Professor
Department of Biology
Marquette University
Milwaukee, Wisconsin

Robert F. Loizzi, Ph.D.
Professor
Department of Physiology and
Biophysics
College of Medicine
University of Illinois at Chicago
Chicago, Illinois

Ravi S. Menon, Ph.D.
Research Associate
MCD Biology
Porter Biosciences
University of Colorado
Boulder, Colorado

Charles S. Nicoll, Ph.D
Professor
Department of Physiology-Anatomy
University of California, Berkeley
Berkeley, California

Takami Oka, Ph.D.
Senior Investigator
Laboratory of Molecular and Cellular
Biology
National Institute of Diabetes,
Digestive and Kidney Diseases
National Institutes of Health
Bethesda, Maryland

James A. Rillema, Ph.D.
Professor
Department of Physiology
Wayne State University
Detroit, Michigan

Sharon M. Russell, Ph.D.
Associate Research Physiologist
Department of Physiology-Anatomy
University of California, Berkeley
Berkeley, California

Michiyoshi Taga, M.D.
Research Fellow
Laboratory of Molecular and Cellular Biology
National Institute of Diabetes, Digestive and Kidney Diseases
National Institutes of Health
Bethesda, Maryland

James L. Voogt, Ph.D.
Professor
Department of Physiology
University of Kansas Medical Center
Kansas City, Kansas

TABLE OF CONTENTS

Chapter 1
The Activation of Molecular Events by Prolactin.. 1
James A. Rillema

Chapter 2
Prolactin Interaction with its Receptors and the Relationship to the
Subsequent Regulation of Metabolic Processes.. 5
Mark E. Costlow

Chapter 3
Actions of Prolactin in the Brain 27
James L. Voogt

Chapter 4
Models of Prolactin Action in Nonmammalian Vertebrates................................41
Nelson D. Horseman

Chapter 5
Prolactin Regulation of Membrane Fluidity and Prostaglandin Formation 69
Jitendra R. Dave

Chapter 6
Role of Calcium Ions and the Phospholipids in the Prolactin Regulation
of its Target Cells... 107
James A. Rillema

Chapter 7
Synergistic Actions of Glucocorticoid and Prolactin in Murine
Milk-Protein Gene Expression... 121
Mihir R. Banerjee and Ravi S. Menon

Chapter 8
Role of Polyamines in Prolactin Actions... 137
Takami Oka and Michiyoshi Taga

Chapter 9
Prolactin and the Regulation of Secretion Including Membrane Flow:
Potential Roles for Tubulin and Microtubules... 153
Robert F. Loizzi

Chapter 10
Protein Phosphorylation of Prolactin Target Tissue: Mammary Gland............... 179
Charles L. Brooks

Chapter 11
Prolactin, Growth Factors, and Cell Growth 199
Charles S. Nicoll, Thomas R. Anderson, Nora J. Hebert, and
Sharon M. Russell

Index.. 213

Chapter 1

THE ACTIVATION OF MOLECULAR EVENTS BY PROLACTIN

James A. Rillema

TABLE OF CONTENTS

I. Introduction 2

II. Initial Interaction of Prolactin with Its Target Cells 2

III. Events Subsequent to the Prolactin-Receptor Interaction 3

References 4

I. INTRODUCTION

The pituitary hormone prolactin was discovered by Stricker and Grueter in 1928.[1] These investigators observed that extracts of bovine pituitary glands induced lactation when injected into pseudopregnant rabbits. Subsequently, prolactin has been shown to have a number of other biological actions in mammalian as well as nonmammalian species.[2] Some of these actions include osmoregulation in teleosts,[4] the regulation of metamorphosis in amphibians, leuteotropic and leuteolytic effects in the ovary, proliferative actions on male sex accessory glands, the regulation of parental behavior, and the stimulation of "crop milk" formation in pigeons and doves. The effect of prolactin on the Columbid crop was first reported by Riddle and Braucher in 1931.[3] This effect of prolactin served for many years as the method of choice for the bioassay of this hormone. It is thus clear that prolactin has a number of proliferative and differentiative actions on many cell types derived from a number of animal species. This book will focus on the molecular processes involved in the expression of the effects of prolactin on its target cells. Most experimental work concerned with the molecular processes by which prolactin carries out its effects has been accomplished employing the mammary gland as the target organ. Accordingly, most of this book will focus on how prolactin carries out its proliferative and differentiative actions on this organ. One chapter, however, deals exclusively with how prolactin functions in nonmammalian species.[4]

At the outset, it can be clearly stated that the sequence of molecular events by which prolactin expresses its actions on target cells is not totally understood. A unifying molecular mechanism by which prolactin acts on target cells has not evolved from the multitude of laboratory studies concerned with this topic. It is not unlikely, however, that the initial or primary molecular perturbations that occur immediately subsequent to the interaction of prolactin with its receptor are the same in all prolactin target cells. The current laboratory studies that focus on the identification of the primary molecular events that occur in response to prolactin will be highlighted in this book.

II. INITIAL INTERACTION OF PROLACTIN WITH ITS TARGET CELLS

The molecular structures of prolactins from many species are quite similar. Ovine prolactin, which was one of the first prolactins to be chemically characterized, consists of 198 amino acids arranged as a single polypeptide chain; it contains 3 internal disulfide bridges and has a molecular weight of about 24,000.[4] Human prolactin, which was identified and isolated more recently,[5-7] has a structure similar to that of ovine prolactin, with several amino acid substitutions in the peptide chain.[8] Other hormones that possess lactogenic activities include the placental lactogens and human growth hormone; each of these hormones is also a single-chain peptide with a molecular weight in the low 20-dalton range.

Being relatively large protein molecules, the lactogenic hormones are believed to initially interact with their target cells by binding to "receptors" located on the outer surface of the plasma membrane (Figure 1). The lactogenic hormones interact with both dissociable and nondissociable binding sites on prolactin target cells. The dissociable binding sites are discussed in detail by Costlow in Chapter 2 and in a recent review by Kelly et al.[9] The interaction of prolactin with the dissociable binding sites appears to initiate at least certain effects of prolactin in target cells. Evidence supporting this conclusion is that antibodies to the prolactin-dissociable "receptors" cause prolactin-like responses on several molecular processes[9,10] in prolactin target cells.

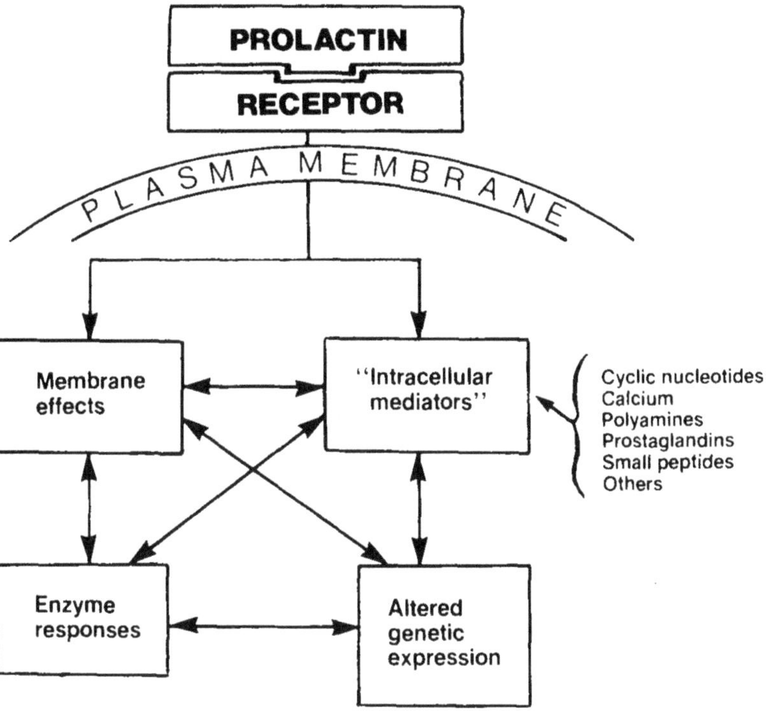

FIGURE 1. Mechanism of prolactin action on its target cells.

Whether the binding of prolactin to nondissociable binding sites on prolactin target cells is involved in the expression of prolactin effects remains to be established.

III. EVENTS SUBSEQUENT TO THE PROLACTIN-RECEPTOR INTERACTION

Subsequent to the interaction of prolactin with cell surface receptors, it is of primary interest to know what occurs on a molecular basis that culminates in cellular responses. It has been shown by several laboratories that prolactin is rapidly taken up into target cells, "internalized", and is bound to specific intracellular organelles.[11-15] The purpose for which prolactin is "internalized" is not clear at the present time. It has been postulated that the "internalization" process could be involved in the mechanism for the inactivation of prolactin. It is also possible that prolactin may have effects inside its target cells. To date, however, there is no convincing information in the literature which supports the concept of intracellular actions of prolactin. In fact, experiments by Serveley et al.[16] suggest that prolactin firmly bound to prolactin receptors on the cell surface is responsible for the stimulation of casein mRNA accumulation.

Assuming that prolactin does not elicit effects intracellularly, one is then left with the question as to how intracellular molecular changes are generated in response to the prolactin-receptor interaction at the cell surface. One would then have to postulate that either prolactin alters metabolic processes at the level of the plasma membrane and/or it generates "intracellular mediators" which then carry out the intracellular responses. "Intracellular mediators" for other hormones are known to include (1) the cyclic nucleotides, (2) calcium ions, (3) the polyamines, (4) the prostaglandins, or (5) low molecular weight peptides (see Figure 1). Each of these substances has been implicated regarding its involvement in prolactin-elicited responses. Most will be discussed in detail in various chapters of this book. A detailed account, however, concerning the cyclic nucleotides and low molecular weight peptides will not be presented. The cyclic nucleotide studies were reviewed earlier[17] and will only be discussed where appropriate. Studies concerning the existence of a putative low molecular weight peptide which mediates the effect of prolactin on gene expression in rabbit mammary cells were recently summarized by Kelly et al.[9] Unfortunately, the existence of this low molecular weight peptide is suspect, since several laboratories could not reproduce the original experiments.[18] A discussion of the low molecular weight peptide mediator will therefore not be included in this volume.

REFERENCES

1. Stricker, P. and Grueter, R., Action du lobe anterieur de l'hypophyse sur la mont'ee laiteuse, *C. R. Soc. Biol.*, 99, p. 1978, 1928.
2. Nicoll, C. S., Physiological actions of prolactin, in *Handbook of Physiology, Section* 7, Vol. 4 (Part 2), Greep, R. O., and Astwood, E. B., Eds., American Physiological Society, Washington, D. C., 1974, 253.
3. Riddle, O. and Braucher, P. F., Studies on the physiology of reproduction in birds; control of special secretion of the crop-glands of pigeons by anterior pituitary hormone, *Am. J. Physiol.*, 97, 617, 1931.
4. Li, C. H., Chemistry of ovine prolactin, in *Handbook of Physiology, Section 7,* Vol. 4 (Part 2), Greep, R. O. and Astwood, E. B., Eds., American Physiological Society, Washington, D.C., 1974, 103.
5. Frantz, A. G. and Kleinberg, D. L., Prolactin: evidence that it is separate from growth hormone in human blood, *Science*, 170, 745, 1970.
6. Lewis, V. J., Singh, N. P., and Seavey, B. K., Human prolactin: isolation and some properties, *Biochem. Biophys. Res. Commun.*, 44, 1169, 1971.
7. Hwang, P., Guyda, H., and Friesen, H. G., Purification of human prolactin, *J. Biol. Chem.*, 247, 1955, 1972.
8. Shome, B. and Parlow, A. F., Human pituitary prolactin: the entire linear amino acid sequence, *J. Clin. Endocrinol. Metab.*, 45, 1112, 1977.
9. Kelly, P. A., Djiane, J., Katoh, M., Ferland, L. H., Houdebine, L.-M., Teyssot, B., and Dusanter-Fourt, I., The interaction of prolactin with its receptors in target tissues and its mechanism of action, *Rec. Prog. Horm. Res.*, 40, 379, 1984.
10. Shiu, R. P. C. and Friesen, H. G., Blockade of prolactin action by antiserum to its receptors, *Science*, 192, 259, 1976.
11. Nolin, J. M. and Witorsch, R. J., Detection of endogenous immunoreactive prolactin in rat mammary epithelial cells during lactation, *Endocrinology*, 99, 949, 1976.
12. Nolin, J. M., Intracellular prolactin in rat corpus luteum and adrenal cortex, *Endocrinology*, 102, 402, 1978.
13. Nolin, J. M., Molecular homology between prolactin and ovarian peptides: evidence for physiologic modification of the parent molecule by the target, *Peptides*, 3, 823, 1982.
14. Josefsberg, Z., Posner, B. I., Patel, B., and Bergeron, J. M., The uptake of prolactin into female rat liver, *J. Biol. Chem.*, 254, 209, 1979.
15. Posner, B. I., Josefsberg, Z., and Bergeron, J. M., Intracellular polypeptide hormone receptors, *J. Biol. Chem.*, 254, 12494, 1979.
16. Serveley, J. L., Teyssot, B., Houdebine, L. M., Delouis, C., and Djiane, J., Evidence that the activation of casein gene expression in the rabbit mammary gland can be elicited by a low amount of prolactin firmly retained on its receptors, *Biochemie*, 64, 133, 1982.
17. Rillema, J. A., Mechanism of action of prolactin, *Fed. Proc.*, 39, 127, 1980.
18. Houdabine, L. M., Djiane, J., Kelly, P. A., Katoh, M., Dusanter-Fourte, I., and Martel, P., The mechanism of action of prolactin on casein gene expression, in *Endocrinology*(Int. Congr. Ser. 655), Labrie, E. and Proulx, L., Eds., Elsevier, New York, 1984, 203.

Chapter 2

PROLACTIN INTERACTION WITH ITS RECEPTORS AND THE RELATIONSHIP TO THE SUBSEQUENT REGULATION OF METABOLIC PROCESSES

Mark E. Costlow

TABLE OF CONTENTS

I. Introduction ... 6

II. Identification of Receptors ... 6

III. Isolation and Characterization of Prolactin Receptors ... 9

IV. Endocrine Control of Prolactin Receptors ... 10

V. Mechanisms of Prolactin Receptor Regulation ... 15

VI. Conclusions ... 19

Acknowledgments ... 20

References ... 20

I. INTRODUCTION

Prolactin interaction with its receptor is the first and requisite step in its action. The first reports identifying these receptors were published in 1972 to 1973,[1,2] and since then, more than 400 papers have appeared on the subject. The data in Table 1, which is not intended to be complete, provide a sampling of the wide variety of tissues and species where the receptor has been identified. These findings indicate that prolactin must have a biological function in each of these tissues, yet the role of prolactin has only been firmly established in relatively few instances. Considering the volume of literature on prolactin receptors (PR) and current directions in receptor research in general, this chapter will deal mainly with the area of receptor regulation. Reports fall into two categories: (1) the effects of hormones, endocrine status, and endocrine organ ablation on receptor levels and (2) possible control mechanisms involved in regulating receptor levels. In spite of the wide selection of tissues and species available for study, most reports have focused on mammary gland, neoplastic mammary tissue, and reproductive tissues of rabbits, mice, and rats.

It will become clear in this review that PR are regulated in a complex way by many hormones, including prolactin. Although implying that such regulation is involved in governing prolactin responsiveness, few studies have dealt with this issue directly. Finally, I should emphasize that prolactin action is initiated by binding to its receptor, but this alone is inadequate to produce a biological response. Since prolactin target tissues are also targets for many other hormones, the regulation of receptor is but one of a series of interrelated cellular activities which ultimately control prolactin action.

II. IDENTIFICATION OF RECEPTORS

In most reports, the identification of receptor-positive tissue is based on competitive radiolabeled prolactin-binding assays of crude or partially purified plasma membranes or tissue homogenates. Autoradiographic localization following in vivo injection or in vitro incubation of cells or tissue slices with radiolabeled prolactin has also been used. To detect endogenously bound or added prolactin, antibodies to prolactin have been appllied to tissue sections or fixed cultured cells and hormone location determined by immunocytochemistry (IC).

Each of these approaches has both advantages and disadvantages. Radio-receptor assays of whole cells or membranes can characterize and quantitate specific, saturable prolactin binding, but fail to identify receptor association with a particular cell type in a mixed population. Autoradiography is useful to identify the cell type containing receptors, provided that the specificity of localization has been determined with coadministered excess unlabeled hormone.[3-8] Not all studies have included such controls,[9-13] but selectivity in localization among different cell types suggests that the hormone is receptor associated. Finally, IC can be used to identify the presence of prolactin in a particular cell type,[14,15] but to determine if it is receptor associated, specificity and saturability of localization must be demonstrated. Specificity can be established only by using an antigenically unrelated prolactin to displace the hormone to which the antisera was raised. To my knowledge, there are no reports of using this approach to determine the specificity of prolactin localization by IC. Additionally, it is imperative that physiologic levels of prolactin be used to determine hormone localization at presumptive receptor sites. This problem is illustrated by the results in a report on prolactin localization by IC to cultured human breast cancer cells in culture.[16] The concentration of prolactin used (1 $\mu g/m\ell$) was well beyond saturation of receptor sites.[17] Nonspecific binding increases linearly with prolactin concentration, and receptor sites were

Table 1
PROLACTIN RECEPTORS IN VARIOUS SPECIES AND TISSUES

Species	Ref.
Rat	
Mammary gland	14, 85, 142
Liver	39, 125, 138, 139, 142, 143
Uterus	144
Ovary	65, 69, 70, 136, 145
Adrenals	15, 76, 140, 146, 147
Kidney	146, 147, 148, 149
Prostate	78, 80, 150, 151
Lung	36
Testes	71, 75, 76, 77, 78
Seminal vesicles	5, 71, 141, 151, 152
Pancreas	153
Brain	6, 8
Pituitary gland	154
Hamster — testes, ovary	64, 73, 155
Frog, tadpole — liver, tail, kidney, bladder	156, 157, 158, 159
Marsupial — corpus luteum	160
Porcine — ovary	28, 161
Mink — uterus	162
Cow — mammary gland, liver, ovary	63, 163, 164
Pigeon — crop sac, liver	159, 165, 166, 167
Human	
Testes	168
Ovary	169, 170
Chorion	171, 172
Amnion	10
Prostate	173, 174
Sheep — uterus, ovary, liver	159
Mouse	
Mammary gland	11, 175, 176
Liver	11, 59, 177, 178, 179
Ovary	11
Adrenals	180
Kidney	11, 181
Brain	181
Prostate	11
Seminal vesicles	11
Rabbit	
Mammary gland	2, 24, 37, 44, 183, 184
Kidney	24, 46, 154
Adrenal	24, 154, 185, 188
Ovary	24, 154
Liver	24, 48, 154
Testes	185
Prostate	185
Seminal vesicles	185
Uterus	188
Brain	186, 187, 188
Adipose tissue	154, 188
Pancreas	154
Milk	189
Epididymus	185

Note: The list of references is far from complete. Those given are intended to provide a sampling of observations on receptor regulation, methods for identifying receptors, and receptor characterization for a particular tissue.

saturated at 50 ng/mℓ. As a result, approximately 99% of the hormone detected was nonspecifically bound.

Autoradiographic localization, with the controls mentioned, has been used successfully to define prolactin binding sites in subcellular fractions of liver.[18] Because subcellular fractions always contain more than one organelle, autoradiography is a powerful, sensitive method to define receptor-containing compartments. With all these approaches, the receptor is identified by its biologic function, i.e., to specifically and selectively recognize prolactin. It must be emphasized that in certain circumstances prolactin may be associated with organelles that are devoid of receptor. Recent evidence indicates that receptors can transfer their ligands to other cellular compartments while the free receptors are recycled to the cell surface.[19] It is not known if such a phenomenon occurs in prolactin target cells.

Most studies of PR have been done using crude membrane fractions that have been incubated with ^{125}I-labeled prolactin in the presence or absence of excess unlabeled prolactin. The latter is used to distinguish specific (displaced by unlabeled prolactin) from nonspecific (not displaced by unlabeled prolactin) interactions. Accurate determination of PR quantity depends of the amount of ^{125}I-labeled prolactin present. Single-point assays using a subsaturating concentration of tracer are often employed for convenience when many samples are involved and to reduce the level of nonspecific binding. This approach, however, results in underestimation of relative receptor content. Using a proportional method to calculate specific binding overcomes this difficulty.[20,21] To measure the total number of receptors in a preparation, saturating concentrations ($>5 \times K_D$) must be used. This is usually accomplished by incubating small quantities of labeled hormone with increasing concentrations of unlabeled prolactin until nonspecific binding is unchanged. After correction for nonspecific binding by a proportional methods,[22] the data are plotted according to Scatchard to obtain the total number of sites and receptor affinity (K_A or K_D). The reasons for this approach are practical ones. First, the tedious preparation of labeled prolactin is done with relatively small quantities of hormone which are rapidly depleted if increasing concentrations of labeled prolactin are used. Second, nonspecific binding increases linearly with concentration and at saturation can become a high proportion of total binding, which results in a less precise calculation of specific binding. To reduce nonspecific binding of radiolabeled prolactin to incubation vessels and membranes, bovine serum albumin from 0.1 to 1% is included in buffers. In addition, even gentle iodination using lactoperoxidase or low concentrations of chloramine T reduces the binding activity of prolactin.[20,23-26] Ion exchange chromatography can be used to remove unreactive labeled hormone, thereby reducing nonspecific binding by about four-fold.[1,27-29]

Prolactin affinity for its receptor measured by Scatchard analysis ranges from 1 to 10×10^{-10} *M*. The plot yields a straight line, indicating that the hormone binds to a single class of high-affinity receptors. The physiologic relevance of these values is questionable because of Scatchard analysis, it is assumed that the interaction of ligand with its receptor is a reversible process. This is not the case for prolactin. Dissociation of the hormone from its receptor decreases with incubation time[30-32] and at steady state (usually 4 to 18 hr at 23°C); only about half of the bound labeled prolactin can be displaced by excess unlabeled hormone. This tight irreversible binding is not covalent because chaotropic salts[31] or brief exposure to low pH[33-35] releases the prolactin. With careful attention to time of exposure to these treatments, neither receptor nor hormone is damaged. Occupation of binding sites by prolactin in vivo has been shown to interfere with estimates of total receptor, but treatment with 5 M $MgCl_2$[31] or low pH can be used effectively to desaturate receptors. That prolactin can interact with receptor irreversibly in in vitro assays does not necessarily mean that this occurs in vivo since the complex is internalized. Although prolactin is degraded after internalization, this tight

Table 2
MOLECULAR WEIGHT ESTIMATES FOR PROLACTIN RECEPTORS

Tissue	Detergent/method	Mol wt	Comments	Ref.
Rat liver	Affinity label	43,000 and 28,000	Reduction, no effect	45
Rabbit MG	Zwittergent® 3-12	42,000	21,000 on reduced SDS/PAGE does not bind ConA	44
Rat Liver, nonpreg.	Triton X-100®	247,000	Binds ConA, pI 7.0—8.5	43
Rat liver, preg.	Triton X-100®	307,000	pI 4.2—4.7 increase with neuraminidase	43
Rabbit kidney	Triton X-100®	—	↑# sites, ↓ Stokes rad. with Prl	46
Rat liver	Triton X-100®	77,000	Hydrodynamic determination	38
Rat liver	Triton X-100®	73,000	Hydrodynamic determination	40
Rabbit mammary gland	Triton X-100®	220,000	—	37
Rat liver	Affinity label	36,000	—	139
Rat liver	CHAPS®	320,000	—	135
Human ovary	Triton X-100®	127,000	Affinity increase with detergent	170
Rabbit mammary gland	Zwittergent® 3-12	297,000	—	181
Liver; lactating mouse	CHAPS®	37,000	Same affinity with detergent	41
Tadpole tailfin	Triton X-100®	47,000	—	39
Lung, liver; E_2, PrL-injected Rats	Aqueous soluble	317,000	OPrl antibody also present	36
Rabbit liver	Triton X-100®	35,000	Multimers nonreduced	182
Rabbit mammary gland	Triton X-100®	35,000	Multimers nonreduced	182

association possibly plays a role in its intracellular routing. Intact prolactin has been found in certain target cells.

III. ISOLATION AND CHARACTERIZATION OF PROLACTIN RECEPTORS

Several properties of solubilized prolactin receptors have been determined in a variety of species and tissues. Table 2 summarizes these reports and indicates, where appropriate, interesting observations. In all but one report,[36] receptors have been rendered soluble and active by treating membranes with nonionic detergents. Triton X-100® has been used frequently, but has the disadvantage of causing ovine prolactin to aggregate, making the separation of free and bound hormone by gel filtration difficult.[37] This problem has been alleviated by the more recently available Zwittergent® and CHAPS® detergents. As is readily apparent, there is little consensus on the molecular weight of prolactin receptors within species or tissues. Values encompass a greater than tenfold range. Some of these differences can be accounted for by detergent binding to receptors. Triton X-100® binding to receptor has been reported to account for as much as one third of its apparent molecular weight.[38-40] Receptor aggregation with other proteins or with itself (a problem with Triton X-100®, see Reference 41) may also account for high molecular weight estimates, although a molecular weight of 32,000 has been observed with Zwittergent® and CHAPS®. Because the plant lectin concanavalin A (ConA) can block prolactin binding, and ConA affinity columns can retain solubilized receptor,[42] it has been suggested that the receptor is a glycoprotein. Neuraminidase treatment of membranes also increases the isoelectric point (pI) of PR.[43] That the carbohydrate may be unrelated to the receptor is suggested by the finding that Zwittergent®-solubilized mammary gland receptor (42,000 mol wt) does not bind ConA,[44] although it is possible that the 42,000-mol wt species may represent a binding subunit.

High molecular weight estimates of receptors might also be due to disulfide-linked subunits, but affinity-labeled receptors (43,000 and 28,000 mol wt) are unaffected by reducing agents,[45] as are Zwittergent®-solubilized rabbit mammary gland receptors.[44] Aggregation of receptors that was reversed by dithiothreitol has been observed in Triton X-100®-solubilized rabbit mammary gland receptors.[182] Proteolytic cleavage of receptors following cell breakage could be responsible for lower molecular weight receptors, but it would be expected that multiple molecular weight forms would have been observed in some studies. Further complicating the interpretation of these reports is the recent observation that PR can exist in a soluble form isolated form cytosols of estrogen (E_2) or prolactin-primed rat lung and liver.[36] In that the soluble PR appears when total PR levels are induced, the authors suggest that the soluble PR represents an intermediate in *de novo* synthesis prior to incorporation into the plasma membrane. Such a conclusion, however, conflicts with accepted information on the membrane-bound routes of integral membrane protein synthesis. Another unusual observation is that prolactin binding to Triton X-100®-solubilized PR decreases the Stokes radius of the complex,[46] suggesting that binding induces a conformational change in the receptor.

Shiu and Friesen were the first to attempt to purify PR from Triton X-100® extracts.[37] Using a prolactin affinity column and 5 *M* $MgCl_2$ elution of receptor, they achieved a 1100-to 1600-fold purification with about 8% recovery. SDS gel electrophoresis, however, indicated that the receptor was not homogeneous. More recently, using Zwittergent®-solubilized receptor from rabbit mammary gland and acidic elution from a prolactin affinity column, a 37,500-fold purification of receptors has been achieved with a 54% yield.[44] Iodinated purified receptor yielded a single radioactive peak on sodium dodecyl sulfate (SDS) gels (21,000 mol wt). In addition to reporting the highest purification and recovery of PR to date, they also addressed the question of varied molecular weight estimates. With purified receptor, molecular weights of 55,000 and 17,000 were obtained with CHAPS® detergent. Using Zwittergent®, a molecular weight of 37,000 was obtained, while in Triton X-100®, a molecular weight of 180,000 was obtained. Using unpurified receptor, the molecular weight in CHAPS® was 55,000 and 350,000 in Triton X-100®. Thus, much of the variation reported earlier may be due to the effects of detergents on receptor aggregation, association with other proteins, or the effect of detergents on the molecular weight standards used to calibrate columns.

IV. ENDOCRINE CONTROL OF PROLACTIN RECEPTORS

PR in rat liver are low or absent in males and present in females. Castration of males or E_2 administration increases PR in liver from male rats. Testosterone administration to castrated male rats or female rats lowers PR levels. The steroid-induced increases in receptor levels require the presence of the pituitary gland, and hypophysectomy (hx) of females results in liver PR loss. Pregnancy increases PR levels.[47-49] Several conflicting reports have examined the role of prolactin on increasing the level of PR in rat and mouse liver.[50-53] Receptor levels were increased in some instances, but not in others. In many of these studies, heterologous prolactin (usually ovine) was administered for long periods (>1 week). As recently reported, the up regulation of PR observed under these conditions is likely due to the generation of anti-ovine prolactin antibodies which bind to liver membranes.[51] Short-term (1 to 2 day) administration of ovine prolactin also increases mouse and rat liver receptor levels,[54-56] but by itself does not result in complete restoration to pre-hx levels. In addition, rat prolactin does not increase receptors, suggesting that homologous and heterologous prolactin may not act similarly. Rat growth hormone, however, does increase the level of PR in rat[57,58] and mouse liver.[59] The loss of PR in short-term cultures of hepatocytes is prevented by prolactin,[60] indicating that prolactin plays a direct role in maintaining receptors. The exact role of the pituitary gland in concert with E_2 is uncertain. It has been proposed that the pituitary may produce an E_2-dependent receptor-producing factor.[61]

PR levels in ovaries change throughout the estrus cycle and during pregnancy. Using ^{125}I-labeled prolactin binding to crude membranes or homogenates of ovarian tissue, it has been shown that receptor levels peak at proestrus and are lowest at estrus.[62,63] There is no obvious correlation of PR levels with serum levels of lutropin (LH), follicle-stimulating hormone (FSH), or prolactin,[62] but it has been suggested that rising E_2 levels might be responsible for the proestrus increase.[63] Autoradiographic studies have shown that receptors are present in hamster ovary interstitial cells, theca, granulosa cells in developing follicles, and in the corpus luteum.[64] In both cycling and primed hx rats, PR have been localized in follicles and corpus luteum;[5,65] in cycling rats they are also present in the theca.[5]

Using this semiquantitative method, hamster receptors are highest in follicles in early proestrus prior to the LH/FSH surge. In the corpus luteum, they peak at diestrus.[64] Blocking the rise in LH/FSH at early proestrus prevents a rapid decrease in follicular PR which normally occurs at late proestrus. Because human chorionic gonadotropin (HCG) decreases PR levels in luetinized rat ovaries,[66] it has been suggested that LH might be responsible for the decrease in binding seen at proestrus.[67] In contrast, LH has been shown to increase PR in granulosa cells and leutinized ovaries in hx immature-E_2, FSH, and LH-primed rats.[65] LH also increases PR in granulosa cells prepared from immature hx-E_2 injected rats, provided they are primed for 2 days with FSH. This increase occurs whether FSH priming is performed in vivo or in vitro.[68,69] The reason for these discrepancies is unresolved, but may be related to the schedule of hormone injection(s) or the time PR were assayed after hormone injection.[67] It is clear, nonetheless, that E_2, FSH, and LH are al ultimately necessary to induce PR in the ovary. Because a number of cell types contain PR and their relative binding is altered throughout the cycle, it is possible that cell-cell interaction may also play a role. In a recent study, the subcellular location of prolactin and PR within luteal cells was found to change from cytoplasmic to cell surface.[70] How such shifts might be related to prolactin action and apparent receptor levels and their control is unknown.

There have been several reports that PR are regulated by sex steroids and gonadotropins in testes and prostate. Testosterone administration to intact rats causes a decrease in testicular PR.[71,72] Injecting prolactin results in a time-dependent loss of PR and testosterone acts synergistically with prolactin.[72] Estradiol also causes a decrease in receptors. This effect is not enhanced by injected prolactin, so that the estradiol effect may be a result of increased prolactin secretion. The decreases in receptor level are unlikely due to occupation by bound hormone, because receptor assays were performed 2 days after the last prolactin injection.[72] In contrast, prolactin reportedly increases the level of PR in hamster testes,[73] although this may be due to antiprolactin antibodies bound to membranes. Lowering serum prolactin levels with the ergot alkaloid CB154 produces a modest decrease[73] or has no effect on receptor levels.[74] Of interest is the effect of LH/HCG on PR levels.[75-77] In neonatal rats, LH causes a rapid and transient increase in PR, while in adult rats, LH administration results in a rapid decrease in PR. Because LH receptors show the same pattern of response to LH, it is possible that both LH and PR are somehow closely associated with each other in the plasma membrane and are cointernalized in adult tissue. Supporting this notion is the observation that increasing serum prolactin levels results in a decrease in both LH and

PR. Clearly other factors are involved, as the overall kinetics of receptor loss differ. LH receptor loss is more sustained thn that of PR.

That LH and prolactin receptors are not down regulated in neonatal rat testes suggests that the transition from fetal to adult Leydig cells is accompanied by the acquisition of cellular mechanisms for receptor-mediate endocytosis. Further studies with this model may provide important clues for understanding this phenomenon.

Although testosterone decreases receptor levels in testes, it is necessary for the maintenance of PR in the ventral prostate. Castration results in an 80 to 90% loss of PR, and subsequent testosterone administration restores PR to precastration levels.[71,78-80] Testosterone also restores PR levels after hx, supporting the finding that pituitary hormones are not involved in regulating PR levels in castrated rats.[80]

The influence of pregnancy and lactation on the level of PR in mammary glands has been investigated in the rabbit, rat, and mouse. Various sources of receptor have been used, including tissue slices, microsomal membranes, and enzymatically dispersed whole cells. For mouse and rabbit tissue, prolactin binding in dispersed cells is higher than during pregnancy,[81,82] whereas it is undetectable in microsomal preparations of rabbit mammary gland.[83] Because mammary tissue contains a variety of cell types (fat, connective tissue, myoepithelial and epithelial cells), which change in relative proportion during pregnancy and lactation, purified epithelial cells are the most desirable source of receptor. Precautions must be taken[82] because commercial collagenase contains proteolytic enzymes that can destroy PR.[84] In addition, the influence of placental lactogen (mouse, rat) and prolactin on the measurement of PR must be considered. Reducing serum prolactin with the ergot alkaloid CB154 has been used by some investigators, but because PR levels can be dramatically influenced by prolactin, this maneuver may affect interpretation of results. Brief treatment of cells at pH 3 is an effective alternative which removes surface-bound prolactin on while cells.[34] Although these precautions have not been employed in all studies, a general pattern of PR changes during pregnancy has emerged in all species studied. Receptor levels are high in virgin tissue, decline during pregnancy, and increase again near or after parturition. Various hormones have been investigated for their role in these changes, and it seems that prolactin increases PR levels,[85,86] while progesterone antagonizes the effect of prolactin.[86] Hysterectomy and/or ovariectomy (ovex) during pregnancy in rats and mice results in increased receptor levels[27,87] in 1 day, but in one report, it resulted in no change.[85] Receptor levels fall to pretreatment values in 3 to 4 days.[88] The increase in PR following ovex is prevented by simultaneous adrenalectomy,[89] suggesting that glucocorticoid is required to maintain PR or is required for the Ovex-induced increase. The positive effect of glucocorticoid is supported by the finding that it increases PR levels in cultured mammary cells.[90] That serum progesterone levels fall following ovex[89] supports the notion that the increase in PR following ovex is due to lowered progesterone levels. A similar fall occurs at late pregnancy where PR levels increased dramatically.[82]

These findings suggest that the level of PR plays a role in regulating mammary gland growth and differentiation, but at present cannot be dissociated from the possible direct effects of glucocorticoids in supporting and progesterone-inhibiting lactogenesis.

The presence of PR in rat mammary tumors was first reported by Turkington,[91] who found receptors in amounts ranging from 30 to 80% of that in normal mammary tissue. Less than 15% of the normal complement of receptors was present in an autonomous transplantable mammary tumor, R3230AC, and autonomous mouse mammary tumors bound far less prolactin than did normal mouse mammary gland. These initial findings suggested a correlation between PR and hormone-dependent growth — receptors being present in hormone-dependent tumors, but low or absent in autonomous tumors. In one subline of the prolactin-dependent MTW9 transplantable mammary tumor, PR were present in amounts equal to those previously measured in lactating rat mammary tissue, but in an autonomous subline, only 15% of the number of sites were present.[92]

Estrone-progesterone-induced mammary tumors in the GR strain of mice are initially hormone dependent; however, when transplanted, they eventually lose their hormone-dependent growth characteristics. Prolactin binding was highest in primary hormone-dependent tumors and declined progressively in transplanted hormone-dependent and hormone-responsive tumors. The decrease in bound hormone is due to a loss in the number of receptor sites, as binding affinity of the receptor was the same for all tumor types.[93]

These results support a causal relationship between loss of PR and loss of hormone dependence in experimental breast cancer, but are complicated by the following observations. First, the R3230AC transplantable rat mammary carcinoma does *not* depend on prolactin for growth, although it *does* respond to prolactin with distinct changes in enzyme activities. In contrast to the autonomous MTW9 subline, the R3230AC tumor contains nearly the same number of PR as normal rat mammary tissue.[94] Studies of DMBA-induced rat mammary tumors have shown that receptor levels were slightly lower in autonomous tumors, but because individual dependent and autonomous tumors varied widely in receptor content, it was impossible to correlate receptor level and hormone dependence. Moreover, ovex alone resulted in a mean decrease of approximately 30%, regardless of response.[95] If tumor prolactin dependence was assessed by response to drugs that either increased or decreased circulating levels of prolactin,[96] receptor levels were higher in tumors that responded to prolactin by increased growth, but response could not be predicted by the PR level in pretreatment biopsies. If tumors were classified based on response to decreased circulating prolactin, receptor levels were higher in responsive tumors both before and after biopsy, but again, this correlation was not predictive of growth behavior for individual tumors. When growing tumors were assayed for PR content and subsequently their hormone dependence was determined by transplantation, PR levels were 50% lower in autonomous tumors. Overlap in receptor concentrations, however, did not allow correlations to be made between hormone-dependent growth and PR levels.[97]

Because the number of receptors in both hormone-dependent and autonomous DMBA-induced mammary tumors covers a broad spectrum, it is possible that the degree of dependence is related to the relative level of receptor. Kelly et al.[98] approached this question by measuring the number of PR in DMBA-induced rat mammary tumors after ranking them according to their growth rate in response to prolactin administration. Receptor levels were highest in tumors showing the greatest increase in tumor mass in response to prolactin, suggesting that the number of receptor sites in a tumor indicated the degree of responsiveness to prolactin. Alternatively, if tumors consisted of a heterogenous cell population with respect to PR, growth response would be directly related to the proportion of tumor cells that contained receptor sites. To distinguish between these alternatives, PR were assessed in DMBA-induced rat mammary tumors using autoradiography. In some tumors, all cells contained receptors while in others up to one half of the tumor cells remained unlabeled. These results suggest that reported variations in receptor content in DMBA-induced mammary tumors may be due to the presence of a heterogenous population of cells — some containing many PR and others containing very few receptor sites or none at all.[4]

Target tissue function is regulated not only by the circulating levels of hormones, but also by target cells, whose response to hormonal stimulation is controlled by hormone receptor content. Hilf and co-workers studied the effect of E_2, prolactin, lergotrile mesylate (which lowers circulating prolactin), and streptozotocin-induced diabetes on PR in DMBA-induced rat mammary tumors and the transplantable R3230AC rat mammary carcinoma.[99-101] Prolactin administration decreased prolactin binding (down regulation ?) in both tumor types, but increased growth only in DMBA-induced tumors. Lergotrile mesylate administration reduced circulating prolactin levels in animals bearing DMBA-induced tumors and caused a decrease in tumor growth, while estradiol(2 μg/day) had the opposite effect; neither treatment altered prolactin binding. In diabetic rats, the growth of R3230AC tumors was slightly enhanced, whereas up to 50% of DMBA-induced tumors regressed. Prolactin binding decreased slightly in R3230AC tumors (30%) and in DMBA-induced tumors, binding fell by 50%. Because of the important role of insulin in prolactin-induced lactogenesis in normal mammary tissue in vitro and growth of DMBA-induced mammary tumors in vivo, these findings suggest that insulin may, in part, control tissue responsiveness to prolactin by regulating the number of prolactin-binding sites. The effect of prolactin on its own receptor is less clear. As mentioned earlier, other investigations have shown that raising prolactin levels increases the number of receptors in DMBA-induced rat mammary tumors, but that this effect may be indirect by increasing the number of PR-containing cells. The binding inhibition they observed might reflect masking of sites or an actual reduction of receptors in response to homologous hormone (so-called down regulation). This latter possibility is supported by the observation that, in the R3230AC tumor, the tumor growth rte has no apparent effect on the apparent number of binding sites. Prolactin, however, does increase the level of its receptors in cultured DMBA-induced rat mammary tumors.[34]

Administration of pharmacologic doses of steroid hormones can cause regression of DMBA-induced rat mammary tumors. The effect of high doses of androgen is counteracted by prolactin coadministration, suggesting that androgens may block the peripheral effect of prolactin at the level of the tumor cell.[102] Similarly, high doses of E_2 cause regression of DMBA-induced rat mammary tumors, and coadministration of prolactin with E_2 overcomes steroid-induced regression, possibly by a mechanism similar to the one proposed for the effects of androgen.[103] In androgen-responsive tumors, PR levels averaged 63% lower than in tissue biopsied before androgen treatment, but receptor levels did not fall in all cases. Compared to biopsy samples, tumors that did not regress following androgen administration showed no change in the number of receptor sites, indicating that the reduction in binding sites in responsive tumors was not due to increased circulating prolactin in response to testosterone proprionate. Prolactin binding in pretreatment biopsies from both groups varied considerably and was not predictive of which tumors were likely to respond.[104] These findings, as well as a report by Kelly et al.[105] showing that androgen can reduce prolactin binding in rat liver, support the hypothesis that androgen reduces the level of PR. There may be alternative mechanisms, however, as not all tumors which regressed showed a reduction in binding sites. If PR were reduced so that a critical level of signal was not obtained, tumor growth would be retarded, but prolactin injection could overcome an inadequate signal by increasing receptor occupancy. Because all tumors initially grew at physiologic levels of prolactin, each tumor would require a unique amount of prolactin-induced signal to support growth.

The observation that tumor regression due to pharmacologic doses of E_2 is also reversed by prolactin prompted Kledzik et al.[106] and Smith et al.[100] to consider whether alterations in prolactin binding might account for these effects. PR were measured in DMBA-induced rat mammary tumors after E_2 injections which either stimulate or inhibit tumor growth. To eliminate possible interference of in vitro prolactin binding by high levels of circulating prolactin, ergocornine was injected before sacrifice. Stimulation of tumor growth by estradiol benzoate, at 0.2 or 2 μg/day, did not alter receptor levels, while dosages of 10, 20, or 25 μg/day resulted in tumor regression and a 26 to 44% decrease in prolactin binding. It is not known if all tumors which regressed lost receptor, since pretreatment receptor levels were not determined. Because PR were decreased and prolactin reversed the effect of *either* steroid on tumor growth, these findings suggest a similar mode of action. However, because PR were not completely abolished in either instance, the same alternatives presented to account for androgen-induced regression may also apply to E_2-mediated regression.

V. MECHANISMS OF PROLACTIN RECEPTOR REGULATION

Although some actions of prolactin may be mediated by its initial interaction with cell surface receptors, both prolactin and its receptor are found intracellularly in a number of target tissues. Prolactin-receptor complexes, like those for other polypeptide hormones, are internalized by receptor-mediated endocytosis. The presence of prolactin within target tissues has been taken by some investigators as evidence that the intracellular hormone itself may mediate certain responses.[14] There is only one report where this notion has been tested directly.[108] When prolactin was incubated with crude mammary cell nuclei, an increase in RNA synthesis was observed. The purity of the nuclear preparation, however, was only assessed by phase microscopy. Because receptors are present on intracellular membranes which could have been present in the reaction mixture, it is possible that prolactin-receptor interactions mediated the increase in ^{3}H-uridine incorporation.

It has been shown that PR are present in internal membranes of the female rat liver. Subcellular fractionation combined with analysis of marker enzymes, electron microscopy, and EM autoradiography of ^{125}I-labeled prolactin-receptor complexes has unequivocally demonstrated that about 70% of female rat liver receptors are located on the internal (cisternal) side of Golgi-like vesicles.[18] In addition, receptors are also present in unique vesicles found in "lysosomal" fractions of rat liver.[109] These vesicles lack Golgi or lysosomal enzyme activities, but are enriched in NADH-pyrophosphatase activity.[109,110] Recent findings also indicate that PR are present in an intracellular vesicle fraction distinct from the plasma membrane in cultured rat mammary tumor cells.[111] The function of these receptor sites is unclear. One possibility is that they may serve as a reservoir for receptors which, under certain conditions, replace those removed from the cell surface. If this migration is a regulated process, it might explain why not all liver receptors are occupied in the presence of high plasma levels of prolactin.[52] Another possibility is that internal receptors may represent the major portion of receptors which are constantly cycling between the cell surface and the cell interior. Several studies have shown that the rate of removal of receptors from the cell surface is considerably more rapid than the rate of receptor synthesis.[33,112,113] At present, there is no direct evidence for PR recycling.

Prolactin, initially presented to cells at their surface, is internalized by an energy-dependent uptake process. Uptake proceeds within minutes, leading to the accumulation of hormone in the Golgi.[7] In vivo and in vitro studies indicate that radiolabeled prolactin is then lost from the cell. In vitro experiments indicate that this loss is due to hormone degradation as the released radioactivity is of low molecular weight[33] and can be prevented by lysosomotropic amines.[33,114,115]

Immunocytochemical studies of prolactin localization have demonstrated the presence of hormone within mammary cells, ventral prostate, adrenal, and ovary.[14,15,70,116] Nuclear localization of prolactin has been observed in mammary gland by both immunocytochemical means[14] and autoradiography.[82] In the latter, grains were localized over the nucleus within 15 min after ^{125}I-prolactin addition. In contrast, following in vivo administration of ^{125}I-prolactin to rats, no nuclear localization was observed in rat liver.[7] In all instances, prolactin was also found in the cytoplasm, but whether the hormone was bound to its receptor is not known.

In a recent study, both prolactin and its receptor were localized in rat ovary using immunocytochemical staining.[70] The pattern of cellular staining for both receptor and prolactin changed in a concerted manner from the cytoplasm to cell surface in post-partum ovaries. These results suggest that the localization of prolactin along with its receptor can change, depending on the endocrine state of the tissue. Prolactin-receptor complexes within cells may be capable of producing signals necessary for prolactin responses. Uptake of prolactin-receptor complexes occurs within minutes, while many prolactin responses require the presence of prolactin for many hours. All of the information necessary to produce prolactin effects, however, resides in the receptor molecule, because antibody to receptor can mimic prolactin action. Bivalent and monovalent antibodies to the receptor can cause receptor down regulation,[117] but only bivalent antibody mimics prolactin action,[114,117] suggesting that receptor cross-linking may be necessary for prolactin action. These results might also be taken as evidence that internalization, which precedes down regulation, is not required for prolactin action. Unfortunately, in these experiments, receptor level was determined 1 day after prolactin addition, but the rate of down regulation or internalization was not measured. Thus, although both forms of antibody caused a reduction of receptor levels, the routing of the receptor or possible effects of either antibody form on PR synthesis is uncertain.

The regulation of PR by prolactin has been investigated in pseudopregnant rabbit mammary gland in vivo and in vitro. Following injection of large doses (3 mg per rabbit) of prolactin, mammary gland receptors are maximally occupied within 15 min. This is followed by a decrease in PR levels at 6 hr after prolactin administration. Receptor levels return to control values at 30 hr. The serum levels of prolactin were 4 μg/mℓ at 6 hr.[118] In vitro studies using explant cultures have shown that PR are in a state of rapid flux, as inhibiting protein synthesis with cycloheximide results in a 60% decrease in receptors within 6 to 12 hr. PR levels return to near control values 18 to 24 hr after cyclohexamide removal.[112] This rapid turnover of PR is similar to that obtained in vivo in rat liver.[113] More extensive in vitro studies have disclosed that during down regulation, it is likely that PR are internalized, delivered to lysosomes, and are then degraded. Incubating explants with prolactin in the presence of lysosomotropic agents prevents down regulation in the presence of 1 μg prolactin per milliliter. That exposing cultures to chloroquine, NH_4Cl, or methylamine alone results in a two- to threefold increase in PR supports the notion that under these in vitro conditions, unoccupied receptors are also delivered to lysosomes. In the above in vitro experiments, extremely high (1 to 5 μg/mℓ) concentrations of prolactin were used to effect receptor down regulation. It is interesting that, even with such high levels of prolactin, complete occupancy of free receptors was not achieved. Saturation of receptors in membrane preparations occurs at 50 to 100 ng/mℓ. Possibly not all sites were occupied because the hormone was unable to penetrate to the explants. Alternatively, some receptors may not have been accessible to prolactin because they were located intracellularly. In addition, it appears that not all occupied PR are down regulated, in vivo or in vitro. The level of total receptor at maximum down regulation is higher than the free level at maximum occupancy. This implies that some recycling of receptor may have occurred or that peripheral receptors are more sensitive to down regulation than possible internal sites.[119]

In a recent report, Djiane et al.[120] investigated the relationship between PR down regulation and prolactin action in rabbit mammary gland organ culture. In contrast to results obtained at supraphysiologic concentrations of prolactin, there was a direct correlation between receptor occupancy and down regulation in the range of 0 to 100 ng prolactin per milliliter. Unfortunately, all measurements were made 24 hr after prolactin addition, so it is not known if a similar delay between occupancy and receptor loss occurred as observed in vivo.[118] The extent of down regulation was inversely related to prolactin-induced casein, casein mRNA, and DNA synthesis. With a large excess of prolactin (1 to 20 μg/mℓ), responses decreased, while there were only modest changes in total receptor. It must be noted that because casein synthesis occurs when hormone and receptor degradation (down regulation) are blocked, the loss of these molecules is unrelated to prolactin action.[115,121,122] That internalized prolactin-receptor complexes are not involved in prolactin action, however, cannot be excluded. In these studies, receptors were measured in broken cell preparations. Thus, their location was not determined. Since PR complexes in rabbit mammary gland cell suspensions are rapidly internalized,[82] it is possible that internal complexes may have elicited the necessary signals to stimulate casein and DNA synthesis. That extremely high prolactin concentration blunted prolactin responsiveness also suggests that there may be alternative intracellular fates of prolactin. Possibly at extremely high prolactin concentration, the hormone is rapidly degraded, terminating a response, while at lower doses, it remains internalized in association with its receptor.

In cultured rat mammary tumor cells, prolactin both up and down regulates the level of its receptors. This regulation is dependent on both the concentration of hormone and the duration of exposure. The majority of receptors in these cells are cryptic under normal conditions and thus are inaccessible to prolactin. There is evidence that a steady state exists between internal (cryptic) and surface receptors and that receptor occupancy causes internal sites to appear at the cell surface.[34] Acute down regulation in the presence of very high prolactin concentrations (1 to 5 μg/mℓ) results in a rapid depletion of all receptor sites, including those in the internal receptor pool. Saturating but lower prolactin concentrations (100 to 500 ng/mℓ) decrease the steady-state level of receptors, but do not eliminate them; prolactin at 50 ng/mℓ which is also a saturating concentration, does not change receptor levels. Because these receptors are rapidly and constantly removed from the cell surface,[33] these results are consistent with the notion that tumor cells regulate the appearance of receptors at the cell surface, depending on the hormonal environment. Once prolactin-receptor complexes are internalized, the hormone is degraded. Lysosomotropic amines block hormone degradation, but not uptake of complexes. Uptake does not require protein synthesis, but is energy dependent.[33] Although prolactin is degraded, the fate of receptor is uncertain. It is not immediately recycled to the cell surface, but may be retained within the cell. The presence of a pool of internal sites that can be shuttled to the cell surface may explain the observed capacity of these tumor cells to recognize changes in prolactin concentrations well beyond the point predicted by the dissociation constant for binding.

When tumor cells are incubated with low (0.1 to 0.5 ng/mℓ) prolactin concentrations, the level of receptors increases (up regulation). Prolactin concentrations causing up regulation are well below the normal basal serum concentration in rats (10 to 20 ng/mℓ) and at 0.5 ng/mℓ, occupy only 5 to 15% of the total receptor sites. There is no direct evidence that this increase in receptors is due to accelerated receptor synthesis or a decreased rate of receptor degradation; however, there is indirect evidence that increased synthesis is involved.[33] Receptors in these cells have a relatively short half-life ($t_{1/2}$, 60 min), and occupancy causes an increased rate of receptor uptake. Furthermore, the rate of uptake of hormone-receptor complexes is independent of hormone concentration between 5 μg and 5 ng of prolactin per milliliter,[33] and the extent of down regulation (about 85%) is independent on the basal level of receptors, which can vary by more than eightfold among tumors.[84] Since the steady-state level of receptors is determined by the rate of synthesis vs. the rate of destruction, an increased rate of synthesis may be maintained at all prolactin concentrations. At higher prolactin levels, the steady-state receptor level could decrease due to increased occupancy, uptake, and degradation. The net effect would be the lower steady-state receptor level observed at 100 to 500 ng prolactin per milliliter. When prolactin levels are increased to very high levels (1 to 5 μg/mℓ), rapid and nearly complete receptor down regulation likely reduces the rate of receptor synthesis by preventing the generation of sufficient cellular signals to induce receptor synthesis. This is supported by the observation that levels of prolactin that down regulate receptors (1 μg/mℓ) in normal mammary gland and tumor cells decrease prolactin-induced DNA synthesis in normal tissue 22 to 24 hr after its addition,[120] but these concentrations do not affect ornithine decarboxylase activity in tumor cells, which is maximally stimulated 3 hr after prolactin addition.[123] Thus, receptor down regulation may blunt a late response, but have no effect on an earlier one. Since saturation of receptor sites alone is insufficient to account for the effect of prolactin on down regulation, it is likely that the steady-state receptor level is controlled by the rate of prolactin-receptor complex internalization.

Down regulation in mammary tumor cells requires energy, but not the integrity of cytoskeletal elements, which is also true for normal mammary tissue.[118] In addition, NH_4Cl, which prevents hormone degradation in both normal and neoplastic tissue, does not inhibit down regulation in tumor cells. This indicates that hormone degradation is not essential for down regulation, and it does not appear to be necessary for biologic responsiveness.[115,121] It has been suggested that NH_4Cl prevents prolactin receptor down regulation in normal mammary gland.[124] Initially, this might appear to be a potentially important difference in normal vs. neoplastic mammary cell receptor processing. This apparent difference, however, can be explained by the different methods employed. In tumor cells, down regulation was defined as a decrease in cell surface receptors.[34] By contrast, in normal mammary tissue,[124] receptors were quantitated in membrane preparations having both surface and internalized receptors. Thus, NH_4Cl blocks hormone degradation without preventing uptake of prolactin-receptor complexes, but also inhibits degradation of internalized receptors. Preliminary experiments in our laboratory indicate that prolactin binding increases in membranes prepared from cells treated with NH_4Cl. Thus, NH_4Cl may also prevent receptor degradation in tumor cells.

Tumor cells appear to be less sensitive to down regulation 1 day after prolactin addition than normal mammary tissue. About 50 ng of prolactin per milliliter does not decrease tumor receptor levels, even after 6 days. In contrast, this concentration causes a 40% decrease in normal mammary cell receptors in 1 day.[120] Although it is possible that this difference stems from the experimental system (organ vs. primary cultures) and/or species (rabbit vs. rat), it nonetheless suggests a mechanism for an increased growth advantage of prolactin-responsive tumors at physiologic prolactin levels.

In further attempts to define possible mechanisms for rapid modulation of PR-binding activity, some investigators have focused on the effect of modifying the cell surface membrane on receptor-binding capacity.

It has been demonstrated repeatedly that treating membranes with phospholipase C causes a reduction in prolactin binding,[24,27,125] by decreasing the movement of receptor within the membrane, altering receptor conformation, or by masking the binding site. Consistent with some of these possible mechanisms are the following observations. When mouse mammary gland or liver membrane phospholipids are methylated in vitro with *S*-adenosyl-L-methionine, there is a dose-dependent increase in lactogen binding.[126,127] Preincubation of rat ventral prostate membranes with ethanol, propanol, or butanol causes a 37 to 77% increase in prolactin binding and increases membrane fluidity.[128] In vivo modification of membrane lipids also results in altered prolactin binding. Knasak and Liu[129] demonstrated that in mouse livers made deficient in essential fatty acids, prolactin binding capacity decreased. In addition, prolactin itself modifies the fluidity of liver membranes. Prolactin doses which increased fluidity increased prolactin binding up to fivefold, while doses that did not alter fluidity had no effect on binding.[107]

A possible mechanism for prolactin induction of its own receptors emerges from the findings that

1. Prolactin stimulates prostaglandin synthesis in liver,[56] while inhibiting prostaglandin synthesis decreases receptor levels.[130,131]
2. Prostaglandins, arachidonic acid, and phospholipase A_2 all modify prolactin-binding capacity and membrane fluidity,[132,133]
3. T_3, which is lipolytic, increases prolactin binding in vivo and in organ cultures of mammary gland in the absence of protein synthesis.[134]
4. Detergent solubilization of membranes, which removes lipids, results in a two- to eightfold increase in prolactin binding in rat liver and ovary.[135,136]

These findings suggest that a portion of PR in certain targets exists in a cryptic form in membranes and becomes exposed as a result of membrane phospholipid perturbation. This could provide a means for cells to respond rapidly to changes in prolactin concentrations without altering the receptor-synthetic machinery of the cells. In mammary tumor cells there are receptors which, under normal conditions, are masked or cryptic.[137] When tumor cells are depleted of ATP using uncouplers of oxidative phosphorylation or inhibitors of electron transport, there is a tenfold increase in surface membrane prolactin-binding capacity. These cryptic receptors are restricted to mammary tumor cells (cultured or in tissue slices) and proliferating mammary cells, but are absent in other targets including lactating mammary gland.[137] While unmasked sites are located on the cell surface, the location of cryptic receptors is uncertain. One possibility is that ATP depletion may result in modification of membrane lipids, thereby unmasking receptors. In this regard, phosphatidyl inositol, but not phosphatidyl choline, is rapidly lost from ATP-depleted cells.[190] Alternatively, cryptic receptors may be located intracellularly in vesicles as in rat liver. ATP depletion could cause fusion of these vesicles with the cell surface. Although exocytosis requires ATP, membrane fusion does not. Supporting the latter mechanism are recent observations showing that tumor cell cryptic receptors are localized in a subcellular vesicle fraction distinct from the plasma membrane. ATP depletion shifts receptors from vesicles to plasma membrane-rich fractions.[111] Permeabilization of tumor cells with saponin results in increased binding.[190] This could be taken to support the role of lipid modification in unmasking receptors already present on the surface membrane. Freeze thawing cells in liquid N_2, however, also unmasks receptors, supporting the contention that cryptic sites are internal. At present, neither mechanism for unmasking receptors can be excluded. Crude membrane fractions used in most studies contain both surface membranes and internal vesicles. Thus, it is possible that lipid modification could cause fusion of vesicles with membranes, the permeabilization of vesicles, or unmasking of receptors within the cell-surface membrane.

VI. CONCLUSIONS

The availability of PR at the cell surface is determined by the rate of PR synthesis, internalization, degradation, and possibly recycling. Changes in the composition of surface membrane components, especially lipids, also alter receptor binding capacity. Besides prolactin, other polypeptide and steroid hormones cause changes in PR levels. The direction and magnitude of change are dependent on the tissue, state of differentiation, hormone concentration, and endocrine history. Although some details of the biochemical and cellular mechanism responsible for prolactin-induced receptor regulation are known, little information is available on the mechanism by which other hormones participate in PR regulation. Future studies, especially with in vitro models, should help to define critical points in prolactin and receptor processing that govern the expression of prolactin responsiveness.

ACKNOWLEDGMENTS

I thank Patricia Powell for typing and Linda Daniels for editing. Preparation of this manuscript was supported by the American Lebanese Syrian Associated Charities and USPHS grant CA 38190.

REFERENCES

1. Frantz, W. L. and Turkington, R. W., Formation of biologically active ^{125}I-prolactin by enzymatic radioiodination, *Endocrinology*, 91, 1545, 1972.
2. Shiu, R. P., Kelly, P. A., and Friesen, H. G., Radioreceptor assay for prolactin and other lactogenic hormones, *Science*, 180, 968, 1973.
3. Bergeron, J. J., Resch, L., Rachubinski, R., Patel, B. A., and Posner, B. I., Effect of colchicine on internalization of prolactin in female rat liver: an in vitro radioautographic study, *J. Cell Biol.*, 96, 875, 1983.
4. Costlow, M. E. and McGuire, W. L., Autoradiographic localization of prolactin receptors in 7,12-dimethylbenz(a)anthracene-induced rat mammary carcinoma, *J. Natl. Cancer Inst.*, 58, 1173, 1977.
5. Costlow, M. E. and McGuire, W. L., Autoradiographic localization of the binding of ^{125}I-labelled prolactin to rat tissues in vitro, *J. Endocrinol.*, 75, 221, 1977.
6. Walsh, R. J., Posner, B. I., Kopriwa, B. M., and Brawer, J. R., Prolactin binding sites in the rat brain, *Science*, 201, 1041, 1978.
7. Josefsberg, Z., Posner, B. I., Patel, B., and Bergeron, J. J. M., The uptake of prolactin into female rat liver. Concentration of intact hormone in the golgi apparatus, *J. Biol. Chem.*, 254, 209, 1979.
8. Walsh, R. J., Posner, B. A., and Patel, B., Binding and uptake of [^{125}I]-iodoprolactin by epithelial cells of the rat choroid plexus: an in vivo autoradiographic analysis, *Endocrinology*, 114, 1496, 1984.
9. Birkinshaw, M. and Falconer, I. R., The localization of prolactin labeled with radioactive iodine in rabbit mammary tissue, *J. Endocrinol.*, 55, 323, 1972.
10. Salih, H., Murthy, G. S., and Friesen, H. G., Stability of hormone receptors with fixation: implications for immunocytochemical localization of receptors, *Endocrinology*, 105, 21, 1979.
11. Rajaniemi, H., Oksanen, A., and Vanha-Perttula, T., Distribution of ^{125}I-prolactin in mice and rats. Studies with whole-body and microautoradiography, *Horm. Res.*, 5, 6, 1974.
12. Reddy, S. and Watkins, W. B., Uptake of ^{125}I-labelled human placental lactogen and human placental lactogen by the tissues of normal and lactating rats, *J. Endocrinol.*, 65, 183, 1975.
13. Shani (Mishkinsky), J., Givant, Y., Sulman, F. G., Eshkol, A., and Lunenfeld, B., Uptake of ^{125}I-labelled prolactin by rat mammary gland and pigeon crop mucosa, *J. Endocrinol.*, 52, 397, 1972.
14. Nolin, J. M. and Witorsch, R. J., Detection of endogenous immunoreactive prolactin in rat mammary epithelial cells during lactation, *Endocrinology*, 99, 949, 1976.
15. Nolin, J. M., Intracellular prolactin in rat corpus luteum and adrenal cortex, *Endocrinology*, 102, 402, 1978.
16. Paterson, J. A., Salih, H., and Shiu, R. P., Immunocytochemical and autoradiographic demonstration of prolactin binding to human breast cancer cells in tissue culture, *J. Histochem. Cytochem.*, 30, 153, 1982.
17. Shiu, R. P., Prolactin receptors in human breast cancer cells in long-term tissue culture, *Cancer Res.*, 39, 4381, 1979.
18. Bergeron, J. J. M., Posner, B. I., Josefsberg, Z., and Sikstrom, R., Intracellular polypeptide hormone receptors. The demonstration of specific binding sites for insulin and human growth hormone in golgi fractions isolated from the liver of female rats, *J. Biol. Chem.*, 253, 4058, 1978.
19. Geuze, H. J., Slot, J. W., Stross, G. J. A. M., Lodish, H. F., and Schwartz, A. L., Intracellular site of asialoglycoprotein-receptor ligand uncoupling: double-label immunoelectron microscopy during receptor-mediated endocytosis, *Cell*, 32, 277, 1983.
20. Brooks, C. L., DeSombre, E. R., and Jensen, E. V., Characterization of [^{125}I]-iodo-ovine prolactin and evaluation of prolactin receptor assay methods, *Horm. Res.*, 16, 365, 1982.
21. Brooks, C. L., Leinen, J. G., DeSombre, E. R., and Jensen, E. V., Calculating specific binding for prolactin receptor assays, *Mol. Cell. Endocrinol.*, 26, 81, 1982.

22. Chamness, G. C. and McGuire, W. L., Scatchard plots: common errors in correction and interpretation, *Steroids*, 26, 538, 1975.
23. Barkey, R. J., Shani, J., Goldhaber, G., and Barzilai, D., Binding of iodinated rat and ovine prolactins to prolactin receptors and to its antibodies, *Mol. Cell. Endocrinol.*, 12, 105, 1981.
24. Shiu, R. P. and Friesen, H. G., Properties of a prolactin receptor from the rabbit mammary gland, *Biochem. J.*, 140, 301, 1974.
25. Hughes, J. P., Tanaka, T., Gout, P. W., Beer, C. T., Noble, R. L., and Friesen, H. G., Effect of iodination on human growth hormone and prolactin: characterized by bioassay, radioimmunoassay, radioreceptor assay, and electrophoresis, *Endocrinology*, 111, 827, 1982.
26. Rhodes, S. B., Frantz, W. L., and Holland, J., The effects of iodination on the receptor binding affinity of ovine prolactin, *Biochem. Biophys. Res. Commun.*, 121, 686, 1984.
27. Holcomb, H. H., Costlow, M. E., Buschow, R. A., and McGuire, W. L., Prolactin binding in rat mammary gland during pregnancy and lactation, *Biochim. Biophys. Acta*, 428, 104, 1976.
28. Rolland, R. and Hammond, J. M., Demonstration of a specific receptor for prolactin in porcine granulosa cells, *Endocrinol. Res. Commun.*, 2, 281, 1975.
29. Costlow, M. E. and Gallagher, P. E., Concanavalin A-induced alterations in ^{125}I-labeled prolactin binding, *Biochem. Biophys. Res. Commun.*, 77, 905, 1977.
30. Brooks, C. L., Leinen, J. G., DeSombre, E. R., and Jensen, E. V., Calculating specific binding for prolactin receptor assays, *Mol. Cell. Endocrinol.*, 26, 81, 1982.
31. Kelly, P. A., LeBlanc, G., and Djiane, J., Estimation of total prolactin-binding sites after in vitro desaturation, *Endocrinology*, 104, 1631, 1979.
32. Powell, B. L. and Hollander, V. P., Studies on the dissociation of prolactin from mammary tumor membranes, *Endocrinology*, 103, 710, 1978.
33. Costlow, M. E. and Hample, A., Prolactin receptors in cultured rat mammary tumor cells. Energy-dependent uptake and degradation of hormone receptors, *J. Biol. Chem.*, 257, 9330, 1982.
34. Costlow, M. E. and Hample, A., Prolactin regulation of cryptic prolactin receptors in cultured rat mammary tumor cells, *J. Cell Physiol.*, 118, 247, 1984.
35. Necessary, P. C. and Ebner, K. E., pH dependence of dissociation of the ovine prolactin rabbit mammary receptor complex, *Biochem. Biophys. Res. Commun.*, 111, 224, 1983.
36. Amit, T., Barkey, R. J., Gavish, M., and Youdim, M. B., Induction of prolactin (PRL) receptors by PRL in the rat lung and liver. Demonstration and characterization of a soluble receptor, *Endocrinology*, 114, 545, 1984.
37. Shiu, R. P. and Friesen, H. G., Solubilization and purification of a prolactin receptor from the rabbit mammary gland, *J. Biol. Chem.*, 249, 7902, 1974.
38. Jaffe, R. C., Hydrodynamic characterization of the Triton X-100® solubilized lactogenic hormone receptor of rat liver, *Biochemistry*, 21, 2936, 1982.
39. Jaffe, R. C., Rana catasbeina tail fin lactogenic hormone receptor: hydrodynamic characterization, *Comp. Biochem. Physiol.*, 76, 529, 1983.
40. Rae-Venter, B. and Dao, T. L., Hydrodynamic properties of rat hepatic prolactin receptors, *Arch. Biochem. Biophys.*, 222, 12, 1983.
41. Liscia, D. S., Alhadi, T., and Vonderhaar, B. K., Solubilization of active prolactin receptors by a nondenaturing zwitterionic detergent, *J. Biol. Chem.*, 257, 9401, 1982.
42. Costlow, M. E. and Gallagher, P. E., Relationship of prolactin receptors to concanavalin A binding, *Biochim. Biophys. Acta*, 587, 192, 1979.
43. Sasaki, N., Tanaka, Y., Imai, Y., Tsushima, T., and Matsuzaki, F., Different characteristics of solubilized lactogen receptors from livers of pregnant and non-pregnant female rats, *Biochem. J.*, 203, 653, 1982.
44. Necessary, P. C., Humphrey, P. A., Mahajan, P. B., and Ebner, K. E., Purification of rabbit mammary prolactin receptor by acidic elution from a prolactin affinity column, *J. Biol. Chem.*, 259, 6942, 1984.
45. Yamada, K. and Donner, D. B., Structures of the somatotropin receptor and prolactin receptor on rat hepatocytes characterized by affinity labeling, *Biochem. J.*, 220, 361, 1984.
46. Roguin, L. P., Bonifacino, J. S., and Paladini, A. C., Properties of human growth hormone binding sites solublized from female rabbit kidney, *Biochim. Biophys. Acta*, 715, 222, 1982.
47. Aragona, C., Bohnet, H. G., and Friesen, H. G., Prolactin binding sites in the male rat liver following castration, *Endocrinology*, 99, 1017, 1976.
48. Kelly, P. A., Posner, B. I., Fjushima, T., and Friesen, H. G., Studies of insulin, growth hormone and prolactin binding: ontogenesis, effects of sex and pregnancy, *Endocrinology*, 95, 535, 1974.
49. Posner, B. I., Kelly, P. A., and Friesen, H. G., Induction of a lactogenic receptor in rat liver: influence of estrogen and the pituitary, *Proc. Natl. Acad. Sci. U.S.A.*, 71, 2407, 1974.
50. Furuhashi, N. and Fang, V. S., Hormone receptors in livers of GH3 tumor-bearing rats: the predominant effect of growth hormone and testosterone, *Endocrinology*, 103, 2053, 1978.

51. Hughes, J. P., Elsholtz, H. P., and Friesen, H. G., Up-regulation of lactogenic receptors — an immunological artifact?, *Endocrinology,* 111, 702, 1982.
52. Posner, B. I., Regulation of lactogen specific binding sites in rat liver: studies on the role of lactogens and estrogen, *Endocrinology,* 99, 1168, 1976.
53. Frantz, W. L., Mann, L. C., and Welsch, C. W., Induction of specific binding sites for prolactin in the livers of Dw/J dwarf mice by injected prolactin, *IRCS Medical Science,* 5, 32, 1977.
54. Bohnet, H. G., Aragona, C., and Friesen, H. G., Induction of lactogenic receptors. I. In the liver of hypophysectomized female rats, *Endocrinol. Res. Commun.,* 3, 187, 1976.
55. Costlow, M. E., Buschow, R. A., and McGuire, W. L., Prolactin stimulation of prolactin receptors in rat liver, *Life Sci.,* 17, 1457, 1975.
56. Dave, J. R., Brown, N. V., and Knazek, R. A., Prolactin modifies the prostaglandin synthesis, prolactin binding and fluidity of mouse liver membranes, *Biochem. Biophys. Res. Commun.,* 108, 193, 1982.
57. Baxter, R. C., Zaltsman, Z., and Turtle, J. R., Rat growth hormone (GH) but not prolactin (PRL) induces both GH and PRL recepotrs in female rat liver, *Endocrinolgoy,* 114, 1893, 1984.
58. Norstedt, G., Mode, A., Eneroth, P., and Gustafsson, J. A., Induction of prolactin receptors in rat liver after the administration of growth hormone, *Endocrinology,* 108, 1855, 1981.
59. Knazek, R. A., Liu, S. C., Graeter, R. L., Wright, P. C., Mayer, J. R., Lewis, R. H., Gould, E. B., and Keller, J. A., Growth hormone causes rapid induction of lactogenic receptor activity in the Snell dwarf mouse liver, *Endocrinology,* 103, 1590, 1978.
60. Rosa, A. A., Djiane, J., Houdebine, L. M., and Kelly, P. A., Stimulatory effects of prolactin and anti-prolactin receptor serum on prolactin binding sites in rat liver cells in suspension culture, *Biochem. Biophys. Res. Commun.,* 106, 243, 1982.
61. Norstedt, G. and Mode, A., On the primary site of action of estrogens and androgens in the regulation of hepatic prolactin receptors, *Endocrinology,* 111, 645, 1982.
62. Cheng, K. W., Changes in rat ovaries of specific binding for LH, FSH and prolactin during the oestrous cycle and pregnancy, *J. Reprod. Fertil.,* 48, 129, 1976.
63. Saito, T. and Saxena, B. B., Specific receptors for prolactin in the ovary, *Acta Endocrinol.,* 80, 126, 1975.
64. Oxberry, B. and Greenwald, G. S., Autoradiographic analysis of proestrous changes in the binding of ^{125}I-labeled prolactin in the hamster ovary, *Biol. Reprod.,* 29, 1255, 1983.
65. Richards, J. S. and Williams, J. J., Luteal cell receptor content for prolactin and luteinizing hormone (LH) regulation by LH and prolactin, *Endocrinology,* 99, 1571, 1986.
66. Davies, T. F., Katikineni, M., Chan, V., Harwood, J. P., Dufau, M. L., and Catt, K. J., Lactogenic receptor regulation in hormone-stimulated steroidogenic cells, *Nature (London),* 283, 863, 1980.
67. Oxberry, B. A. and Greenwald, G. S., Autoradiography analysis of proestrus changes in the binding of ^{125}I-labeled prolactin to the hamster ovary, *Biol. Reprod.,* 29, 1255, 1983.
68. Navickis, R. J., Jones, P. B., and Hsueh, A. J., Modulation of prolactin receptors in cultured rat granulosa cells by FSH, LH and GnRH, *Mol. Cell. Endocrinol.,* 27, 77, 1982.
69. Wang, C., Hsueh, A. J., and Erickson, G. F., Induction of functional prolactin receptors by follicle-stimulating hormone in rat granulosa cells *in vivo* and *in vitro, J. Biol. Chem.,* 254, 11330, 1979.
70. Dunaif, A. E., Zimmerman, E. A., Friesen, H. G., and Frantz, A. G., Intracellular localization of prolactin receptor and prolactin in the rat ovary by immunocytochemistry, *Endocrinology,* 110, 1465, 1982.
71. Barkey, R. J., Shani, J., and Barzilai, D., Regulation of prolactin binding sites in the seminal vesicle, prostate gland, testis and liver of intact and castrated adult rats: effect of administration of testosterone, 2-bromo-α-ergocryptine and fluphenazine, *J. Endocrinol.,* 81, 11, 1979.
72. Amit, T., Barkey, R. J., and Youdim, M. G. H., Effect of prolactin, testosterone and estrogen on prolactin binding in the rat testis, prostate, seminal vesicle and liver, *Mol. Cell. Endocrinol.,* 30, 179, 1983.
73. Klemcke, H. G., Bartke, A., and Borer, K. T., Regulation of testicular prolactin and luteinizing hormone receptors in golden hamsters, *Endocrinology,* 114, 594, 1984.
74. Aragona, C., Bohnet, H. G., and Friesen, H. G., Localization of prolactin binding in prostate and testis: the role of serum prolactin concentration on the testicular LH receptor, *Acta Endocrinol.,* 84, 402, 1977.
75. Chan, V., Katikineni, M., Davies, T. F., and Catt, K. J., Hormonal regulation of testicular luteinizing hormone and prolactin receptors, *Endocrinology,* 108, 1607, 1981.
76. Katikineni, M., Davies, T. F., and Catt, K. J., Regulation of adrenal and testicular prolactin receptors by adrenocorticortopin and luteinizing hormone, *Endocrinology,* 108, 2367, 1981.
77. Huhtaniemi, I. T., Warren, D. W., and Catt, K. J., Development of heterologous down-regulation of lactogen receptors in the rat testis, *Mol. Cell. Endocrinol.,* 29, 287, 1983.
78. Aragona, C. and Friesen, H. G., Specific prolactin binding sites in the prostate and testis of rats, *Endocrinology,* 97, 677, 1975.

79. Manlin, M. L. and Yount, A. P., Prolactin binding in the rat ventral prostate, *Endocrinol. Res. Commun.*, 2, 489, 1975.
80. Kledzik, G. S., Marshall, S., Campbell, G. A., Gelato, M., and Meites, J., Effects of castration, testosterone, estradiol and prolactin on specific prolactin-binding activity in ventral prostate of male rats, *Endocrinology*, 98, 373, 1976.
81. Sakai, S., Enami, J., Nandi, S., and Banerjee, M. R., Prolactin receptor on dissociated mammary epithelial cells at different stages of development, *Mol. Cell. Endocrinol.*, 12, 285, 1978.
82. Suard, Y. M. L., Kraehenbuhl, J.-P., and Aubert, M. L., Dispersed mammary epithelial cells. Receptors of lactogenic hormones in virgin, pregnant, and lactating rabbits, *J. Biol. Chem..*, 254, 10466, 1979.
83. McNeilly, A. S. and Friesen, H. G., Binding of prolactin to the rabbit mammary gland during pregnancy, *J. Endocrinol.*, 74, 507, 1977.
84. Costlow, M. E., Gallagher, P. E., and Koseki, Y., Prolactin receptors in primary cultures of carcinogen-induced rat-mammary tumors, *Mol. Cell. Endocrinol.*, 14, 81, 1979.
85. Bohnet, G. H., Gomez, F., and Friesen, H. G., Prolactin and estrogen binding sites in the mammary gland of the lactating and non-lactating rat, *Endocrinology*, 101, 1111, 1977.
86. Djiane, J. and Durand, P., Prolactin-progesterone antagonism in self regulation of prolactin receptors in the mammary gland, *Nature (London)*, 266, 641, 1977.
87. Harigaya, T., Sakai, S., and Kohmoto, K., Induction of mammary prolactin receptors and lactose synthesis after ovariectomy in the pregnant mouse, *Endocrinol. Jpn.*, 25, 157, 1978.
88. Djiane, J., Houdebine, L. M., and Kelly, P. A., Correlation between prolactin-receptor interaction, down-regulation of receptors, and stimulation of casein and deoxyribonucleic acid biosynthesis in rabbit mammary gland explants, *Endocrinology*, 110, 791, 1982.
89. Harigaya, T., Sakai, S., Kohmoto, K., and Shoda, Y., Influence of glucocorticoids on mammary prolactin receptors in pregnant mice after ovariectomy, *J. Endocrinol.*, 94, 149, 1982.
90. Sakai, S., Bowman, P. D., Yang, J., McCormick, K., and Nandi, S., Glucocorticoid regulation of prolactin receptors on mammary cells in culture, *Endocrinology*, 104, 1447, 1979.
91. Turkington, R. W., Prolactin receptors in mammary carcinoma cells, *Cancer Res.*, 34, 758, 1974.
92. Costlow, M. E., Buschow, R. A., Richert, N. J., and McGuire, W. L., Prolactin and estrogen binding in transplantable hormone-dependent and autonomous rat mammary carcinoma, *Cancer Res.*, 35, 970, 1975.
93. Costlow, M. E., Sluyser, M., and Gallagher, P. E., Prolactin receptors in mammary tumors of GR mice, *Endocrinol. Res. Commun.*, 4, 285, 1977.
94. Costlow, M. E., Buschow, R. A., and McGuire, W. L., Prolactin receptors in estrogen receptor-deficient mammary carcinoma, *Science*, 184, 85, 1974.
95. Costlow, M. E., Buschow, R. A., and McGuire, W. L., Prolactin receptors in 7,12-dimethylbenz(a)anthracene-induced mammary tumors following endocrine ablation, *Cancer Res.*, 36, 3941, 1976.
96. Holdaway, I. M. and Friesen, H. G., Correlation between hormone binding and growth response of mammary tumor, *Cancer Res.*, 36, 1562, 1976.
97. DeSombre, E. R., Kledzik, G., Marshall, S., and Meites, J., Estrogen and prolactin receptor concentrations in rat mammary tumors and response to endocrine ablation, *Cancer Res.*, 36, 354, 1976.
98. Kelly, P. A., Bradley, C., Shiu, R. P. C., Meites, J., and Friesen, H. G., Prolactin binding to rat mammary tumor tissue (38197), *Proc. Soc. Exp. Biol. Med.*, 146, 816, 1974.
99. Smith, R. D., Hilf, R., and Senior, A. E., Prolactin binding to R3230AC mammary carcinoma and liver in hormone-treated and diabetic rats, *Cancer Res.*, 37, 595, 1977.
100. Smith, R. D., Hilf, R., and Senior, A. E., Prolactin binding to mammary gland, 7,12-dimethylbenz(a)anthracene-induced mammary tumors, and liver in rats, *Cancer Res.*, 36, 3726, 1976.
101. Smith, R. D., Hilf, R., and Senior, A. E., Prolactin binding to 7,12-dimethylbenz(a)anthracene-induced mammary tumors and liver in diabetic rats, *Cancer Res.*, 37, 4070, 1977.
102. Quadri, S. K., Kledzik, G. S., and Meites, J., Counteraction by prolactin of androgen-induced inhibition of mammary tumor growth in rats, *J. Natl. Cancer Inst.*, 52, 875, 1984.
103. Meites, J., Cassell, E., and Clark, J., Estrogen inhibition of mammary tumor growth in rats; counteraction by prolactin, *Proc. Soc. Exp. Biol. Med.*, 137, 1225, 1971.
104. Costlow, M. E., Buschow, R. A., and McGuire, W. L., Prolactin receptors and androgen-induced regression of 7,12-dimethylbenz(a)anthracene-induced mammary carcinoma, *Cancer Res.*, 36, 3324, 1976.
105. Kelly, P. A., LeBlanc, G., Ferland, L., Labrie, F., and DeLean, A., Androgen inhibition of basal and estrogen-stimulated prolactin binding in rat liver, *Mol. Cell. Endocrinol.*, 9, 195, 1977.
106. Kledzik, G. S., Bradley, C. J., Marshall, S., Campbell, G. A., and Meites, J., Effects of high doses of estrogen on prolactin-binding activity and growth of carcinogen-induced mammary cancers in rats, *Cancer Res.*, 36, 3265, 1976.

107. Dave, J. R., Knazek, R. A., and Liu, S. C., Prolactin modifies the fluidity of rat liver membranes, *Biochem. Biophys. Res. Commun.*, 100, 45, 1981.
108. Chomczynski, P. and Topper, Y. J., A direct effect of prolactin and placental lactogen on mammary epithelial nuclei, *Biochem. Biophys. Res. Commun.*, 60, 56, 1974.
109. Kahn, M. N., Posner, B. I., Verma, A. K., Kahn, R. J., and Bergeron, J. J. M., Intracellular hormone receptors: evidence for insulin and lactogen receptors in a unique vesicle fraction sedimenting in lysosome fractions of rat liver, *Proc. Natl. Acad. Sci. U.S.A.*, 78, 4980, 1981.
110. Basset, M., Smith, G. D., Pease, R., and Peters, T. J., Uptake and processing of prolactin by alternative pathways in rat liver, *Biochim. Biophys. Acta*, 769, 79, 1984.
111. Costlow, M. E. and Rodgers, Q. E., Subcellular localization of cryptic prolactin receptors in mammary tumor cells, *Exp. Cell Res.*, 163, 159, 1986.
112. Djiane, J., Delouis, C., and Kelly, P. A., Prolactin receptors in organ culture of rabbit mammary gland: effect of cycloheximide and prolactin, *Proc. Soc. Exp. Biol. Med.*, 162, 342, 1979.
113. Kelly, P. A., Posner, B. I., and Friesen, H. G., Effect of hypophysectomy, ovariectomy and cycloheximide on specific binding sites for lactogenic hormones in rat liver, *Endocrinology*, 97, 1408, 1975.
114. Shiu, R. P., Elsholtz, H. P., Tanaka, T., Friesen, H. G., Gout, P. W., Beer, C. T., and Noble, R. L., Receptor-mediated mitogenic action of prolactin in a rat lymphoma cell line, *Endocrinology*, 113, 159, 1983.
115. Houdebine, L. M., Effect of various lysosomotropic agents and microtubule disrupting drugs on the lactogenic and the mammogenication of prolactin, *Eur. J. Cell Biol.*, 22, 755, 1980.
116. Nolin, J. M., The prolactin incorporation cycle of the milk secretory cell. An integral component of the prolactin response cycle, *J. Histochem. Cytochem.*, 27, 1203, 1979.
117. Dusanter-Fourt, I., Djiane, J., Kelly, P. A., Houdebine, L. M., and Teyssot, B., Differential biological activities between mono- and bivalent fragments of anti-prolactin receptor antibodies, *Endocrinology*, 114, 1021, 1984.
118. Djiane, J., Clauser, H., and Kelly, P. A., Rapid down-regulation of prolactin receptors in mammary gland and liver, *Biochem. Biophys. Res. Commun.*, 90, 1371, 1979.
119. Djiane, J., Houdebine, L. M., and Kelly, P. A., Down-regulation of prolactin receptors in rabbit mammary gland: differential subcellular localization, *Proc. Soc. Exp. Biol. Med.*, 168, 378, 1981.
120. Djiane, J., Houdebine, L. M., and Kelly, P. A., Correlation between prolactin-receptor interaction, down-regulation of receptors, and stimulation of casein and deoxyribonucleic acid biosynthesis in rabbit mammary gland explants, *Endocrinology*, 110, 791, 1982.
121. Houdebine, L. M. and Djiane, J., Effects of lysosomotropic agents, and of microfilament- and microtubule-disrupting drugs on the activation of casein-gene expression by prolactin in the mammary gland, *Mol. Cell. Endocrinol.*, 17, 1, 1980.
122. Houdebine, L. M., Djiane, J., and Clauser, M., Role of lysosomes, microtubules and microfilaments in the mechanism of the lactogenic action of prolactin in the rabbit mammary gland, *C.R. Acad. Sci.*, 289, 697, 1979.
123. Frazier, R. P. and Costlow, M. E., Prolactin stimulation of ornithine decarboxylase activity in cultured rat mammary tumors, *Exp. Cell Res.*, 138, 39, 1982.
124. Djiane, J., Kelly, P. A., and Houdebine, L. M., Effects of lysosomotropic agents, cytochalasin b and colchicine on the "down-regulation" of prolactin receptors in mammary gland explants, *Mol. Cell. Endocrinol.*, 18, 87, 1980.
125. Silverstein, A. M. and Richards, J. F., Characterization of prolactin binding by membrane preparations from rat liver, *Biochem. J.*, 178, 743, 1979.
126. Bhattacharya, A. and Vonderhaar, B. K., Membrane modification differentially affects the binding of the lactogenic hormones human growth hormone and ovine prolactin, *Proc. Natl. Acad. Sci. U.S.A.*, 78, 5704, 1981.
127. Bhattacharya, A. and Vonderhaar, B. K., Phospholipid methylation stimulates lactogenic binding in mouse mammary gland membranes, *Proc. Natl. Acad. Sci. U.S.A.*, 76, 4489, 1979.
128. Dave, J. R. and Witorsch, R. J., Modulation of prolactin binding sites in vitro by membrane fluidizers. I. Effects on adult rat ventral prostatic membranes, *Biochem. Biophys. Res. Commun.*, 113, 220, 1983.
129. Knazek, R. A. and Liu, S. C., Dietary essential fatty acids are required for maintenance and induction of prolactin receptors, *Proc. Soc. Exp. Biol. Med.*, 162, 346, 1979.
130. Knazek, R. A., Liu, S. C., Dave, J. R., Christy, R. J., and Keller, J. A., Indomethacin causes a simultaneous decrease of both prolactin binding and fluidity of mouse liver membranes, *Prostaglandins Med.*, 6, 403, 1981.
131. Muccioli, G., Bellussi, G., Lando, D., and DiCarlo, R., Modifications in prolactin binding capacity in the rat liver induced by non-steroidal anti-inflammatory drugs, *J. Pharm. Pharmacol.*, 36, 95, 1984.
132. Dave, J. R. and Knazek, R. A., Prostaglandin I2 modifies both prolactin binding capacity and fluidity of mouse liver membranes, *Proc. Natl. Acad. Sci. U.S.A.*, 77, 6597, 1980.

133. Dave, J. R., Knazek, R. A., and Liu, S. C., Arachidonic acid, bradykinin and phospholipase A_2 modify both prolactin binding capacity and fluidity of mouse hepatic membranes, *Biochem. Biophys. Res. Commun.*, 130, 727, 1981.
134. Bhattacharya, A. and Vonderhaar, B. K., Thyroid hormone regulation of prolactin binding, *Biochem. Biophys. Res. Commun.*, 88, 1405, 1979.
135. Gavish, M., Amit, T., Barkey, R. J., and Youdim, M. B., Characterization of CHAPS-solubilized prolactin receptors induced by estradiol in liver of male rats. Unmasking of cryptic sites by CHAPS, *FEBS Lett.*, 154, 81, 1983.
136. Koppelman, M. C. and Dufau, M. L., Prolactin receptors in luteinized rat ovaries: unmasking of specific binding sites with detergent treatment, *Endocrinology*, 111, 1350, 1982.
137. Costlow, M. E. and Hample, A., Prolactin receptors in cultured rat mammary tumor cells. Unmasking of cell surface receptors by energy depletion, *J. Biol. Chem.*, 257, 6971, 1982.
138. Aragona, C., Bohnet, H. G., and Friesen, H. G., Prolactin binding sites in the male rat liver following castration, *Endocrinology*, 99, 1017, 1976.
139. Borst, D. W., and Sayare, M., Photoactivated crosslinking of prolactin to hepatic membrane binding sites, *Biochem. Biophys. Res. Commun.*, 105, 194, 1982.
140. Calvo, J. C., Finocchiaro, L., Luthy, I., Charreau, E. H., Calandra, R. S., Engstrom, B., and Hansson, V., Specific prolactin binding in the rat adrenal gland: its characterization and hormonal regulation, *J. Endocrinol.*, 89, 317, 1981.
141. Barkey, R. J., Shani, J., Amit, T., and Barzilai, D., Characterization of the specific binding of prolactin to binding sites in the seminal vesicle of the rat, *J. Endocrinol.*, 80, 181, 1979.
142. Hayden, T. J., Bonney, R. C., and Forsyth, I. A., Ontogeny and control of prolactin receptors in the mammary gland and liver of virgin, pregnant and lactating rats, *J. Endocrinol.*, 80, 259, 1979.
143. Gustafsson, J. A., Mode, A., Norstedt, G., Eneroth, P., and Hokfelt, T., Central control of prolactin and estrogen receptors in rat liver — expression of a novel endocrine system, the hypothalamo-pituitary-liver axis, *Annu. Rev. Pharmacol. Toxicol.*, 23, 259, 1983.
144. Williams, G. H., Hammond, J. M., Weisz, J., and Mortel, R., Binding sites for lactogenic hormone in the rat uterus, *Biol. Reprod.*, 18, 697, 1978.
145. Midgley, A. R., Autoradiographic analysis of gonadotropin binding to rat ovarian tissue section, *Adv. Exp. Med. Biol.*, 36, 365, 1973.
146. Marshall, S., Bruni, J. F., and Meites, J., Effects of hypophysectomy, thyroidectomy, and thyroxine on specific prolactin receptor sites in kidneys and adrenals of male rats, *Endocrinology*, 104, 390, 1979.
147. Marshall, S., Huang, H. H., Kledzik, G. S., Campbell, G. A., and Meites, J., Glucocorticoid regulation of prolactin receptors in kidneys and adrenals of male rats, *Endocrinology*, 102, 869, 1978.
148. Mountjoy, K., Cowden, E. A., Dobbie, J. W., and Ratcliffe, J. G., Prolactin receptors in the rat kidney, *J. Endocrinol.*, 87, 47, 1980.
149. Donatsch, P. and Richardson, B., Localization of prolactin in rat kidney tissue using a double-antibody technique, *J. Endocrinol.*, 66, 101, 1975.
150. Charreau, E. H., Attramadal, A., Torjese, P. A., Calandra, R., Purvis, K., and Hansson, V., Androgen stimulation of prolactin receptors in rat prostate, *Mol. Cell. Endocrinol.*, 7, 1, 1977.
151. Witorsch, R. J., Immunohistochemical studies of prolactin binding in sex accessory organs of the male rat, *J. Histochem. Cytochem.*, 26, 565, 1978.
152. Barkey, R. J., Shani, J., Amit, T., and Barzilai, D., Specific binding of prolactin to seminal vesicle, prostate and testicular homogenates of immature, mature and aged rats, *J. Endocrinol.*, 74, 163, 1977.
153. Tesone, M., Oliveira-Filho, R. M., and Charreau, E. H., Prolactin binding in rat Langerhans islets, *J. Recept. Res.*, 1, 355, 1980.
154. Frantz, W. L., Payne, P., and Dombroske, O., Binding of ovine ^{125}I-prolactin to cultured anterior pituitary tumour cells and normal cells, *Nature (London)*, 255, 636, 1975.
155. Klemcke, H. G., Bartke, A., and Borer, K. T., Regulation of testicular prolactin and luteinizing hormone receptors in golden hamsters, *Endocrinology*, 114, 594, 1984.
156. Carr. F. E. and Jaffe, R. C., Characterization of solubilized prolactin receptors from *Rana catesbeiana* tadpole tissues, *Mol. Cell. Endocrinol.*, 25, 317, 1982.
157. Jaffe, R. C., *Rana catesbeiana* tail fin lactogenic hormone receptor: hydrodynamic characterization, *Comp. Biochem. Physiol.*, 76, 529, 1983.
158. White, B. A., Occurrence and binding affinity of prolactin receptors in amphibian tissues, *Gen. Comp. Endocrinol.*, 45, 153, 1981.
159. Posner, B. I., Kelly, P. A., Shiu, R. P., and Friesen, H. G., Studies of insulin, growth hormone and prolactin binding; tissue distribution, species variation and characterization, *Endocrinology*, 95, 521, 1974.
160. Stewart, F. and Tyndale-Biscoe, C. H., Prolactin and luteinizing hormone receptors in marsupial corpora lutea: relationship to control of luteal function, *J. Endocrinol.*, 92, 63, 1982.

161. Rolland, R., Gunsalus, G. L., and Hammond, J. M., Demonstration of specific binding of prolactin by porcine corpora lutea, *Endocrinology,* 98, 1083, 1976.
162. Rose, J., Stormshak, F., Adair, J., and Oldfield, J. E., Prolactin binding sites in the uterus of the mink, *Mol. Cell.Endocrinol.,* 31, 131, 1983.
163. Gertler, A., Ashkenazi, A., and Madar, Z., Binding sites of human growth hormone and ovine and bovine prolactins in the mammary gland and the liver of lactating dairy cow, *Mol. Cell. Endocrinol.,* 34, 51, 1984.
164. Poindexter, A. N., Buttram, V. C., Jr., Besch, P. K., and Smith, R. G., Prolactin receptors in the ovary, *Fertil. Steril.,* 31, 273, 1979.
165. Kledzik, G. S., Marshall, S., Gelato, M., Campbell, G., and Meites, J., Prolactin binding activity on the crop sacs of juvenile, mature, parent and prolactin-injected pigeons, *Endocrinol. Res. Commun.,* 2, 345, 1975.
166. Shani, J., Barkey, R. J., and Amit, T., Endogenous prolactin maintains its own binding sites in the pigeon crop sac mucosa, *J. Recept. Res.,* 82, 407, 1981.
167. Forsyth, I. A., Buntin, J. D., and Nicoll, C. S., A pigeon crop sac radio-receptor assay for prolactin, *J. Endocrinol.,* 79, 349, 1978.
168. Wahlstrom, T., Huhtaniemi, I., Hovatta, O., and Seppala, M., Localization of luteinizing hormone, follicle-stimulating hormone, prolactin and their receptors in human and rat testis using immunohistochemistry and radio-receptor assay, *J. Clin. Endocrinol. Metab.,* 57, 825, 1983.
169. Ben-David, M. and Schenker, J. G., Human ovarian receptors to human prolactin: implications in infertility, *Fertil. Steril.,* 38, 182, 1982.
170. Kelly, P. A., Posner, B. I., and Friesen, H. G., Effects of hypophysectomy, ovariectomy and cycloheximide on specific binding sites for lactogenic hormones in rat liver, *Endocrinology,* 97, 1408, 1975.
171. Herington, A. C., Graham, J., and Healy, D. L., The presence of lactogen receptors in human chorion laeve, *J. Clin. Endocrinol. Metab.,* 51, 1466, 1980.
172. McWey, L. A., Singhas, C. A., and Rogol, A. D., Prolactin binding sites on human chorion-decidua tissue, *Am. J. Obstetr. Gynecol.,* 144, 283, 1982.
173. Keenan, E. J., Kemp, E. D., Ramsey, E. E., Garrison, L. B., Pearse, H. D., and Hodges, C. V., Specific binding of prolactin by the prostate gland of the rat and man, *J. Urol.,* 122, 43, 1979.
174. Leake, A., Chisholm, G. D., and Habib, F. K., Characterization of the prolactin receptor in human prostate, *J. Endocrinol.,* 99, 321, 1983.
175. Nagasawa, H. and Yanai, R., Mammary gland prolactin receptor and pituitary prolactin secretion in lactating mice with different lactational performance, *Acta Endocrinol.,* 88, 94, 1978.
176. Sakai, S. and Banerjee, M. R., Glucocorticoid modulation of prolactin receptors on mammary cells of lactating mice, *Biochim. Biophys. Acta,* 582, 79, 1979.
177. Dave, J. R. and Knazek, R. A., Changes in the prolactin-binding capacity of mouse hepatic membranes with development and aging, *Mech. Ageing Dev.,* 23, 235, 1983.
178. Marshall, S., Bruni, J. F., and Meites, J., Prolactin receptors in mouse liver; species differences in response to estrogenic stimulation, *Proc. Soc. Exp. Biol. Med.,* 159, 256, 1978.
179. Posner, B. I., Characterization and modulation of growth hormone and prolactin binding in mouse liver, *Endocrinology,* 98, 645, 1976.
180. Frantz, W. L., MacIndoe, J. H., and Turkington, R. W., Prolactin receptors: characteristics of the particulate fraction binding activity, *J. Endocrinol.,* 60, 485, 1974.
181. Church, W. R. and Ebner, K. E., Solubilization of prolactin receptor by a Zwitterionic detergent, *Experientia,* 38, 434, 1982.
182. Haeuptle, M.-T., Aubet, M. L., Djiane, J., and Kraebenbuht, J.-P., Binding sites for lactogenic and somatogenic hormones from rabbit mammary gland and liver, *J. Biol. Chem.,* 258, 305, 1983.
183. Djiane, J., Durand, P., and Kelly, P. A., Evolution of prolactin receptors in rabbit mammary gland during pregnancy and lactation, *Endocrinology,* 100, 1348, 1977.
184. Richards, S. R., Malarkey, W. B., Nicol, S. J., and Matthews, R. H., Assessment of mammary lactogenic receptor changes in pregnant rabbits, *Am. J. Obstetr. Gynecol.,* 149, 159, 1984.
185. Orgebin-Crist, M. C. and Djiane, J., Properties of a prolactin receptor from the rabbit epididymis, *Biol. Reprod.,* 21, 135, 1979.
186. DiCarlo, R. and Muccioli, G., Presence of specific prolactin binding sites in the rabbit hypothalamus, *Life Sci.,* 28, 2299, 1981.
187. DiCarlo, R. and Muccioli, G., Changes in prolactin binding sites in the rabbit hypothalamus induced by physiological and pharmacological variations of prolactin serum levels, *Brain Res.,* 230, 445, 1981.
188. Bolander, F. F., Jr., Hurley, T. W., Handwerger, S., and Fellows, R. E., Localization and specificity of binding of subprimate placental lactogen in rabbit tissues, *Proc. Natl. Acad. Sci., U.S.A.,* 73, 2932, 1976.
189. Waters, M. J., McNeilly, A. S., Ohgo, S., and Friesen, H. G., Prolactin receptor content of rabbit milk, *Endocrinology,* 107, 16, 1980.
190. Costlow, M. E., Unpublished observations.

Chapter 3

ACTIONS OF PROLACTIN IN THE BRAIN

James L. Voogt

TABLE OF CONTENTS

I. Introduction 28

II. Prolactin in the Brain and CSF 28

III. Electrical Activity of the Hypothalamus Following Prolactin 30

IV. Brain Dopamine: Mechanism for Prolactin Regulation of Its Own Secretion 31

V. Prolactin Alters Hypothalamic Luteinizing Hormone-Releasing Hormone 35

VI. Prolactin and Male Sex Behavior 36

References 37

INTRODUCTION

The early observations in the late 1960s and early 1970s that placement of prolactin into the median eminence of the rat hypothalamus dramatically altered prolactin secretion[1-4] has led to hundreds of studies appearing in the literature describing the influence of prolactin on the brain. This chapter begins with an accounting of the evidence that prolactin is indeed found in the brain and cerebrospinal fluid (CSF). Next, the effect of prolactin on the electrical activity of neurons is described. A section follows which is devoted to the physiological importance of prolactin short-loop feedback. The effect of prolactin on the dopaminergic system has been studied more than any other area, particularly the tuberoinfundibular dopamine pathway (TIDA), which extends from the arcuate nucleus to the median eminence of the hypothalamus. Since dopamine is very important in the regulation of pituitary prolactin, the mechanism for feedback of prolactin on itself probably involves dopaminergic neurons in the hypothalamus. Prolactin also has effects on other systems in the brain, particularly neurons containing luteinizing hormone release hormone (LHRH). Finally, a section dealing with the influence of prolactin on male sex behavior is included.

II. PROLACTIN IN THE BRAIN AND CSF

Numerous reports have appeared which clearly show that prolactin is found in the CSF, especially in that of humans and other primates.[5-8]Rats with very high serum prolactin due to pituitary tumors also have elevated prolactin levels in the CSF.[9] It is not known for certain what route(s) prolactin uses to reach the CSF. It has been suggested that prolactin may be filtered across the blood-CSF barrier, originating in the peripheral blood.[7,10] Others found indirect evidence to suggest that prolactin reaches the CSF via a retrograde flow of blood into the pituitary portal vessels, with prolactin reaching the CSF following diffusion across the median eminence.[11] However, this route may only allow a finite amount of prolactin to reach the CSF, as suggested by the study in which rat prolactin levels in excess of several thousand nanograms per milliliter did not elevate CSF prolactin concentrations any further than that found in rats with much lower prolactin levels.[9]

There is a significant amount of evidence showing a correlation between peripheral and CSF blood levels of prolactin in monkeys and humans.[7,12-14] Several experiments have been done in rhesus monkeys in which a stimulus to release prolactin was given and the CSF and blood levels of prolactin monitored over time.[12-14] In general, most reports indicate the rise in CSF prolactin lags somewhat behind that of blood prolactin. Intravenous injection of morphine sulfate caused a brisk, marked rise in plasma prolactin in monkeys, followed by a later, smaller rise in CSF prolactin. Infusion of human prolactin into monkeys resulted in a pattern of plasma and CSF prolactin similar to that following morphine.[13] The ratio of plasma to CSF prolactin was reported to be about five in rhesus monkeys, and this ratio remained constant with steady state or persistent hyperprolactinemia induced by sulpiride. Acute elevations in plasma prolactin were followed by a similar, slow increase in prolactin in the CSF.[14] Injection of thyrotropin-releasing hormone (TRH) into rhesus monkeys increased both plasma and lumbar CSF prolactin levels (Figure 1); however, the rise in CSF prolactin was of lower magnitude and delayed time compared to plasma prolactin.[12] Acute rises in plasma prolactin induced by haloperidol were correlated with smaller and delayed increases in CSF prolactin.[15] In summary, the rhesus monkey has prolactin levels in the CSF that reflect the dynamic changes occurring in the peripheral circulation, albeit at a slower rate and lower magnitude. Thus, there must be some limit to the rate at which prolactin diffuses into the CSF.

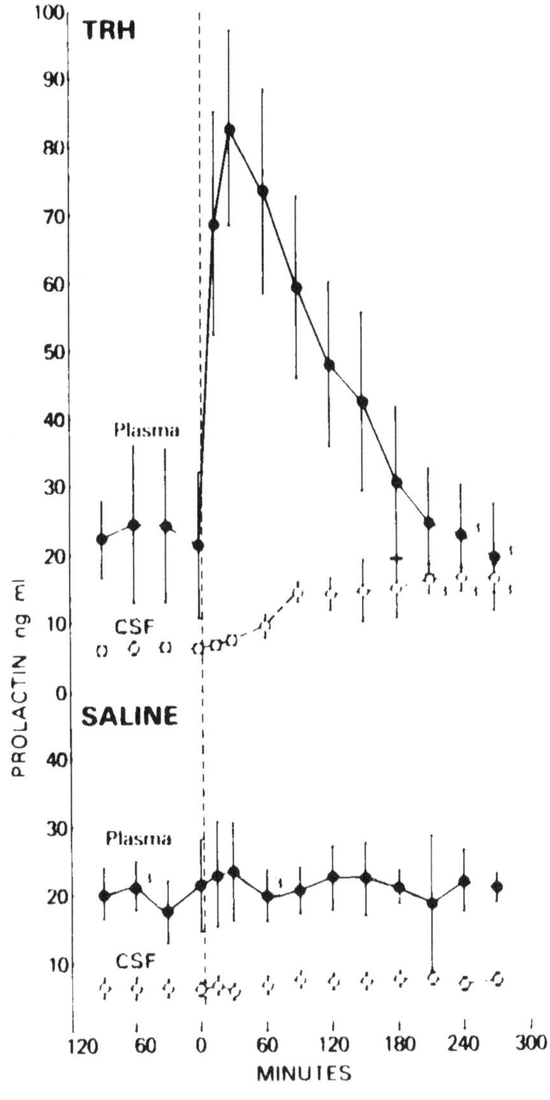

FIGURE 1. The simultaneous comparison of prolactin in plasma and lumbar CSF after i.v. injection of either TRH (100 μg) or saline at 0 time. Each point represents the means ± S.E.M. for four, or where indicated, three monkeys. (From Kalin, N. H. et al., *Life Sci.*, 31, 159, 1982. With permission.)

In the human, much less is known about the relationship of CSF to blood prolactin under physiological conditions. One of the questions asked is what mechanisms exist for the entry of prolactin into the CSF? A study was recently reported in which serum to CSF ratios of pituitary-derived prolactin were compared to placentally derived placental lactogen (hPL) and chorionic gonadotropin (hCG) in term pregnant women.[16] Since hPL shares structural homology with prolactin, serum to CSF ratios of hPL compared to prolactin should give some insight into whether prolactin has an access to CSF via a mechanism in addition to the blood-CSF barrier. The serum to CSF ratio for hCG was 289, with a highly significant correlation coefficient between serum and CSF levels. Similar results were found for hPL, although the ratio of serum to CSF hPL was only 23. The serum to CSF ratio for prolactin in these subjects was even lower, 11, and there was no correlation between prolactin levels in serum and CSF. It would appear that these results indicating a lack of correlation of serum and CSF prolactin levels are contrary to what was found in the rhesus monkey and what has been found in patients with neurological and psychiatric diseases.[7,11] However, in this study by Peake et al.,[16] serum prolactin levels were much higher, in general, than found in patients in other reports. It is possible that there is a limit to how much prolactin can enter the CSF, regardless of the serum levels. This has been seen in pituitary tumor-bearing rats.[9] The lower ratio of serum to CSF prolactin compared to hPL or hCG suggests that there are additional mechanisms for transporting prolactin into the CSF than passage across the blood-CSF barrier. Receptors for prolactin have been found in the rat choroid plexus and may play a role in transporting prolactin from the blood to the CSF.[17,18] Many different peptidergic pathways in the brain have been described in recent years. There now is some evidence that prolactin is present in CNS neurons. Fuxe et al.[20] were the first to describe, using antibodies to rat prolactin in immunohistochemical studies, the presence of a prolactin-like protein in nerve terminals in many hypothalamic areas. In addition, hypophysectomy did not affect the neuronal storage of this protein. Similar findings were reported by Toubeau et al.[21] using immunohistochemical techniques. They also found no difference in labeling between female and male rats, nor following hypophysectomy. Prolactin-like material was also demonstrated in rat brain[22] and brain and pituitary of the African freshwater fish *Calamoichthys calabaricus*.[23] In these studies, anti-ovine prolactin as the primary antibody was used. Its binding was completely blocked by absorption with highly purified rat prolactin, suggesting that the antibody is directed against some portion of the prolactin molecule. Immunoreactive perikarya were found in the supraoptic nucleus, paraventricular nucleus, anterior hypothalamus, anterior commissural nucleus, and preoptic nucleus. The authors of the study in the rat[22] suggest that only a segment of the prolactin molecule is present in these neurons, and may function as a neuromodulator or neurotransmitter. Data from a study by Harlan et al.[24] showed prolactin-like immunoreactivity in cells in the mediobasal hypothalamus and fibers in the dorsal midbrain, suggesting that the hypothalamus produces a prolactin-like substance which travels via nerve fibers to the dorsal midbrain. If the prolactin-like molecule found in the brain is indeed prolactin, and hypophysectomy does not eliminate it, as these studies have shown, then prolactin may be synthesized in the brain. To support this hypothesis, Schachter et al.[25] searched for prolactin mRNA. They found a hybridizable RNA in both male and female hypothalami which was the same size as pituitary prolactin mRNA, although its concentration was less.

Thus, it would appear that a source of prolactin in the brain can be either the peripheral blood via the CSF or synthesis of prolactin by neurons or other brain tissue. The questions remaining concern the regulation and function of prolactin found in the brain. The remainder of this chapter will describe some parameters in the brain affected by prolactin originating from outside the brain.

III. ELECTRICAL ACTIVITY OF THE HYPOTHALAMUS FOLLOWING PROLACTIN

Only three studies have been reported on the effect of exogenous prolactin on the unit activity of neurons in the hypothalamus.[26-28] Clemens et al.[26] were the first to show that prolactin injected intravenously altered the electrical activity of neurons in the hypothalamus of unanesthetized rabbits. Out of 40 neurons tested with prolactin, 14 decreased, 11 increased, and 15 showed no change in activity. To rule out the possibility that prolactin may have affected activity of neurons in other locations, which in turn induced excitation or inhibition of hypothalamic neurons, Yamada[27] applied prolactin iontophoretically to neurons in various areas of the brain of the rat. Similar to what Clemens et al.[26] found, some hypothalamic neurons were activated, while others were inhibited by prolactin. Almost none of the neurons in the preoptic area were affected, and no neurons of the cerebral cortex responded to prolactin. More recently, Chan et al.[28] investigated the effect of prolactin on the electrical activity of neurons of the ventromedial hypothalamus of ovariectomized rats. A dose-related increase in firing rate with iontophorectically applied prolactin was found in about half the neurons. About 9% showed a decrease and 42% showed no change following prolactin. Priming the ovariectomized rats with estrogen-progesterone did not alter this response.

A recent report by Ramaswamy et al.[29] indicated that prolactin is capable of producing a dose-dependent analgesic effect in mice. Because the effect of prolactin was antagonized by naltrexone, opiate mechanisms may be involved in this effect of prolactin.

There are many other peptidergic or monoaminergic systems in the brain that may be the mechanism by which prolactin influences some aspect of brain function. No one has shown any link between altered electrical activity in response to prolactin and a change in synthetic enzyme or neurotransmitter activity, even though prolactin has been shown to alter many substances found in neurons. This very difficult problem will require simultaneous measurement of electrical activity and neurotransmitter function following prolactin uptake or secretion by brain tissue.

IV. BRAIN DOPAMINE: MECHANISM FOR PROLACTIN REGULATION OF ITS OWN SECRETION

Considerable evidence has emerged indicating that hypothalamic dopamine is a primary inhibitor of prolactin secretion by acting directly on the prolactin cell in the anterior pituitary.[30-32] Somewhat preceding these observations were experiments in which prolactin was shown to inhibit its own secretion, primarily via an action in the hypothalamus.[1-4,33] Placement of small amounts of ovine prolactin into the median eminence of the hypothalamus resulted in decreased circulating prolactin levels associated with pregnancy,[34] pseudopregnancy,[35] suckling, and the proestrous surge.[4] Systemic injection of various prolactin preparations also have been effective in the inhibition of prolactin in response to various stimuli. TRH-induced,[36] suckling-induced,[37] and stress-induced[36,37] prolactin secretion as well as basal secretion[38] have been shown to be inhibited by prolactin. The minimum exposure time that appears to be necessary for systemically injected prolactin to have an effect is 3 to 4 hr.[38] By injecting microgram amounts of ovine or bovine prolactin into the third ventricle of the ovariectomized rat, prolactin was lowered significantly in 15 to 30 min.[39]

The question remains as to what is the most physiological route for administration of prolactin in order to demonstrate feedback. Prolactin in the peripheral circulation has ready access to hypothalamic neurons, including the TIDA of the arcuate nucleus, which extend to the median eminence. It is the TIDA neurons which are believed to be important in controlling prolactin release. Thus, prolactin could be taken up into the median eminence from the peripheral circulation. Alternatively, prolactin could reach the median eminence by way of retrograde blood flow in the portal vessels.[40] Large amounts (4900 ng/mℓ) of prolactin are found in portal blood delivered to the median eminence in the adult male rat.[41] This prolactin could then perfuse the arcuate nucleus via capillaries which branch off the primary portal plexus,[42] exposing the TIDA neurons to prolactin. The discussion earlier, which reported possible prolactin-secreting neurons in the hypothalamus,[20-24] suggests an alternative delivery of prolactin. Thus, the physiological route prolactin takes in regulating its own secretions is not clear. However, there is considerable evidence that the mechanism by which the prolactin short-loop feedback system works involves hypothalamic dopamine.

Given the evidence cited above, that hypothalamic dopamine is a chronic inhibitor of prolactin and prolactin inhibits its own secretion by an action in the hypothalamus, it was obvious to many researchers that one possible mechanism for this effect of prolactin is via an increase in release of dopamine from TIDA neurons. Hence, the hypothesis was developed and tested to show that there was, in fact, an increase in activity of TIDA neurons and dopamine in portal blood in response to prolactin. There have been several excellent reviews recently which thoroughly discussed the anatomical distribution of dopamine neurons, methods for estimating dopaminergic activity, influence of pharmacological and endocrinological manipulation, and physiological states on dopamine neurons.[43,44] Therefore, this chapter will only focus on the most critical findings regarding prolactin feedback on dopamine which were reported prior to the reviews, and will give more attention to those papers appearing after 1981.

Various methods to increase prolactin have been used to show an effect on dopamine-containing neurons. These include systemic on intraventricular injections, use of pituitary transplants or pituitary tumors secreting large amounts of prolactin, or administration of neuroleptic drugs that elevate or depress serum prolactin levels.

An increase in dopamine turnover in the median eminence following injection of ovine or rat prolactin was first reported using histochemical techniques,[45] and later confirmed biochemically.[46] Intracerebroventricular (i.c.v.) injection of prolactin increased dopamine turnover[47,48] and release into hypophysial portal blood.[49] The increase in activity of tyrosine hydroxylase, the rate-limiting enzyme for catecholamine synthesis, following i.c.v. injection of prolactin[50] suggests that part of the mechanism of action of prolactin involves stimulation of protein synthesis in the neurons. This is supported by the finding that cycloheximide, an inhibitor of protein synthesis, prevented the increase in dopamine synthesis rate in the median eminence following prolactin.[48]

Several studies have investigated the effect of chronic elevation of prolactin in the peripheral circulation on the dopamine system. Transplantation of anterior pituitaries resulted in elevated serum prolactin, which was correlated with increased activity of tyrosine hydroxylase in the arcuate nucleus-median eminence area[50-52] and synthesis of hypothalamic dopamine, but not norepinephrine.[53] If extremely high prolactin levels were attained, using rats with prolactin-secreting tumors, there were increased amounts of dopamine in portal blood[54] and an increased rate of conversion of ^{3}H-tyrosine to dopamine in the hypothalamus.[55]

There is considerable interest in changes that occur in the TIDA neurons after long-term exposure to high levels of prolactin. This may be important in patients who have prolactin-secreting tumors or in individuals who use neuroleptic drugs which influence central dopaminergic systems, and thereby have elevated prolactin levels. Two commonly employed methods to elevate prolactin chronically include administration of estrogens and the use of rats which have pituitary tumors transplanted under their skin producing prolactin. A study by Demarest et al.[56] used a moderate time period of 18 days of prolactin elevation induced by estrogen administration. They found that turnover of dopamine in the median eminence was increased at 6 days, not at 12 days, and decreased at 18 days, even though prolactin was continuously elevated. Initially, these results suggested that chronic exposure to high circulating levels of prolactin decreased the ability of the TIDA neurons to respond to prolactin. This effect of estrogen was reversible. and, in fact, there was increased dopamine sensitivity to prolactin 18 days after estrogen removal. However, it was not the high levels of prolactin for 18 days that decreased the responsiveness of the TIDA neurons initially, since elevation of prolactin for 11 or 22 days by giving haloperidol resulted in increased dopamine turnover. Thus, estrogen may have a direct effect on the prolactin-TIDA feedback mechanism.

A second method to develop chronic high levels of prolactin is to transplant a pituitary tumor secreting prolactin to a recipient rat or to inject estrogen chronically to create a pituitary tumor *in situ* that secretes a large amount of prolactin. Casanueva et al.,[57] using the estrogen-induced pituitary tumor model, found pathological changes in the arcuate nucleus region after 25 weeks of elevated prolactin. This was accompanied by a reduction in dopamine concentrations in the median eminence. Using the prolactin-secreting MtTW15 tumor, Simpkins et al.[58] found marked reductions in dopamine in the median-eminence-arcuate nucleus region and the medial basal hypothalamus. In agreement with this is a study[59] in which either estrogen-induced pituitary tumors or MtTW15 tumor-bearing rats had markedly elevated prolactin, reduced dopamine in the median eminence, and TIDA neurons which showed neuronal degeneration. Taken together, these observations suggest that prolactin and estrogen may exert neurotoxic effects on TIDA neurons when present for long periods of time. As TIDA neurons lose their responsiveness to prolactin, there is a loss in the feedback control of prolactin via dopamine. This may lead to development of prolactin-secreting tumors, which are found in old rats.[59]

There has been a considerable amount of discussion in the literature concerning the importance of prolactin feedback via dopamine in situations when prolactin is elevated for only a short period of time. In other words, what are the acute effects of prolactin on the TIDA system? Several studies have shown the systemic or intraventricular administration of prolactin caused a delayed (12 to 24 hr) increase in median eminence dopamine synthesis and turnover.[46,47] Injection of dopamine antagonists such as haloperidol, which raise prolactin levels, also caused a delayed increase in dopamine turnover[60] and dopamine secretion into hypophysial portal blood.[49] Just the opposite occurs when the pituitary is removed or the rats are treated with a dopamine agonist, such as bromocriptine.[61] In both of these cases, serum prolactin levels were very low or nondetectable and dopamine turnover decreased. These studies suggest that prolactin regulates its own secretion via the TIDA system in a very sluggish fashion.

Contrary to these in vivo studies are two in vitro studies in which prolactin was shown to cause a rapid synthesis and release of dopamine from fragments or synaptosomes of median eminence tissue.[62,63] To rectify these two seemingly opposite results, a well-conceived study[64] was done using either hypophysectomized or bromocriptine-treated rats to reduce prolactin levels. In these animals, intracerebroventricular injection of prolactin increased dopamine synthesis in the median eminence within 4 hr and caused an even further increase at 12 hr (Figures 2 and 3). Additional evidence also was presented suggesting that the circulating concentration of prolactin at the time of dopamine synthesis measurement (rapid component) and at an earlier time (delayed component) influenced the activity of TIDA neurons.[64]

From these experiments, it can be concluded that prolactin exerts a tonic inhibitory effect on its own secretion by way of activating the tuberoinfundibular neuronal system. There are physiological states in which there are prolonged (several hours) elevated levels of prolactin in the serum, including pregnancy, pseudopregnancy, and proestrus. It is possible that termination of these prolactin surges involves the dopamine system. The difficulty encountered in trying to solve this problem is that dopamine neurons may be involved in the events leading to the prolactin increases (decreased dopaminergic activity), confounding the possibility of finding dopamine changes several hours later. Results reported recently suggest that the cyclical pattern of TIDA neurons that is evident in early pregnancy is not due to cyclical changes in prolactin at that time.[65] Similarly, the lactating rat does not respond to prolactin with an increase in TIDA neuronal activity.[66] Thus, it may be that prolactin inhibits its own secretion in a tonic manner via the TIDA system, but phasic decreases in prolactin following acute increases are under different controlling mechanisms.

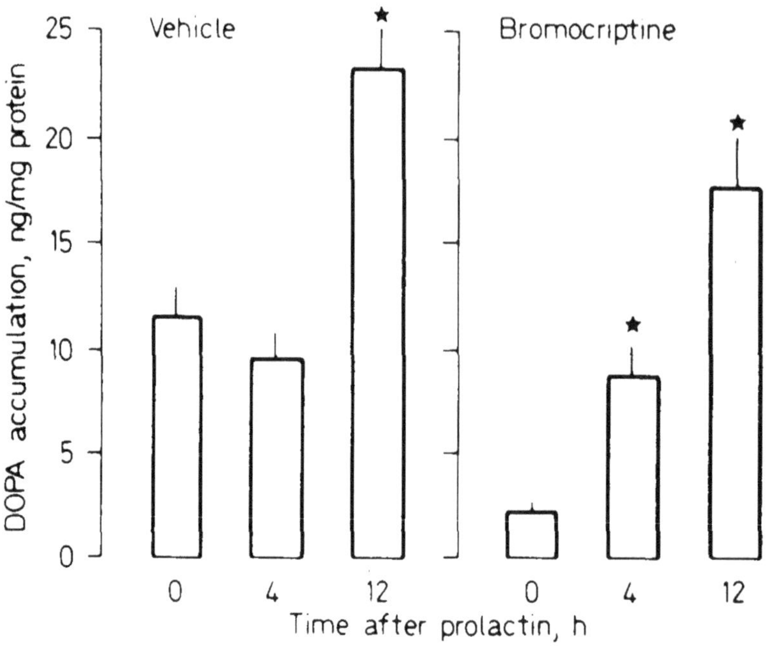

FIGURE 2. Time course for the effect of i.c.v. prolactin on the rate of DOPA accumulation in the median eminence of rats pretreated with bromocriptine (3 mg/kg, s.c.) or its vehicle 24 hr prior to sacrifice and then with prolactin (10 μg/mℓ, i.c.v.) or its saline vehicle 4 or 12 hr prior to sacrifice. Each column represents the mean and vertical line 1 SE of 8 determinations. A star indicates values significantly different from the prolactin vehicle-treated (0 time) animals. (From Demarest, K. T. et al., *Neuroendocrinology*, 38, 467, 1984. With permission.)

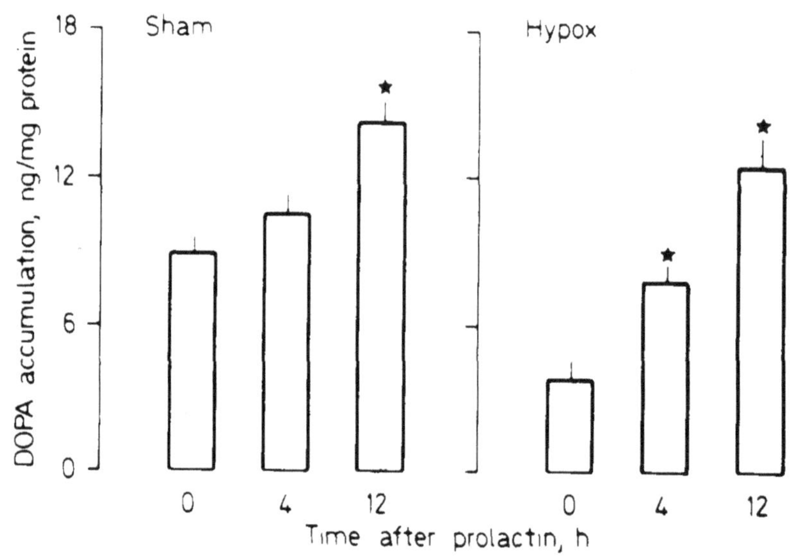

FIGURE 3. Time course for the effect of i.c.v. prolactin on the rate of DOPA accumulation in the median eminence of hypophysectomized rats. Rats were either hypophysectomized (Hypox) or were sham-operated (Sham) 48 hr prior to sacrifice and were treated with prolactin (10 μg/10 mℓ, i.c.v.) or its saline vehicle 4 or 12 hr prior to sacrifice. Each column represents the mean and the vertical line 1 SE of 6 to 8 determinations. A star indicates values significantly different from the prolactin vehicle-treated (0 time) animals. (From Demarest, K. T. et al., *Neuroendocrinology*, 38, 467, 1984. With permission.)

The preceding discussion has concentrated entirely on the effects of prolactin on the tuberinfundibular dopamine system, which is found completely within the hypothalamus. Another major dopamine pathway in the brain is the nigroneostriatal, in which dopaminergic neurons originate in the substantia nigra and terminate in the caudate-putamen and play a major role in control of motor function. Evidence has been reported that prolactin may affect these striated neurons or their dopamine receptors. Injection of prolactin caused an increase in the density of striatal dopamine receptors.[67] Haloperidol, a dopamine antagonist which increases prolactin levels, also has been shown to increase these dopamine receptors[68] and has led to the conclusion that long-term treatment with neuroleptics, such as haloperidol, may affect striatal dopamine via prolactin. However, there is not complete agreement on this conclusion. Chronic estradiol treatment led to elevated prolactin, resulting in increased dopamine receptors in striatal tissue. However, hypophysectomized rats treated with estradiol also had increased striatal dopamine receptors.[69] Secondly, neuroleptic drugs, which raise prolactin but do not penetrate the brain readily, failed to affect striatal dopamine receptors.[70] Thus, it cannot be concluded that estrogen and neuroleptics require prolactin to have their effect on striatal dopamine.

Prolactin has been shown to have other effects on the striatum. Prolactin caused dopamine release from fragments of rat striatal tissue.[71] Striatal tissue from rats treated with prolactin had increased incorporation of dopamine in vitro, although release of preaccumulated dopamine was not affected.[72] Finally, haloperidol increased the striatal dopamine synthesis rate, and antiserum to prolactin blunted this response.[73] Thus, it appears that prolactin can alter the activity of both tuberoinfundibular and nigrostriatal neurons.

V. PROLACTIN ALTERS HYPOTHALAMIC LUTEINIZING HORMONE-RELEASING HORMONE

During periods of increased prolactin secretion, serum levels of gonadotropins, LH and FSH, are usually low. During pregnancy and lactation, there is a cessation of cyclic release of LH and FSH. Women with galactorrhea-amenorrhea often have high prolactin and low gonadotropin secretion. Thus, several investigators have looked at the influence of prolactin on hypothalamic luteinizing hormone-releasing hormone (LHRH). One hypothesis is that the TIDA neurons terminating in the lateral aspects of the median eminence function as inhibitory neuromodulators of LHRH release. Injection of ovine prolactin increased dopamine turnover about fivefold in this lateral area, whereas turnover in striatal, mesolimbic, and incertohypothalamic neurons was unaffected.[74]

The next step was to show that prolactin actually does affect LHRH neurons. When hyperprolactinemia was induced in ovariectomized rats, LH secretion decreased, but there were no changes in medial basal hypothalamic LHRH content.[75] On the other hand, another study reported that hyperprolactinemia resulted in an increase in the medial basal hypothalamic LHRH content, reduced serum LH, and an increase in dopamine in the same hypothalamic area.[76] These same researchers later showed, by inducing hyperprolactinemia with pituitary transplants rather than using neuroleptic drugs, a decreased concentration of LHRH in the hypothalamus and plasma of hypophysial portal vessels, as well as increased turnover of dopamine in the hypothalamus.[77] Finally, rats bearing prolactin-secreting pituitary tumors had reduced gonadotropin secretion and reduced LHRH secretion into hypophysial portal blood, and increased dopamine levels in portal blood.[78] Adrenalectomy at the time of tumor inoculation prevented all of these changes, except for the increased prolactin and dopamine levels. Based on this result, it cannot be concluded that dopamine is solely responsible for the inhibitory effect of prolactin on LHRH and gonadotropin secretion.

VI. PROLACTIN AND MALE SEX BEHAVIOR

A large number of studies has been done on the influence of prolactin on maternal behavior and have been reviewed several times.[79,80] This section will concentrate on the effect of elevated prolactin on male behavior. There are many reports that hyperprolactinemia can depress sexual function in men, which can be counteracted by treatment with bromocryptine, a dopamine agonist which inhibits prolactin.[81,82] Male rats exhibit impaired copulatory behavior if prolactin is elevated due to pituitary transplants.[83,84] To determine whether inhibition of chronically elevated prolactin would reverse both gonadotropin suppression and impaired copulatory behavior, experienced male rats were given bromocryptine. This led to increased sexual activity, reduced prolactin, and increased LH.[85] This treatment also increased sexual activity in control rats (no pituitary transplants), reduced prolactin, but did not affect LH. These experiments show that hyperprolactinemia results in inhibition of sex behavior in experienced male rats and strongly suggest that the reduction in serum prolactin (but not LH elevation) is involved in the return of normal behavior following CB154 treatment. However, a direct effect of CB154 on sexual activity has not been ruled out.

One of the endocrine abnormalities that results from hyperprolactinemia is increased serum levels of corticosterone. Two experiments appeared recently in which the contribution of the hyperactivity of the adrenal cortex during hyperprolactinemia to the inhibition of male copulatory behavior was evaluated. Castrate male rats implanted with testosterone-filled capsules and bearing a pituitary tumor that secreted large amounts of both prolactin and ACTH showed suppressed copulatory behavior compared to similarly treated, but non tumor-bearing castrate control animals.[86] If these animals also were adrenalectomized, there was no difference between tumor-bearing and control rats, suggesting that hyperprolactinemia does not suppress male sexual behavior in the adrenalectomized rat. The second study used the model in which four pituitaries were grafted under the kidney capsule of intact males, thus inducing hyperprolactinemia.[87] This treatment reduced copulatory behavior and increased serum prolactin. Adrenalectomy plus supplemental corticosterone did not alter this response. Thus, it was concluded that the elevated levels of corticosterone that occur during hyperprolactinemia does not mediate the effects of prolactin on copulatory behavior. The conclusions in both studies appear to be valid, since one study used supplemental corticosterone[87] whereas the other did not.[86] Unfortunately, plasma levels of corticosterone were not monitored in either study. Putting the two studies together, it appears that some corticosterone may be necessary for the expression of the behavioral effects of prolactin in the male, but the hyperactivity of the adrenal that usually accompanies hyperprolactinemia is not an important factor.

The sites and systems in the brain upon which prolactin may be acting to impair copulatory behavior have received very little attention. In one recent experiment, rats bearing prolactin-secreting tumors had an increase in dopamine turnover in a part of the incertohypothalamic periventricular system found in the preoptic-anterior hypothalamus. Since androgens normally inhibit dopamine turnover in this area, it was suggested that these dopamine neurons play a role in normal male sex behavior induced by androgens, and that hyperprolactinemia may alter the response of these neurons to androgens.

VII. SUMMARY

The brain is exposed to prolactin even though prolactin is secreted by the anterior pituitary. Prolactin is found in the cerebrospinal fluid and may even be produced by neurons in the hypothalamus, or reach the hypothalamus by way of retrograde blood flow in the hypophysial portal vessels. The electrical activity of hypothalamic neurons is altered by prolactin, although much remains to be done to fully characterize this response. Prolactin has been conclusively shown to increase the activity of the tuberoinfundibular neurons, which contain dopamine and extend from the arcuate nucleus to the median eminence. Experiments have been done to suggest that elevated prolactin results in increased release of dopamine into portal vessels, and this is part of the inhibitory feedback mechanism that prolactin exerts on itself. Prolactin may be inhibitory to LHRH release also, thereby providing a mechanism whereby elevated prolactin results in deficiencies in gonadotropin secretion. Male sex behavior is usually decreased in cases of hyperprolactinemia, and reduction of prolactin often results in a return of normal behavior. Thus, the action of prolactin in the brain affects a number of physiological events, many of which will require further investigation to determine the mechanisms by which prolactin acts.

REFERENCES

1. Clemens, J. A. and Meites, J., Inhibition by hypothalamic prolactin implants of prolactin secretion, mammary growth and luteal function, *Endocrinology,* 82, 878, 1968.
2. Chen, C. L., Voogt, J. L., and Meites, J., Effect of median eminence implants of FSH, LH or prolactin on luteal function in the rat, *Endocrinology,* 83, 1273, 1968.
3. Voogt, J. L. and Meites, J., Effects of an implant of prolactin in median eminence of pseudopregnant rats on serum and pituitary LH, FSH and prolactin, *Endocrinology,* 88, 286, 1971.
4. Voogt, J. L. and Meites, J., Suppression of proestrous and suckling-induced increase in serum prolactin by hypothalamus implant of prolactin, *Proc. Soc. Exp. Biol. Med.,* 142, 1056, 1973.
5. Kendall, J. and Orwoll, E., Anterior pituitary hormones in the brain and other extrapituitary sites, in *Frontiers in Neuroendocrinology,* Vol. 6, Martini, L. and Ganong, W. F., Eds., Raven Press, New York, 1980, 33.
6. Hyppa, M. T. and Liira, J., Neuropeptide hormones in cerebrospinal fluid, *Med. Biol.* 57, 367, 1979.
7. Assies, J., Schellekens, A. P. M., and Touber, J., Prolactin in human cerebrospinal fluid, *J. Clin. Endocrinol. Metab.,* 46, 576, 1978.
8. Schaub, C., Bluet-Pajot, M. T., Szikla, G., Lornet, C., and Talairach, J., Distribution of growth hormone and thyroid-stimulating hormone in cerebrospinal fluid and pathological compartments of the central nervous system, *J. Neurol. Sci.,* 31, 123, 1977.
9. Login, I. S. and MacLeod, R. M., Prolactin in human and rat serum and cerebrospinal fluid, *Brain Res.,* 132, 477, 1977.
10. Schroeder, L. L., Johnson, J. C., and Malarkey, W. B., Cerebrospinal fluid prolactin: a reflection of abnormal prolactin secretion in patients with pituitary tumors, *J. Clin. Endocrinol. Metab.,* 43, 1255, 1976.
11. Assies, J., Schellekens, A. P. M., and Touber, J. L., Protein hormones in cerebrospinal fluid: evidence for retrograde transport of prolactin from the pituitary to the brain in man, *Clin. Endocrinol.,* 8, 487, 1978.
12. Kalin, N. H., Burns, R. S., Risch, S. C., Cosgrove, S. A., Warden, D., and Murphy, D. L., The relationship between blood and cerebrospinal fluid prolactin in nonhuman primates, *Life Sci.,* 31, 159, 1982.
13. Belchetz, P. E., Ridley, R. M., and Baker, H. F., Studies on the accessibility of prolactin and growth hormone to brain: effect of opiate agonists on hormone levels in serial, simultaneous plasma and cerebrospinal fluid samples in the rhesus monkey, *Brain Res.,* 239, 310, 1982.
14. Dubey, A. K., Herbert, J., Martensz, N. D., Beckford, U., and Jones, M. T., Differential penetration of three anterior pituitary peptide hormones into the cerebrospinal fluid of rhesus monkeys, *Life Sci.,* 32, 1857, 1983.

15. Perlow, M. J., Cerebrospinal fluid prolactin: a daily rhythm and response to an acute perturbation, *Brain Res.*, 243, 382, 1982.
16. Peake, G. T., Buckman, M. T., Davis, L. E., and Standefer, J., Pituitary and placentally derived hormones in cerebrospinal fluid during normal human pregnancy, *J. Clin. Endocrinol. Metab.*, 56, 46, 1983.
17. Walsh, R. J., Posner, B. I., Kopriwa, B. M., and Brawer, J. R., Prolactin binding sites in the rat brain, *Science*, 201, 1041, 1978.
18. Posner, B. I., van Houten, M., Patel, B., and Walsh, R. J., Characterization of lactogen binding sites in choroid plexus, *Exp. Brain Res.*, 49, 300, 1983.
19. Landas, S., Uptake of prolactin from cerebral spinal fluid in rat brain, *Neuropeptides*, 3, 173, 1982.
20. Fuxe, K., Hokfelt, T., Eneroth, P., Gustafsson, J., and Skett, P., Prolactin-like immunoreactivity: localization in nerve terminals of rat hypothalamus, *Science*, 196, 899, 1977.
21. Toubeau, J., Desclin, J., Parmeter, M., and Pasteels, J. L., Cellular localization of a prolactin-like antigen in the rat brain, *J. Endocrinol.*, 83, 261, 1979.
22. Hansen, B. L., Hansen, G. N., and Hagen, C., Immunoreactive material resembling ovine prolactin in perikarya and nerve terminals of the rat hypothalamus, *Cell. Tissue Res.*, 226, 121, 1982.
23. Hansen, B. L. and Hansen, G. N., Immunocytochemical demonstration of somatotropin-like and prolactin-like activity in the brain of *Calamoichthys calabarius*, *Cell. Tissue Res.*, 222, 615, 1982.
24. Harlan, R. E., Shivers, B. D., and Pfaff, D. W., Midbrain microinfusions of prolactin increase the estrogen-dependent behavior, lordoris, *Science*, 219, 1451, 1983.
25. Schachter, B. S., Durgerian, S., Harlan, R. E., Pfaff, D. W., and Shivers, B. D., Prolactin in mRNA exists in rat hypothalamus, *Endocrinology*, 114, 1947, 1984.
26. Clemens, J. A., Gallo, R.V., Whitmoyer, D. I., and Sawyer, C. H., Prolactin responsive neurons in the rabbit hypothalamus, *Brain Res.*, 25, 371, 1971.
27. Yamada, Y., Effects of iontophoretically-applied prolactin on unit activity of the rat brain, *Neuroendocrinology*, 18, 263, 1975.
28. Chan. A., Dudley, C. A., and Moss, R. L., Action of prolactin, dopamine and LHRH on ventromedial hypothalamic neurons as a function of ovarian hormones, *Neuroendocrinology*, 36, 397, 1983.
29. Ramaswamy, S., Pillai, N. P., and Bapna, J. S., Analgesic effect of prolactin: possible mechanism of action, *Eur. J. Pharmacol.*, 96, 171, 1983.
30. Weiner, R. I. and Ganong, W. F., Role of brain monoamines and histamine in regulation of anterior pituitary secretion, *Physiol. Rev.*, 58, 905, 1978.
31. Neill, J. D., Neuroendocrine regulation of prolactin secretion, in *Frontiers in Neuroendocrinology*, Vol. 6, Martini, L. and Ganong, W. F., Eds., Raven Press, New York, 1980, 129.
32. Leong, D. A., Frawley, L. S., and Neill, J. D., Neuroendocrine control of prolactin secretion, in *Annual Review of Physiology*, Vol. 45, Berne, R. M., Ed., Annual Reviews, Palo Alto, Calif. 1983, 109.
33. Meites, J., Lu, K. H., Wuttke, W., Welsch, C. W., Nagasawa, H., and Quadri, S. K., Recent studies on functions and control of prolactin secretion in rats, *Recent Prog. Horm. Res.*, 38, 471, 1972.
34. Voogt, J. L., Regulation of nocturnal prolactin surges during pregnancy in the rat, *Endocrinology*, 106, 1670, 1980.
35. Dang, B. T. and Voogt, J. L., Termination of pseudopregnancy following hypothalamic implantation of prolactin, *Endocrinology*, 100, 873, 1977.
36. Bybee, D. E., Nakawatase, C., Szabo, M., and Frohman, L. A., Inhibitory effects of prolactin on its secretion involve central nervous system dopaminergic mediation, *Neuroendocrinology*, 36, 27, 1983.
37. Whitworth, N. S., Grosvenor, C. E., and Mena, F., Autofeedback regulation of prolactin (PRL) secretions: effect of PRL before suckling on the subsequent nursing-induced release of PRL in the lactating rat, *Endocrinology*, 108, 1279, 1981.
38. Selmanoff, M. and Gregerson, K. A., Autofeedback effects of prolactin on basal, suckling-induced, and proestrous secretion of prolactin, *Proc. Soc. Exp. Biol. Med.*, 175, 398, 1984.
39. Mangat, H. K. and McCann, S. M., Acute inhibition of PRL and TSH secretion after intraventricular injection of PRL in ovariectomized rat, *Am. J. Physiol.*, 244, E31, 1983.
40. Mezey, E. and Palkovits, M., Two-way transport in the hypothalamohypophyseal system, in *Frontiers in Neuroendocrinology*, Vol. 7, Ganong, W. F. and Martini, L., Eds., Raven Press, New York, 1982, 1.
41. Oliver, C., Mical, R. S., and Porter, J. C., Hypothalamic-pituitary vasculature: evidence for retrograde blood flow in the pituitary stalk, *Endocrinology*, 101, 598, 1977.
42. Ambach, G. and Palkovits, M., The blood supply of the hypothalamus in the rat, in *Handbook of the Hypothalamus*, Vol. 1, Morgan, P. J., and Panksepp, J., Eds., Marcel Dekker, New York, 1979, 267.
43. Moore, K. E. and Demarest, K. T., Effects of prolactin on central dopaminergic neurons, *Proc. Neuro-Psycholpharmacol. Biol. Psychiatr.*, 6, 645, 1982.

44. Moore, K. E. and Demarest, K. T., Tuberinfundibular and tuberohypophyseal dopaminergic neurons, in *Frontiers in Neuroendocrinology,* Vol. 7, Ganong, W. F. and Martini, L., Eds., Raven Press, New York, 1982, 161.
45. Hokfelt, T. and Fuxe, K., Effects of prolactin and ergot alkaloids on the tuberoinfundibular dopamine neurons, *Neuroendocrinology,* 9, 100, 1972.
46. Gudelsky, G. A., Simpkins, J., Mueller, G. P., Meites, J., and Moore, K. E., Selective actions of prolactin on catecholamine turnover in the hypothalamus and on serum LH and FSH, *Neuroendocrinology,* 22, 206, 1976.
47. Annunziato, L. and Moore, K. E., Prolactin in CSF selectively increases dopamine turnover in the median eminence, *Life Sci.,* 122, 2037, 1978.
48. Johnston, C. A., Demarest, K. T., and Moore, K. E., Cycloheximide disrupts the prolactin-mediated stimulation of dopamine synthesis in tuberoinfundibular neurons, *Brain Res.,* 195, 736, 1980.
49. Gudelsky, G. A. and Porter, J. C., Release of dopamine from tuberoinfundibular neurons into pituitary stalk blood after prolactin or haloperiodol administration, *Endocrinology,* 106, 526, 1980.
50. Nicholson, G., Greeley, G. H., Jr., Humm, J., Youngblood, W. W., and Kizer, J. S., Prolactin in cerebrospinal fluid: a probable site of prolactin autoregulation, *Brain Res.,* 190, 447, 1980.
51. Morgan, W. M. and Herbert, D. C., Early responses of the dopaminergic tuberoinfundibular neurons to anterior pituitary homographs, *Neuroendocrinology,* 31, 215, 1980.
52. Krieger, A. and Wuttke, W., Effects of ovariectomy and hyperprolactinemia on tyrosine hydroxylase and dopamine-β-hydroxylase activity in various limbic and hypothalamic structures, *Brain Res.,* 193, 173, 1980.
53. Bennett, B. A. and Sundberg, D. K., Hypothalamic catecholamine biosynthesis and pituitary gonadotropin secretion in vitro: effect of hyperprolactinemia, *Mol. Cell. Endocrinol.,* 30, 149, 1983.
54. Cramer, O. M., Parker, C. R., and Porter, J. C., Secretion of dopamine into hypophysial portal blood by rats bearing prolactin-secreting tumors of ectopic pituitary glands, *Endocrinology,* 105, 636, 1979.
55. Perkins, N. A., Westfall, T. C., Paul, C. V., MacLead, R. M., and Rogol, A. D., Effect of prolactin on dopamine synthesis in medial basal hypothalamus: evidence for a shortloop feedback, *Brain Res.,* 160, 431, 1979.
56. Demarest, K. T., Riegle, G. D., and Moore, K. E., Long-term treatment with estradiol induces reversible alterations in tuberoinfundibular dopaminergic neurons: a decreased responsiveness to prolactin, *Neuroendocrinology,* 39, 193, 1984.
57. Casanueva, F., Cocchi, D., Locatelli, V., Flauto, C., Zambotti, F., Bestetti, G., Rossi, G. L., and Muller, E., Defective central nervous system dopaminergic function in rats with estrogen-induced pituitary tumors, as assessed by plasma prolactin concentrations, *Endocrinology,* 110, 590, 1982.
58. Simpkins, J. W., Hodson, C. A., Kalra, P. S., and Kalra, S. P., Chronic hyperprolactinemia depletes hypothalamic dopamine concentrations in male rats, *Life Sci.,* 30, 1349, 1982.
59. Sarkar, D. K., Gottschall, P. E., and Meites, J., Damage to hypothalamic dopaminergic neurons is associated with development of prolactin-secreting pituitary tumors, *Science,* 218, 684, 1982.
60. Gudelsky, G. A. and Moore, K. E., A comparison of the effects of haloperidol on dopamine turnover in the striatum, olfactory tubercle and median eminence, *J. Pharmacol. Exp. Ther.,* 202, 149, 1977.
61. Demarest, K. T. and Moore, K. E., Sexual differences in the sensitivity of tuberoinfundibular dopamine neurons to the actions of prolactin, *Neuroendocrinology,* 33, 230, 1981.
62. Forman, M. M. and Porter, J. C., Prolactin augmentation of dopamine and norepinephrine release from superfused medial basal hypothalamic fragments, *Endocrinology,* 108, 800, 1981.
63. Perkins, N. A. and Westfall, T. C., The effect of prolactin on dopamine release from rat striatum and median basal hypothalamus, *Neuroscience,* 3, 59, 1978.
64. Demarest, K. T., Riegle, G. D., and Moore, K. E., Prolactin-induced activation of tuberoinfundibular dopaminergic neurons: evidence for both a rapid 'tonic' and a delayed 'induction' component, *Neuroendocrinology,* 38, 467, 1984.
65. Demarest, K. T., Moore, K. E., and Riegle, G. D., Role of prolactin feedback in the semicircadian rhythm of tuberoinfundibular dopaminergic neuronal activity during early pregnancy in the rat, *Neuroendocrinology,* 36, 371, 1983.
66. Demarest, K. T., McKay, D. W., Riegle, G. D., and Moore, K. E., Biochemical indices of tuberoinfundibular dopaminergic neuronal activity during lactation: a lack of response to prolactin, *Neuroendocrinology,* 36, 130, 1983.
67. Hruska, R. E., Pitman, K. T., Sibergeld, E. K., and Ludmer, L. M., Prolactin increases the density of striatal dopamine receptors in normal and hypophysectomized male rats, *Life Sci.,* 30, 547, 1982.
68. Hruska, R. E., Ludmer, L. M., and Silbergeld, E. K., Hypophysectomy prevents the striatal dopamine receptor supersensitivity produced by chronic haloperidol treatment, *Eur. J. Pharmacol.,* 65, 455, 1980.
69. Paolo, T. Di., Poyet, P., and Labrie, F., Effect of prolactin and estradiol on rat striatal dopamine receptors, *Life Sci.,* 31, 2921, 1982.

70. Rupniak, N., Hong, M., Mansfield, S., Fleminger, S., Dyer, R., Jenner, P., and Marsden, C. D., Elevation of circulating prolactin concentrations may not cause striatal dopamine receptor supersensitivity, *Eur. J. Pharmacol.*, 93, 195, 1983.
71. Chen. Y. F. and Ramirez, V. D., Prolactin stimulates dopamine release from male but not female rat striatal tissue superfused *in vitro, Endocrinology,* 111, 1740, 1982.
72. Gregerson, K. A. and Selmanoff, M. K., Differential effects of prolactin on the uptake and rapid release of dopamine (DA) from median eminence (ME) and striatal (S) synaptosomes, *Proc. 13th Annu. Meet. Soc. Neurosci.*, Abstr. No. 96.9, 1983.
73. Van Loon, G. R., Shum, A., George, S. R., and Shin, S. H., Prolactin increases the activity of tuberoinfundilbular and nigrostiatal dopamine neurons: prolactin antiserum inhibits the haloperidol-induced increase in dopamine synthesis rates in median eminence and striatum of rats, *Brain Res. Bull.*, 10, 539, 1983.
74. Selmanoff, M., The lateral and medial median eminence: distribution of dopamine, norepinephrine, and luteinizing hormone-releasing hormone and the effect of prolactin on catecholamine turnover, *Endocrinology,* 108, 1716, 1981.
75. Carter, D. A., Lakhani, S., and Whitehead, S. A., Characterization of the inhibitory effects of hyperprolactinemia on the mechanisms controlling LH secretion in chronically ovariectomized rats, *J. Reprod. Fertil.*, 69, 57, 1983.
76. Chatani, F., Aono, T., Koike, K., Tasaka, K., and Kurachi, K., Effect of sulpiride-induced hyperprolactinemia on catecholamine turnover and LRH concentration in the medial basal hypothalamus of rats, *Acta Endocrinol.*, 102, 321, 1983.
77. Koike, K., Aono, T., Miyake, A., Tasaka, K., Chatani, F., and Kurachi, K., Effect of pituitary transplants on the LH-RH concentrations in the medial basal hypothalamus and hypophyseal portal blood, *Brain Res.*, 301, 253, 1984.
78. Weber, R. F. A., de Greef, W. J., deKoning, J., and Verburg, J. T. M., LH-RH and dopamine levels in hypophysial stalk plasma and their relationship to plasma gonadotropins and prolactin levels in male rats bearing a prolactin- and adrenocorticotrophin-secreting pituitary tumor, *Neuroendocrinology,* 36, 205, 1983.
79. Nicoll, C. S., Physiological actions of prolactin, in *Handbook of Physiology, Section 7,* Vol. 4 (Part 2), Knobil, E. and Sawyer, W. H., Eds., American Physiological Society, Washington, D. C., 1974, 253.
80. Tindal, J. S., Control of prolactin secretion, in *The Endocrine Hypothalamus,* Jeffcoate, S. L. and Hutchinson, J. S. M., Eds., Academic Press, London, 1978, 253.
81. Tolis, G. and Van Vliet, S., Leydig cell function in hyperprolactinemia, *Clin. Res.*, 24, 179A, 1976.
82. Thorner, M. O. and Besser, G. M., Hyperprolactinemia and gonadal function: results of bromocryptine treatment, in *Prolactin and Human Reproduction,* Crosignani, P. G. and Robyn, C., Eds. Academic Press, New York, 1977, 285.
83. Svare, B., Bartke, A., Doherty, P., Masen, I., Michael, S. D., and Smith, M. S., Hyperprolactinemia suppresses copulatory behavior in male rats and mice, *Biol. Reprod.*, 21, 529, 1979.
84. Bailey, D. J. and Herbert, J., Impaired copulatory behavior of male rats with hyperprolactinemia induced by domperidome or pituitary grafts, *Neuroendocrinology,* 35, 186, 1982.
85. Doherty, P. C., Bartke, A., and Smith, M. S., Differential effects of bromocryptine treatment on LH release and copulatory behavior in hyperprolactinemic male rats, *Horm. Behav.*, 15, 436, 1981.
86. Weber, R. F. A., Ooms, M. P., and Vreeburg, J. T. M., Effects of a prolactin secretory tumor on copulatory behavior in male rats, *J. Endocrinol.*, 93, 223, 1982.
87. Doherty, P. C., Bartke, A., Hogan, M. P., Klemcke, H., and Smith, M. S., Effects of hyperprolactinemia on copulatory behavior and testicular human chorionic gonadotropin binding in adrenalectomized rats, *Endocrinology,* 111, 820, 1982.
88. Kalra, P. S., Simpkins, J. W., Luttge, W. G., and Kalra, S. P., Effects on male sex behavior and preoptic dopamine neurons of hyperprolactinemia induced by MtTW15 pituitary tumors, *Endocrinology,* 113, 2065, 1983.

Chapter 4

MODELS OF PROLACTIN ACTION IN NONMAMMALIAN VERTEBRATES

Nelson D. Horseman

TABLE OF CONTENTS

I. Evolution and Physiology of Prolactin Mechanisms 42

II. Prolactin Mechanisms in Teleost Osmoregulation 45

III. Prolactin Mechanisms in Amphibian Metamorphosis 49

IV. Prolactin Mechanisms in Columbid Crop 51

V. Prospects and Perspectives 58

Acknowledgments 60

References 60

I. EVOLUTION AND PHYSIOLOGY OF PROLACTIN MECHANISMS

During the past two decades, such a large volume of information on the variety of the physiological actions of prolactin has become available that it has become a cliché to recognize that prolactin has the greatest diversity of actions of any vertebrate protein hormone. This obvious diversity demands an eclectic approach toward understanding the mechanisms of prolactin. Hormones with relatively limited target tissue reactions (e.g., the gonadotropins or thyroid-stimulating hormone) might be understood largely by choosing a single or a few target models upon which to focus. In contrast, we should not expect to comprehensively understand prolactin actions if exclusive attention is focused on a very narrow group of models. Thus, despite the inherent attractiveness of mammary gland biology and the historical fact that this organ was the earliest mammalian target described for prolactin,[1] the mammary gland remains only one of a variety of important vertebrate target organs. Even the early history of prolactin studies is noteworthy because of Oscar Riddle's prescient recognition that prolactin performed a variety of functions and was best thought of as "integrative" in the physiology of organisms.[2]

Given the diversity of prolactin actions, what, within the context of this volume, are the lessons regarding the study of the actions of prolactin at the molecular level? I believe there are at least two important points. First, that there is likely to be not one, but several primary pathways by which prolactin actions are mediated within its target cells. The dominances and interactions of the mechanisms must have been determined by evolutionary and developmental differentiation of the appropriate targets. There is recent recognition of the multiplicity of target tissue responses to hormones.[3] Hormonal regulation at the cellular level may be coming to be seen less as unique switching mechanisms for regulated processes and more as comprehensive integration of cellular metabolism. Secondly, the diversity of the effects of prolactin on target tissue offers a uniquely rich abundance of systems in which to study regulatory mechanisms at the molecular level. The molecular biology of cell regulation has thus far been elucidated largely through hormonal studies (i.e., cAMP, steroid regulation of gene expression, prostaglandins, phosphorylation, receptors, etc.). If this history is continued, prolactin represents a ligand which has input into a rich variety of regulatory pathways. Despite these facts, the mammary gland has remained by far the primary focus of studies of prolactin action at the molecular level. This situation is being modified by an expanding number of interesting models being examined at the molecular level. Some such models derived in mammals include the female rat liver,[4,5] a lymphoma cell line,[6-8] and T-lymphocytes.[9] This paper will focus on several nonmammalian systems which are productive and potential models of prolactin action and which can be further exploited.

At the cellular level, prolactin actions may be divisible into several hierarchical domains. These domains include general trophic actions that are shared with several other hormones, prolactin-growth hormone actions that are physiological responses (not dependent on extraordinary doses) which are shared by both prolactin and growth hormone, "conserved" prolactin effects, and "recent" prolactin effects which include several well-known effects in the tetrapods. General trophic actions of prolactin are exemplified by the induction of ornithine decarboxylase, which occurs in response to a wide variety of hormonal and nonhormonal growth stimuli.[10,11] A number of prolactin effects are shared with growth hormone.[12] In some tadpole tissues, receptors bind both prolactin and growth hormone equally well,[13] and both prolactin and growth hormone act partially by stimulating somatomedin release.[14,15] Conserved effects in most, if not all, vertebrate groups include some actions on ion balance and lipid metabolism and the antigonadotropic actions often reported. The most obvious recently evolved prolactin effect is mammary gland growth and development, but other elaborated prolactin actions include crop milk stimulation and the control of "second metamorphosis" in some amphibians. The array of prolactin responses known in various vertebrates offers models which are especially well suited to study any of the various aspects of molecular responsiveness to prolactin. Much of the exploitation of this rich diversity remains. Science has been called the art of the "doable" and it is quite clear that there are animal models in which the study of particular aspects of prolactin actions are most readily "doable". It is hoped that this paper might stimulate either the exploitation of, or search for, models of significant utility.

The molecular divergence of prolactin and growth hormone was apparently initiated by a gene duplication at about the time of the appearance of the first vertebrates 400 million years ago.[16] The family of related polypeptide hormones which also includes the placental lactogens has expanded by means of several gene duplications. The nature and functional significance of the prototypic polypeptide from which this hormone family evolved is unknown. However, recent evidence suggests it might have been a prolactin-like molecule. The protochordate ascidian, *Styela picata,* contains in its neural gland and cerebral ganglion a polypeptide with specific immunological cross-reactivity with mammalian prolactin.[17,18] The functional significance of this polypeptide is unknown.

The evolutionary history of the prolactin-growth hormone family[19] and several physiological considerations, especially tissue and developmental specificities, suggest potentially important considerations about target receptor mechanisms. The first consideration is that prolactin belongs to a reasonably large family of related hormones which have arisen following several gene duplications. This includes not only prolactin, growth hormone, and the placental lactogens, but probably, multiple forms of prolactin in nonmammals, such as recently found in a teleost,[20,21] and a protein expressed in proliferating cells.[22] The two teleost prolactins are highly homologous, but partial sequence analysis shows they are distinct.[184] This family of hormones must be complemented by a family of receptors which probably also arose by gene duplications and possibly other modifications of expression such as alternative RNA splicing or post-translational modification. The initial origin of the hormones and of alternate receptors would have been independent events which would have been coupled by subsequent natural selection. This concept could account for certain heterogeneous effects of prolactin and the partially overlapping, but yet distinct, effects of members of the prolactin-growth hormone gene family. The independent origins of hormone and receptor genes would, in theory, provide impetus for proliferation of an expanding family of related regulatory systems. The extent to which developmental and tissue-specific expression of alternate receptors (and forms of hormones) bears on the physiology and molecular biology of prolactin effects is obviously unknown, but likely to form an important future research direction.

Target cell responsiveness to prolactin has been shown to be highly dependent upon the time of hormone administration during the daily (circadian) cycle.[23,24] All prolactin target responses which have been systematically assayed at several phases of the day are known to be circadian phase dependent. Physiologically, this has been shown to be important in seasonal regulation of metabolic and reproductive conditions and in the maintenance of early pregnancy in rodents.[25,26] In the latter case, a novel nocturnal prolactin surge occurs during the phase of the day when prolactin is capable of maintaining corpus luteum function and arresting the estrous cycle.[26] Because of the association of circadian prolactin responsiveness with fat storage in various vertebrates,[27] the finding of an altered prolactin rhythm in obese women may have pathological significance.[28] The mechanisms by which circadian variations in prolactin responsiveness manifest themselves are unclear. In the case of lipogenic regulation, prolactin administration at particular phases of the day appear to enhance insulin responsiveness.[29] This effect may require induction of new proteins, since in a teleost it takes place with a time course of 10 to 24 hr.[30] The well-known mitogenic action of prolactin in crop tissue is highly circadian phase dependent[31,32] and is associated with a variation in the inducibility of ornithine decarboxylase activity.[33] Since we have come to know that most cellular metabolism undergoes marked circadian variations, it is understandable that prolactin interacts with the circadian organization of its target cells. The specific means by which those interactions take place are likely to inform us about the molecular basis of the actions of prolactin.

A well-known generalization about the evolution of prolactin is its common role in activities concerned with "parental care".[34] These actions range from the stimulation of crop milk in pigeons and mammary gland milk in mammals, which were the first two prolactin actions defined,[1] to effects such as brood patch development,[35,36] discus fish mucous secretion,[37] and numerous behaviors associated with "parenting".[34] These "parenting" responses have independently arisen numerous times by convergent evolution. What then accounts for the repeated theme of prolactin association with parental nurturing? That prolactin positively regulates many influences involved in nurturing young must derive secondarily from the fact that it is generally secreted during the post mating phase of the reproductive cycle, during which parenting occurs in species adapted to utilize this option. The origin of this pattern of prolactin secretion can only be speculated on. This pattern might reflect the hypothetically primitive association of prolactin with osmoregulation, since most euryhaline species enter freshwater to rear their young, and tropical fish species are stimulated to produce their broods by an influx of freshwater. If the direct stimulation of prolactin gene expression by estrogen[38] is phylogenetically general and primitive, the association of prolactin and parental aspects of reproduction might be an outgrowth of that stimulation. Prolactin and its receptors have been found in the brain[39,40] and its association with behaviors such as broodiness,[41] hyperphagia,[42] or sterotyped nesting and nurturing behaviors[34] might have promoted the independent development of the numerous parenting effects in peripheral tissues.

Despite the obvious diversity of prolactin targets and the fact that they must have arisen through independent evolutionary events, there are some interesting emerging generalities regarding prolactin action at the molecular and cellular levels. Nearly all of the well-known effects of prolactin are long-term adaptive changes associated with cell differentiation and growth or long-term shifts in metabolic activity. This is in clear distinction from other protein- or peptide-type hormones which act rapidly within minutes to alter cell activity. Prolactin, like the other protein and peptide hormones, seems to act by way of plasma membrane receptor(s).[43] There have been sporadic reports of possible direct intracellular actions,[44,45] but none have withstood substantiation yet. Good correlations of membrane receptor levels and known physiological responsiveness support the contention that the membrane receptors are physiologically relevant in various vertebrates.[43,46,47] Cyclic AMP, though a mediator for many other hormones with membrane receptors, is apparently not a mediator in any of the several nonmammalian systems tested. These include actions as diverse as teleost kidney and gill effects,[48] inhibition of tadpole metamorphosis,[49] and pigeon crop stimulation.[50] As is true in the mammals, many of the actions of prolactin in nonmammalian vertebrates are dependent on synergisms or antagonisms with other hormones.[34] Common interacting hormones are thyroxine,[51] glucocorticoids,[52] gonadal steroids,[34] and, based on recent information, somatomedins.[15] Thus far, none of these interactions is well understood at a mechanistic level, and it is unlikely that they will be until the mechanisms responsible for discreet prolactin effects are known.

Three systems that I want to deal with in substantial detail include teleost osmoregulation, amphibian metamorphosis, and the crop response of pigeons and doves. Many other nonmammalian models offer potential as attractive models, but are little understood from a molecular point of view. Two systems are somewhat unique in that they react very rapidly to prolactin. One of these is xanthophore pigment dispersion in various fishes[53] in which prolactin from fishes, but not tetrapods, causes yellowing of the skin within minutes; in chick renal cells in vitro, prolactin induces 1α-hydroxylase activity (vitamin D activating), also within a short time course.[54,56] This reaction is specific for hormones with lactogenic activity and, therefore, may depend on the same active structure as is necessary for mammary gland stimulation. The potential of these systems is suggested by the notion that such rapid effects may be directly controlled by some "second messenger" of prolactin action. Another important, but little studied, target is the liver. Although regulation of lipogenesis in mammary gland is a well-known prolactin effect,[57] its mechanisms of action on hepatic lipogenesis[30,58-60] have been largely ignored. It is clear that the liver is a major prolactin target, since it is one of the richest sources of prolactin receptors. The role of prolactin in hepatic metabolism seems to be exaggerated in many nonmammalian species because of its association with seasonal breeding conditions.

The three models I have chosen illustrate salient features of prolactin action and offer experimental values. In both teleost osmoregulation and amphibian metamorphosis, numerous tissues are simultaneously involved, and interactions with cortisol in the former and thyroxine in the latter system are important. The pigeon crop is well known for its utility in prolactin bioassay. The relative simplicity and hormonal specificity of this system are attractive advantages. It is now becoming a potent system for analysis of prolactin actions with new focuses on the mechanisms involved in mitogenic stimulation and application of the tools of molecular biology to identify changes in crop gene expression.

II. PROLACTIN MECHANISMS IN TELEOST OSMOREGULATION

One of the earliest and most potent indications that prolactin was involved in a broad array of physiological regulatory systems was the discovery in the 1950s that prolactin was the necessary hormone for freshwater survival of hypophysectomized killifish (*Fundulus heteroclitus*).[61-63] This discovery touched off a flood of investigations that have led to proposed or proven osmoregulatory effects of prolactin in all the vertebrate groups from fishes through mammals. The literature on the physiology of these effects is vast and has been reviewed extensively.[64-70] My choices are intentionally selective. In the context of molecular mechanisms, the osmoregulatory effects of prolactin are noteworthy in the integrated variety of tissues which are recruited into specific, and in some cases paradoxical, actions that promote osmotic balance. The model systems which have been most informative regarding osmoregulatory influences of prolactin have been freshwater and euryhaline fishes such as the killifish (*Fundulus* sp.), stickleback (*Gillichthyes mirabilis*), molly (*Poecilia latipinna*), and tilapia (*Oreochromis mossambica*). It is because of the exaggerated osmoregulatory problems such species face that the role of prolactin is most readily discernible in them. These fishes exist within a medium which is either always hypotonic, therefore diluting, (freshwater species) or varies from strongly hypertonic to strongly hypotonic (euryhaline species). Gas exchange in the aquatic environment requires a large surface area in proximate contact with the water, and both the lining of the gut and body and gill surfaces are obliged to be osmoregulatory organs.

Mechanisms responsible for the osmoregulatory influences of prolactin include reduced water influx through permeability control and increased conservation of ions (particularly Na^+ and Cl^-) through influences on transport mechanisms and permeability. The relative dominance of prolactin in osmoregulation seems to vary substantially among species, even in close relatives. As only one example, the classical model for osmoregulatory prolactin effects, *F. heteroclitus*, dies rapidly of osmotic failure if kept in freshwater after hypophysectomy.[62] Its congener, *F. kansae*, can survive indefinitely in freshwater after hypophysectomy,[70] although it too shows subnormal osmoregulation which is repaired by prolactin. Given the breadth of environments to which even closely related teleosts have adapted, such variations among species are to be expected. The normal range of *F. kansae* is strictly freshwater, whereas that of *F. heteroclitus* is estuarine. Much of the literature regarding prolactin actions in teleost osmoregulation is devoted to examining the sometimes bewildering array of differences among species regarding hormonal participation, and that literature has been reviewed often enough. The specific molecular processes involved have been examined, at least to some extent, and it has become obvious that prolactin exerts its effects through several or all of the osmoregulatory organs with the gills, kidneys, and urinary bladder being especially important. A common finding is that plasma osmolarity is reduced following hypophysectomy leading to death within several days in freshwater. Plasma concentrations of sodium and chloride ions fall dramatically in most species without concomitant loss of potassium and other electrolytes.[71-73] Mechanisms therefore must account primarily for NaCl conservation. A number of the osmoregulatory effects of prolactin have been questioned because of potential contamination of other hormones and high doses used in experiments. Undoubtedly, some putative effects, such as rapid changes in water permeability in gills,[74,75] can be best attributed to contaminating vasopressin.[76,77] Effects of NaCl fluxes, however, are well documented to be specific for prolactins, and while most studies have used ungulate preparations out of necessity, fish preparations, either unpure[78] or highly purified,[79] have essentially equivalent effects. Early on, Ball and Ensor (1967) examined the specificity of maintenance of plasma [Na^+] by prolactin and found that it was dose dependent over approximately one order of magnitude (0.1 to 1 IU per animal). Of eight other possible hormonal contaminants (ACTH, growth hormone, thyrotropin, α-MSH, oxytocin, isotocin, vasopressin, and arginine vasotocin), none could substitute for prolactin. In the mollyfish *(Poecilia latipinna)*, hypophysectomy-induced osmotic failure in a dilute environment was accompanied by failure of Na^+ conservation. Total body sodium efflux increased by over sevenfold (from 1.6 to 11.8 $\mu g \cdot g^{-1} \cdot hr^{-1}$), whereas Na^+ influx did not change (approximately 0.5 $\mu g \cdot g^{-1} \cdot hr^{-1}$). Prolactin restored Na^+ efflux to normal (low) levels. In the same vein, Maetz and colleagues found that hypophysectomy of *F. heteroclitus* led to elevated Na^+ loss in freshwater, and prolactin acted to limit Na^+ loss.[80,81]

To discern relevant molecular mechanisms for the physiological effects of a hormone, a first question is, "what is the time course through which regulation occurs?" In coping with changes of environmental salinity, teleosts apparently require a relatively long period during which cellular and molecular adaptation occurs. Hypophysectomy does not generally result in immediate osmotic demise, but rather gradual electrolyte loss over several days in many widely varied species.[69,82] Normal adaptation to changes in environmental salinity include either increased or decreased numbers of "chloride cells" in the gill mucosa coincident with saltwater or freshwater adaptation, respectively. Such changes are seen over several days following transfer from one osmotic environment to another. In several teleost osmoregulatory surfaces, including pharyngeal mucosae, intestine, and urinary bladder, Bern and co-workers[77] showed that osmoregulatory effects of PRL took place with time courses of at least greater than several hours.

The relevant tissues for regulating electrolyte concentrations are far more extensive in aquatic than in terrestrial environments. The integument, gills, pharyngeal mucosa, and intestine all are in proximate environmental contact. In the teleost urinary tract, the bladder is an active regulatory organ[83] as well as serving its storage function. Each of the aforementioned organs, as well as the kidney, is regulated by prolactin. However, the gills and the associated mucosa of the pharynx and opercles are probably the most important adaptive organs because of their enormous surface area and direct contact with the external medium. Not surprisingly, they have received extensive experimental attention and are coming to be reasonably well understood.

The primary cellular component involved in branchial electrolyte regulation is the "chloride cell".[84,85] Chloride cells are mitochondria-rich columnar epithelia with highly invaginated cell surfaces and a distinct polarity of anatomy, indicating unidirectional transport. They are present not only in the gill epithelium, but also in the lining of the pharynx and opercles.[85] In individuals adapted to saltwater, they are much more common than in freshwater.[86-88] Electron micrographic analysis[89] of eel's gill epithelium shows chloride cells to be the sites of the bulk of Na^+-K^+-activated ATPase activity in this tissue. Recent tracer and electrophysiological data conclusively show that chloride cells are the sites of active ion extrusion from the gills;[90] thereby they contribute significantly to salt excretion in saltwater adaptation and act counter to necessary salt conservation in freshwater. Differentiated chloride cell numbers are significantly depressed by prolactin in several species studied.[69] Gill Na^+-K^+-ATPase, apparently involved in salt excretion, is elevated by hypophysectomy and repressed by prolactin injections in both killifish[63] and eels.[91] This is in contrast with renal tubular Na^+-K^+ ATPase which is stimulated by prolactin.[63,91,92] The renal Na^+ pump, as in terrestrial vertebrates, is important for reabsorption of salt from renal tubular fluid. Contrary statements in the literature notwithstanding,[68,69] I can find no convincing evidence that there is prolactin-stimulable ion uptake in gill epithelium. It is conceivable that heterogeneity of Na^+-dependent ATPases, such as suggested by Gallis and co-workers,[93] might make possible regulation of both efflux and influx of salt by prolactin. However, all available evidence for gill tissue points only to repression of ion excretion as important. Chloride cell salt secretory activity is under the positive regulation of corticosteroids.[67] Therefore, prolactin is antagonistic to this effect and opposes the differentiation of branchial epithelial cells into the highly characteristic "chloride cell" cytology. An interesting question, apparently not addressed yet, is whether prolactin represses chloride cell responsiveness to cortisol per se, or whether the opposition is accounted for by inhibition of a cortisol-induced parameter (e.g., ATPase) distal to cortisol reactivity.

Another predominant epithelial cell type present in gill, pharynx, and also the integument, is the mucous cell. This cell type is positively regulated by prolactin. In a wide variety of species, including standard models such as the killifish, stickleback, and eels, as well as various more obscure species, branchial and/or integumentary mucous cells are known to be stimulated by prolactin.[61,94,95] Conversely, hypophysectomy generally represses mucous cells.[61] In molecular terms, prolactin injection into eels increases the sialic acid content of the skin,[98] sialic acid being a primary component of secreted mucopolysaccharide. Elaboration of mucus provides an important osmotic "buffer zone" to the surface of epithelia in proximate contact with the environment. Ionic concentrations can be maintained at reasonably high concentrations near the cell surfaces, because the mucous layer turns over only very slowly, whereas the external water is constantly changed.

The stimulation of mucous secretion has been elaborated in at least one species of fish to serve as nourishment for their young. In several of the cichlid fishes, copious mucous secretion along with sloughed cells are grazed by the offspring. Prolactin is responsible for this "milk" secretion.[97]

As previously stated, prolactin positively regulates renal Na^+-K^+ ATPase in *F. heteroclitus*.[60] In freshwater, the induction of this enzyme could conceivably contribute to excretion of a dilute urine. A cytological correlate of this induction is an increase of cell surface particles in freeze-fractured renal tubule cells.[98] The paradox of branchial repression and renal induction of Na^+-K^+ ATPase in a given species points up the integrated, organizational response of individuals to prolactin. Specific uptake of radioiodinated prolactin by teleost kidney has been demonstrated.[99] This apparent binding was found in tubule epithelium and could not be displaced by excess growth hormone or luteinizing hormone. The number of filtration units in the teleost kidney is adaptively regulated by prolactin. Glomerular tufts are increased in number by prolactin injection,[100] and this effect presumably contributes to the elevated urine volume excreted in freshwater.[69]

Prolactin also affects cellular differentiation in the teleost urinary bladder. In vivo[101] or in vitro[87] treatment of teleost bladders with ovine prolactin for several days resulted in decreased osmotic permeability. Co-culture of bladders with rostral pituitary lobes (almost exclusively lactotrophs) also yielded reduced osmotic permeability;[102] therefore, this effect is characteristic of both heterologous mammalian prolactin, as well as homologous teleost prolactin. The reduced water permeability was not accompanied by any changes in ion transport.[102] Presumably, reduced bladder osmotic permeability, with its slow time course, depends on cellular differentiation; corollary with this is prolactin-induced DNA synthesis in flounder urinary bladder as measured by thymidine incorporation.[103]

Prolactin-binding activities in teleost osmoregulatory tissues have been observed. Fryer[104] originally used purified homologous tilapia prolactin to identify binding in tilapia tissues. Iodinated tilapia prolactin bound specifically to kidney microsomal fractions. Neither liver nor gill membranes had significant prolactin-binding activity. The renal prolactin binding appeared to be higher in freshwater- than in seawater-adapted fishes, suggesting that the receptors may be regulated by endogenous prolactin levels. While tilapia renal prolactin receptors bound mammalian (ovine) prolactin at reasonably high affinities, mammalian prolactin receptors (rat liver) had no affinity for either prolactin or growth hormone from tilapia. Binding of radioiodinated ovine prolactin to teleost kidney was also observed autoradiographically by Gona.[99] This binding was displaced by excess unlabeled prolactin, but not growth hormone or other pituitary hormones. In a recent study, binding of ovine prolactin to several teleost target tissues was examined.[105] Osmoregulatory tissues which had measurable receptor activities included gill, intestine, kidney, and urinary bladder, albeit at apparently low levels. Binding, however, was expressed only as percent specific binding (0.5 to 1.6%), so it is not possible to know the apparent numbers of binding sites from these data. High levels of prolactin receptor (207 fmol/mg membrane protein) were found in tilapia liver and equivalently high levels in both ovary and testis.

The primary mechanisms responsible for teleost electrolyte conservation induced by prolactin, which are graphically illustrated in Figure 1, include inhibition of active ion efflux in the gill chloride cells, probably because of Na^+-K^+ ATPase repression, stimulation of active salt reabsorption in kidney (Na^+-K^+-ATPase induction), decreased bladder water permeability leading to excretion of a dilute urine, and stimulation of the protective integumental mucous coat. Little is known regarding the cellular mediation of these effects. Foskett and co-workers[106,107] examined the possible role of cyclic nucleotides. Inhibition of cyclic nucleotide phosphodiesterase by isobutyl-methylxanthine (IBMX) opposed prolactin in the inhibition of gill salt extrusion. This is similar to evidence in mammary gland where cyclic AMP acts contrary to prolactin, and methyl xanthines inhibit prolactin responsiveness. Cyclic AMP then does not appear to mediate prolactin action in the teleost gill. Prostaglandin involvement is implicated by studies of Horseman and Meier.[108] Inhibitors of prostaglandin synthesis (indomethacin, aspirin) depressed plasma chloride concentrations in intact animals and blocked

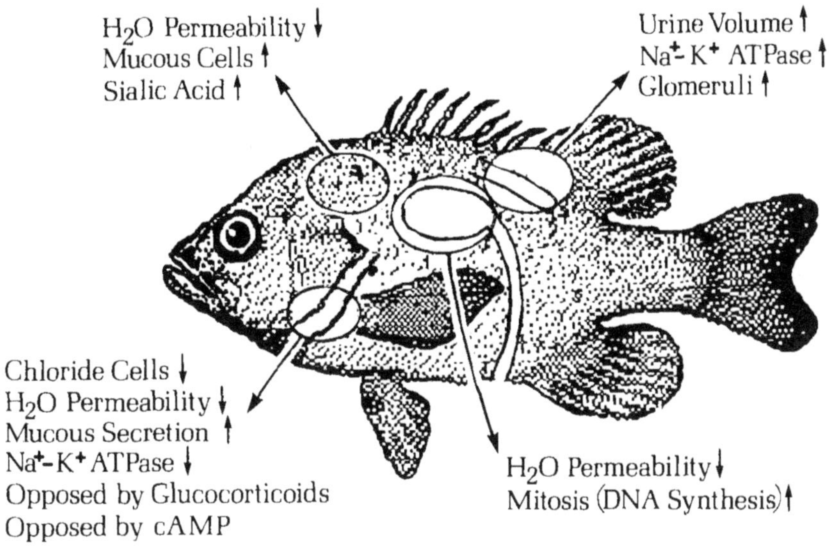

FIGURE 1. The major prolactin targets and specific actions on teleost water and ion balance. The four major targets are (clockwise from upper left): the skin, kidney, urinary bladder, and gill (including opercular lining and pharynx). Note that the actions on nonrenal tissues are all focused on decreasing water gain in a freshwater environment. Renal actions focus on increasing the volume of filtrate and urine and on conserving sodium (and consequently chloride) ions in the face of elevated urine production.

the maintenance of plasma electrolyte by prolactin. While the elevation of chloride concentrations by prolactin was prevented by either aspirin or indomethacin, prostaglandin E_1 induced significant chloride elevation in aspirin or indomethacin-treated animals.[108] These data are consistent with involvement of prostaglandins in at least some of the osmoregulatory actions of prolactin. Determination of precise mediating mechanisms will depend on analysis of specific molecular responses in prolactin-dependent cells, such as branchial chloride cells.

III. PROLACTIN MECHANISMS IN AMPHIBIAN METAMORPHOSIS

Prolactin is an important regulator in the complex and varied life cycles of amphibians. Prolactin apparently acts in at least three distinct developmental capacities. First, it opposes some, but not all, of the metamorphic stimuli of thyroxine.[57] Second, prolactin promotes larval growth independent of any metamorphic changes.[109] Third, prolactin is the hormonal determinant of "second metamorphosis" in certain newts that have a distinct aquatic form for reproductive purposes.[110] The antimetamorphic and thyroxine antagonistic effects of prolactin are apparently not shared with mammalian growth hormones, whereas interestingly, the growth-promoting effects in amphibian larvae are effectively common to both prolactins and growth hormones, although prolactin may be somewhat more potent.[34,109] The opposition of thyroxine-induced metamorphosis takes place in the peripheral target tissues rather than by way of inhibition of thyroxine secretion.[111,112]

The effects of prolactin on tadpole development are highly tissue specific and depend on the stage of development when treatments are made. In premetamorphic tadpoles prior to thyroxine influence, prolactin inhibits hindlimb growth and increases tailfin cell division.[113] Yoshizato and Yasumasu[114-117] defined a number of molecular mechanisms by which prolactin was both anabolic and anticatabolic in the tailfin of premetamorphic bullfrog *(Rana catesbeiana)* tadpoles. The central connective tissue core running through the tailfin of tadpoles was markedly enhanced by prolactin injections, and both collagen synthesis (hydroxyproline incorporation)[114,116] and hyaluronic acid synthesis[115] were stimulated by prolactin. Increases in both these synthetic parameters were dose dependent between 0.1 and 10 units of bovine prolactin per day. Prolactin depressed a number of catabolic processes, including hyaluronidase activity, collagen breakdown, and ribonuclease.[116]

In direct comparisons of molecular responses of tadpole tailfin and liver tissues, both of which have prolactin receptors,[119] marked differences in prolactin responsiveness have been measured.[120,121] In tailfin tissue, prolactin effectively opposed thyroxine induction of lactate dehydrogenase (LDH)[120] and β-glucuronidase,[121] whereas hepatic induction by thyroxine of glutamate dehydrogenase,[120] ornithine transcarbamylase, and tyrosine aminotransferase[121] were not opposed by prolactin.

At metamorphosis, one of the actions of prolactin appears to be the prevention of degradative enzyme release in tissues which undergo involution. In the tail[122,123] and gut,[123] prolactin acts to stabilize lysosomes against enzyme release. Whereas the total concentrations of acid phosphatase were not different in tailfin tissues with or without prolactin, the level of soluble, nonlysosome-bound activity was depressed by 80% following prolactin injections.

Prolactin receptors have been measured in each of four tissues that have been examined in tadpoles: gill, liver, kidney, and tailfin.[47,124,125] It seems likely, given the major organizational changes of metamorphosis, that many tissues are directly responsive to prolactin in amphibian larvae. Receptor levels for prolactin undergo marked change at metamorphosis. White and Nicoll[47] first observed a large increase in renal prolactin receptors at metamorphical climax and their data were substantiated by autoradiographic observations of Gona.[124] Prolactin receptors in both tailfin and liver, in addition to kidney, increase markedly at metamorphosis.[126] The rise of prolactin receptor activity reflects numbers of sites and not affinity.[126] Since receptor activity in the kidney returns to premetamorphic levels in the young froglets, it is unlikely to be primarily reflective of increased renal involvement in ion balance in the terrestrial form as suggested by White and Nicoll,[47] but rather it may related more to the metamorphical growth or differentiation of the kidney.

The characteristics of tadpole receptors for prolactin have several interesting features. Unlike other vertebrate receptor systems examined, tadpole liver and tailfin receptors bind both ovine prolactin and ovine growth hormone with nearly equivalent affinity.[13] In contrast, two different approaches suggest that prolactin receptors in tadpole kidney have little or no crossreactivity with growth hormone. Gona[124] showed that ovine prolactin, but not growth hormone, could displace autoradiographically assayed ^{125}I ovine prolactin from kidney tubules, and White and Nicoll[47] used electrophoretically purified frog prolactin and growth hormone and found that the former, but not the latter, effectively competed for kidney ^{125}I ovine prolactin binding. The tadpole liver and tailfin prolactin receptors, which apparently differ in hormonal specificity from the kidney receptor, are different from one another also. Neuraminidase treatment of either receptor containing membranes or solubilized prolactin receptor decreases the binding affinity of liver, but not tailfin prolactin receptors.[13,125] This enzyme effect indicates that carbohydrate moieties (sialic acid) are important for prolactin binding in liver, but not tailfin. In addition, the solubilized liver receptor displays a higher net charge than the tailfin receptor.[119] Although these studies have used primarily mammalian hormones as ligands, operationally, the finding of biochemical differences is meaningful. Whether the receptor characteristics which differ between liver and tailfin are in any way responsible for the lack of thyroxine antagonistic activity of prolactin in liver compared with tailfin[120,121] is not yet known. Given the presence of at least four receptor-containing tissues (gill, liver, tailfin, and kidney) and the already known biochemical and physiological differences among at least three of these target tissues, the amphibian tadpole offers rich promise.

The possibility that receptors such as those described above, which seem to differ slightly from one another in some characteristics, are truly members of a family of related receptor proteins is exciting. It seems likely that since the tools of molecular biology have shown that protein hormones are represented in moderate to large families of related genes,[127-129] we may also anticipate that the target receptors display similar diversity. It seems self-evident that at least portions of the receptors for prolactin, growth hormone, and other similar hormones diverged from a common genetic origin, as did the hormones themselves. I can think of no more favorable system to analyze such a possibility than the metamorphosing tadpole. If divergent forms of receptors are physiologically relevant, that relevance is likely to be exaggerated in the major developmental changes of metamorphosis.

IV. PROLACTIN MECHANISMS IN COLUMBID CROP

Of all the multifarious effects ascribed to prolactin in nonmammalian vertebrates, none has more experimental and historical prominence than the stimulation of "crop milk" production in Columbid birds (pigeons and doves). Oscar Riddle and co-workers discovered that pituitary glands were the source of a factor which could induce pigeon and dove crops to develop and produce crop milk,[130] and subsequently isolated the factor showing it to be distinct from other pituitary hormones known at the time.[131] Riddle's group also named "prolactin" and showed it to be indistinguishable from the pituitary principal which induced lactation in mammary glands.[1,131] Since then, crop response has been the established international bioassay standard for prolactin.[132] Subsequent work demonstrated that prolactin has other actions including brood patch development, parental behavior stimulation, antigonadal effects, and metabolic effects which are broadly integrative in organismal function.[2,34]

A number of facts make the production of Columbid crop milk one of the most remarkable of all the vertebrate responses to prolactin. Although sharing an analogous function with mammalian milk, the nature of crop milk is quite different. It is produced by the sloughing off of whole, highly differentiated mucosal cells (a holocrine secretion). Although very rich in protein and fat, crop milk is essentially carbohydrate free. The crop itself is a modified outpouching of the esophagus, and therefore of endodermal origin, whereas the mammalian milk glands are integumental, therefore ectodermal, derivatives. Other hormones apparently are not very important for the crop response, in sharp contrast with the usual requirement for several hormones for mammalian lactogenesis (so-called "lactogenic complex"). Prolactin is fully effective in hypophysectomized, thyroidectomized, adrenalectomized, and castrated pigeons.[133-135] Provision of exogenous growth hormone, thyroxine, and glucocorticoid in hypophysectomized pigeons produced only modest enhancement of crop response to prolactin at very high doses (>1 mg/day).[135] This contrasts with several other parameters in the same birds in which there were substantial synergisms and additivities of the several hormones.[135] Insulin requirements for crop response have apparently not been directly examined, but since crop growth is a highly anabolic process, insulin is likely to be necessary. Normal responsiveness of pigeon crop to prolactin is developed in hatchlings by as early as 1 week of age;[130] earlier times have apparently not been tested. One of the most marked contrasts of Columbid crop milk formation is its relative histological simplicity compared with mammalian lactogenesis. At rest, the crop mucosum is a thin membrane (8 to 12 cells thick) of squamous epithelium overlying a dense fibrous layer (lamina propria). Upon stimulation, the basal epithelial cells (stratum basale) begin to divide at a highly accelerated rate. Cells differentiate as they are forced toward the lumen by proliferation in the basal layer.[136] Ultimately the cells, filled with lipidic and proteinaceous deposits, are sloughed off into the lumen in sheets and curds. For the first 3 to 4 days of life, young pigeons and doves are fed exclusively on crop milk, after which their diet includes not only crop milk, but also seeds and other foods.[137] Involution of the parental crops begins slowly within 1 week after hatching as the squabs feed less on crop milk and more on seeds and other foods.

Prolactin levels in the circulation of nesting ring doves have been measured by heterologous radioimmunoassays with either anti-human prolactin,[138] or anti-turkey prolactin sera.[139] While absolute levels in each assay were quite different (lower and less range with anti-human prolactin[138]), some generalizations of importance are observable in these data. Circulating prolactin rises to peak levels in both male and female parents during the last 7 days of the 14-day period during which eggs are incubated. Crop development, measured as total weight, lags behind the increase of prolactin levels by 2 to 3 days and reaches its highest values during the first 4 days of feeding the hatchlings. During the latter stages of brooding, prolactin levels decline to basal levels and crop weight follows as the young develop independent feeding.[138,139] Using the more sensitive anti-turkey prolactin serum, Cheng and Burke[139] observed a small, but significant, prolactin increase in females near the time of their ovulation.

The feeding of the hatchlings is a strong stimulus for prolactin secretion in doves, as in nursing behavior in mammals.[140,141] The feeding of crop milk to squabs is commonly referred to as "regurgitation" behavior; however, since the milk is not food the parents have taken in and is not mixed with food, at least during the first few days, this behavior is quite different from the regurgitation behavior of many birds that return partially digested food for their offspring. When access to squabs is limited experimentally, hungry, food-deprived squabs provide a stronger crop growth stimulus than do well-fed squabs.[141] Much of the stimulus from the young is apparently tactile, since squabs present, but isolated by a clear plastic screen, provided a weaker crop stimulus.[141] Prolactin release from parent dove pituitary glands, measured by radio-receptor assay, is stimulated during exposure to squabs.[142]

The cytological changes in the crop during stimulation by either injected prolactin or incubation were thoroughly described by Dumont.[136] In addition to the obvious induction of cell proliferation in the basal epithelium layer, only two prominent changes in cell structure were seen during stimulation. These included development of an active ribosomal system for protein synthesis and the growth of large lipid (actually lipoprotein) globules within the mucosal cells.[136] The induction of accumulation of these globules is highly specific for tetrapod prolactins. Even though other prolactins (i.e., fish) induce crop growth, they do not induce lipoprotein globule formation.[143] The differentiation of crop cells is noticeable earliest within the layer of cells between the proliferative basal epithelium and the cells lining the lumenal surface. This layer was termed the "stratum spinosum" (or nutritive layer) by Dumont.[136] In unstimulated control animals, the stratum spinosum contained relatively large numbers of free single ribosomes, but few polysomes and little rough endoplasmic reticulum. There were only small lipoprotein globules in occasional cells. By as early as 12 hr after a single prolactin injection, lipoprotein globules, though small, were already common. The protein synthetic apparatus aggregated to form large polysomes which were not bound to endoplasmic reticulum (rough ER). Golgi apparatus was not prominent in these, or any cells of the crop epithelium, under any treatment or brooding conditions.[136] The lack of rough ER and Golgi apparatus correlates with the fact that crop milk is not a secreted product. A prominent, and unusual, feature of prolactin-stimulated crop tissue was the close association of many polysomes with the lipoprotein globules.[127] As will be elaborated subsequently, these globules contain large quantities of stored protein as well as lipid.[144] It is the synthesis of these proteins in which ribosomes coating the lipoprotein globules are presumably engaged. By 4 days of prolactin treatment, the cells which are sloughed off the lumenal surface of the crop are heavily laden with lipoprotein deposits, thereby accounting for the "cheese-like" characteristics of crop milk.

One of the primary responses of crop tissue to prolactin is, obviously, the accelerated mitotic rate of cells in the basal epithelium. It is intentionally stressed that this represents an increased rate of division in an already proliferating population of cells. Since the crop is part of the digestive tract, it normally has a relatively high cell turnover rate. Whether the mechanisms regulating the normal rate of divisions are related to those involved in the prolactin response is unknown. It is possible that other factors may control the basal rate of cell division. Epidermal growth factor is involved in the control of alimentary tract cell division,[145] and it also causes a proliferation in crop tissue which is not morphologically similar to the prolactin response.[15,185] Although generally measured gravimetrically,[146] mitogenesis has been assayed directly as induction of DNA synthesis by Ben-David.[147] Radiolabeled thymidine incorporation was dependent on the dose of prolactin and paralleled crop weight increase over at least three orders of magnitude (0.04 to 4 mg in 4 injections over 2 days). Induction of ornithine decarboxylase (ODC) activity may be an important component of the mitogenic signal of prolactin in crop.[33] ODC induction is generally associated with a variety of chemical and physical stimuli for cell proliferation[148-150] and has been implicated in the action of prolactin in mammary gland and other target tissues.[151,152] Prolactin injected locally over one hemicrop caused approximately a 2.5-fold stimulation of ODC activity. That stimulation was directly correlated with mitogenesis, since no stimulation occurred when injections were made at a circadian time when prolactin did not cause elevated DNA synthesis.[33]

Recent evidence regarding the mitogenic mechanism of prolactin in the crop strongly implicates the participation of a somatomedin-like growth factor.[15,153] Various members of the family of insulin-like growth factors (IGFs — such as relaxin, somatomedin-C, proinsulin, and insulin) synergistically enhanced the stimulation of crop growth by prolactin, although none had an effect by itself.[15] Somatomedin-C was the most potent prolactin synergist. Prolactin locally injected (1 μg) was sixfold more stimulatory to crop mucosal weight when coinjected with somatomedin-C (8 μg). Proinsulin, but not prolactin, stimulated DNA synthesis rapidly (within 6 hr) in vitro in crop tissue which had previously been primed with prolactin, but not in unprimed crop tissue.[15] This latter result suggests that prolactin causes the crop to become sensitized to an insulin-like growth factor. Systemic injection of growth hormone, prolactin, or other preparations which might be expected to stimulate somatomedin secretion increased crop responsiveness to local prolactin more than additively, suggesting that an appropriate growth factor can be induced in the pigeons.[153] Pigeon serum[153] and culture medium conditioned by incubation with pigeon or rat liver[185] contain a factor(s) which synergizes with prolactin to cause crop mitogenesis. These results together have been taken to indicate that crop mitogenesis is dependent not only on prolactin, but also on an insulin-like growth factor for which Nicoll's group has coined the name "synlactin".[15] The particular molecular responses to prolactin which may be dependent on the presence of IGFs are unknown. Prolactin is known to cause secretion of insulin-like growth factors from mammalian liver.[154-156] Hyperplastic growth is a component of the full spectrum of prolactin effects including, as examples, integumentary mucous cell proliferation in fishes,[95] teleost bladder mitogenesis,[103] larval amphibian growth,[109] brood patch development in birds,[157] and mammary gland growth.[158,159] The role of growth factor mediation, therefore, requires substantial future attention to elaborate the mechanisms of prolactin. If a liver-derived IGF is of importance in mediating some prolactin effects, this would rationalize the common finding of relatively high levels of prolactin receptors in liver tissue of a number of species.[119,160-162]

Binding of prolactin in crop was first evidenced by uptake of radioiodinated prolactin by crop in vivo.[163,164] Isolated crude microsomes were a source of specific, high-affinity prolactin-binding sites.[165,166] The crop microsomal preparation had a prolactin dissociation constant of 7×10^{-10} mol · ℓ^{-1}. Preparations of four mammalian prolactins had binding potencies directly correlated with their relative crop stimulating activity. Low levels of binding of growth hormone preparations were consistent with the known levels of prolactin contamination in those preparations.[165,166] Kledzick and co-workers[165] found that prolactin stimulation elevated binding by severalfold and that juvenile birds had higher levels of binding than unstimulated adults. Apparent "upregulation" of prolactin receptor sites by prolactin stimulation has been studied by Shani and colleagues.[167] The concentration of binding sites for prolactin was increased sevenfold by injection of low doses (1 μg twice per day) of ovine prolactin locally. Systemic injections of antiserum to prolactin caused a 63% reduction of prolactin receptors in crop.

Induction of crop mucosal proliferation is accompanied by a dramatic and specific phenotypic differentiation. Cytologically, the primary elements of crop differentiation are assembly of an active polysomal protein synthesizing apparatus and the accrual of stored lipoprotein globules.[136,144] The role of gene expression in crop stimulation was experimentally tested in studies of Sherry and Nicoll.[168] Blockage of either RNA (and/or DNA) synthesis by actinomycin D or 80S ribosomal protein synthesis by puromycin inhibited prolactin-induced crop mucosal proliferation. Although nonspecific effects of either drug are contingencies, these results indicate a likely role for prolactin-induced gene expression. Amino acid uptake and incorporation into protein was stimulated within 16 to 20 hr after a single prolactin injection over the crop.[169] In a further study, Sherry and Nicoll[168] isolated partially pure RNA fractions from either induced crop or pigeon liver tissue and injected them locally over crop tissue. The RNA isolated from prolactin-induced crop stimulated crop growth, and that stimulation was eliminated by prior RNase treatment. Liver RNA had no stimulatory effect. The capacity of crop RNA to induce growth in a prolactin-like manner suggests that crop mucosal proliferation is secondary to alteration of RNAs and points to the pivitol role of prolactin actions on gene expression.

The most dramatic phenotypic correlate of altered gene expression in crop is the growth of large globules of lipid and protein.[136,144] The lipid in the globules is more similar to mammalian milk fat (unsaturated, medium chain length) than to adipose fat (saturated, long chain). However, crop milk lipids are apparently not synthesized in the crop per se. Neither of the two necessary fatty acid synthesizing enzymes, fatty acid synthetase[144] or acetyl CoA carboxylase,[186] are present in either control or stimulated crop tissue. Sequestering of lipid at least partially depends on lipoprotein lipase, which is induced by prolactin approximately tenfold in crop tissue and two fold in adipose.[170] The growth of lipid-containing globules has an earlier onset during stimulation (12 hr) than does lipoprotein lipase induction (2 days);[136,170] therefore, increased lipid uptake by this route probably is not a primary determinant of globule formation. The so-called "lipid globules" of the crop actually contain an unusually high protein concentration of 35% by weight,[144] and given the characteristic association of polysomes with them,[136] it is likely that protein is of major importance in lipoprotein globule growth. Upon isolation of protein from crop globules, we found that essentially all of this protein consisted of two polypeptides having molecular weights of 58,000 and 50,500.[144] The synthesis of these crop milk polypeptides (CMP58, CMP50.5) was stimulated over fivefold by prolactin injection, whereas total protein synthesis increased only approximately twofold. CMP58 is polymerized by disulfide linkage and occupies

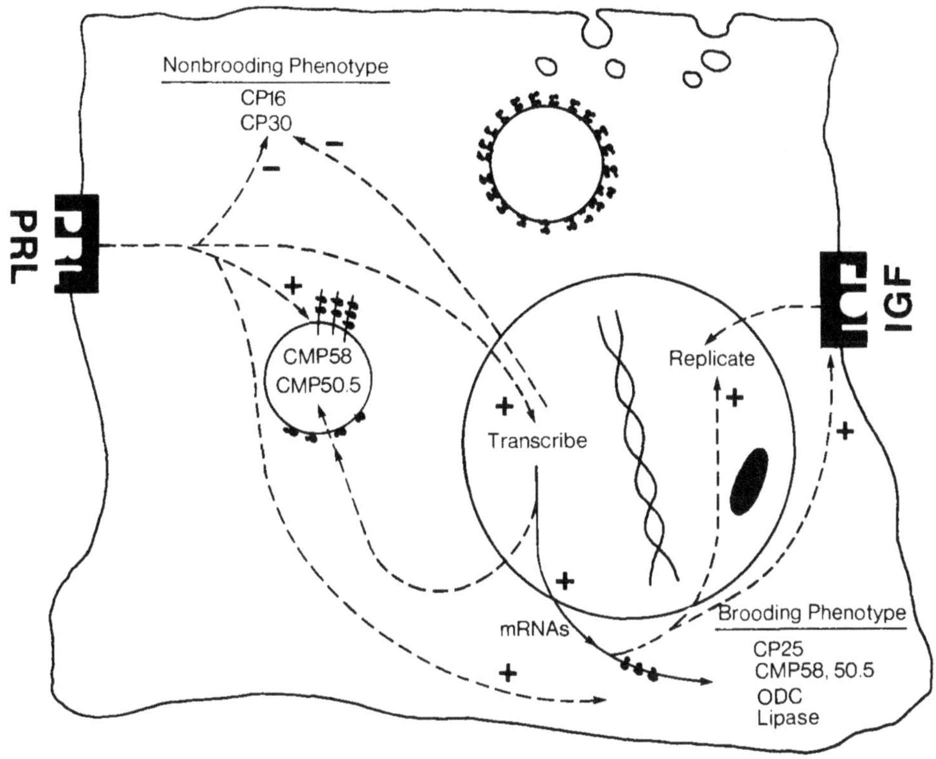

FIGURE 2. Schematic model of the control of crop cell proliferation and differentiation. The major differentiative change includes growth of large lipid globules which are surrounded by polysomes and contain large quantities of the proteins designated CMP58 and CMP50.5. A number of other proteins are included in the changes to the "brooding phenotype". The possible mediating pathways are indicated by broken lines. Insulin-like growth factor (IGF) is needed for mitogenic stimulation and prolactin apparently controls responsiveness to it. The regulation by prolactin not only includes induction of prolactin-regulated products, but also includes the selective repression of at least two proteins (CP16, CP30) present in the "nonbrooding" (unstimulated) crop tissue.

primarily the core of the lipoprotein globules, since it was not recognized when whole globules were used as an antigen, but was readily antigenic if the lipid globules were dispersed in an emulsion. CMP50.5 is exposed on the lipid globules surface and is sensitive to protease degradation, whereas CMP58 is largely intrinsic to the globule and only an 8000- to 9000-mol wt portion of it was digested when globules were treated with protease.[144] The mechanism by which CMP58 and CMP50.5 are induced might be either transcriptional or post-transcriptional. We have isolated mRNA from crop and translated in the cell-free reticulocyte lysate system. No identifiable CMP58 or CMP50.5 translation product has ever been detected in this system either by electrophoretic or immunoreactive criteria.[171] Since these polypeptides are highly hydrophobic, and reticulocyte lysate contains large amounts of a protein which binds to hydrophobic proteins and inhibits their translation[172] (SRP, signal recognition protein), it is possible that competent mRNA for CMP58 and CMP50.5 cannot be detected in this system. We have begun analysis of crop gene expression by recombinant DNA techniques, and it is anticipated that mRNA for these polypeptides may be discovered by this approach. Probes of this sort will make possible direct analysis of regulation of these interesting proteins involved in lipid storage and nourishment of young.

Differentiation of the crop mucosal epithelium is observed at the molecular level as specific changes at both the protein and RNA levels.[148,171] The differentiation state induced by prolactin includes several gene products distinctive for the prolactin response (Figure 2). I refer to the two differentiation states as "brooding phenotype" (prolactin dependent) and "nonbrooding phenotype" (unstimulated). Gradient sodium dodecyl sulfate (SDS) polyacrylamide gel electrophoresis yielded over 80 distinct resolvable polypeptide bands in cytosol extracts from crop tissue.[171] In these protein extracts two specific bands were present only in control samples (designated CP16, CP30), whereas two bands were present only after prolactin treatment (designated CP25, CP154). Protein isolates prepared after in vivo labeling with tritiated amino acids reflected appropriate changes in synthesis of proteins comigrating with CP16, CP30, and CP25. With the exception of these proteins, a pattern of general similarity of 95% of the profile is maintained, in spite of the dramatic change from the relatively inactive nonbrooding phenotype to the highly differentiated and metabolically active brooding phenotype. It is likely, then, that the abundant gene products regulated by prolactin are of essential importance in the induction of the brooding phenotype.

By direct extraction of crop tissue with phenol-SDS followed by oligo-dt cellulose chromatography, we isolated high-quality mRNA, which we then translated in the cell-free reticulocyte lysate system.[171] After systemic (intramuscular) prolactin injections, mRNA for CP25 was induced (see Figure 2) and mRNA for CP16 was repressed.[171] Local intradermal injection of prolactin over one hemicrop caused a dose-dependent induction of translatable mRNA for CP25 ($mRNA_{CP25}$) (Figure 3), and no other changes in translatable mRNAs were observed.[171] The CP25 product might be imagined to be related to caseins since it is of the same size, is controlled by prolactin, and used to nourish offspring. However, there is no evidence it is secreted, and phosphorus labeling shows it not to be phosphorylated as are the caseins.[186] It is noteworthy that the electrophoretic pattern of protein bands in isolates from crop tissues is very faithfully replicated in the pattern of translation products from crop mRNAs. At least up to a molecular weight of over 60,000, the relative proportions of the major protein bands is similar in the protein directly isolated from crop and that synthesized from crop RNA. This includes an increase in the relative amount of both protein and translation product at a molecular weight of approximately 40,000. Therefore, much of the regulatory action of prolactin is accounted for by changes in mRNA levels. $mRNA_{CP25}$ appears to be the predominant gene product regulated by prolactin since it is absent in the nonbrooding phenotype and highly abundant (14% of translation profile)[171] in the brooding phenotype.

The differentiation and proliferation of crop mucosal cells provides a model of prolactin mechanisms with great potential value. The essential features known and hypothesized regarding the crop response are summarized in Figure 2. The model proposes that the primary elements of the crop response focus on transcriptional, and possibly post-transcriptional regulation of gene expression, so as to promote the "brooding phenotype" and repress the "nonbrooding phenotype". The brooding phenotype includes molecular changes associated with the synthesis of characteristic proteins and lipid sequestration. It also includes cell proliferation, which is the result of prolactin-dependent sensitization to an insulin-like growth factor. That sensitization might include expression of growth factor receptors and/or synergism along the pathway(s) mediating the effect of the growth factor.

Several studies have examined the potential mediating mechanisms for prolactin action in the crop. As is true for other prolactin targets, no satisfactory mediator is known; however, some of the components likely to be involved have been identified. Shani and co-workers[173,174] strongly proposed cAMP as the prolactin mediator in the crop. They stated that crop mucosal proliferation was stimulated by dibutyryl cyclic AMP, theophylline, or aminophylline, but not by cAMP,[173] and they reported that four intradermal doses of prolactin injected over 2 days resulted in elevation of crop

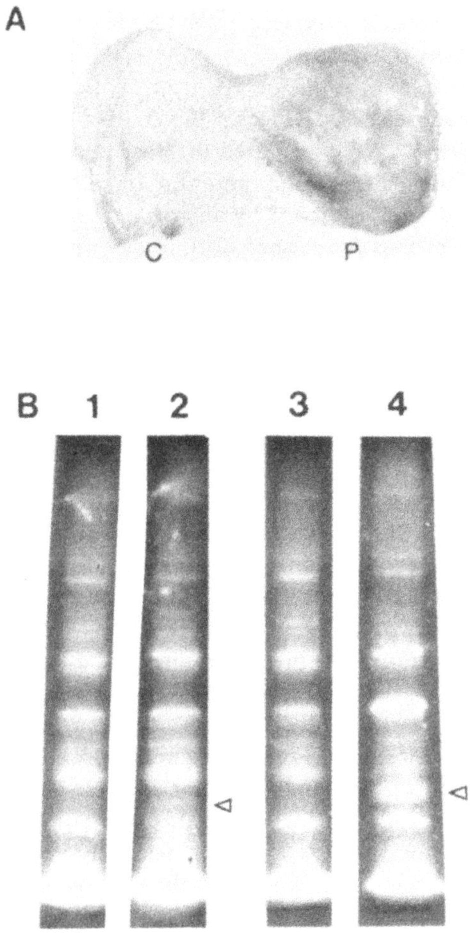

FIGURE 3. The induction by local intradermal prolactin injections of morphological (A) and molecular (B) differentiation in the pigeon crop. Three daily injections of even relatively high prolactin doses (50 μg/day) over one hemicrop do not cause any observable stimulation of the contralateral lobe of the crop. In section (B), lanes 1 and 2 are left and right hemicrops of one bird, and lanes 3 and 4 are from another. Messenger RNA was isolated from each tissue, translated in a cell-free system with radioactive amino acids, and the proteins synthesized were electrophoresed and visualized by autoradiography. Each of lanes 1 and 3 are controls with no prolactin. Lane 2 represents crop tissue treated with 16 μg/day of prolactin and lane 4 represents crop tissue treated with 25 μg/day. The major prolactin-induced protein "CP25" is marked by the arrow. (From Pukac, L. A. and Horseman, N. D., *Endocrinolgoy*, 114, 718, 1984. With permission.)

cAMP levels measured by radioimmunoassay.[174] Their results, however, are not consistent with findings of Sinha and Schmidt[175] and also Anderson and colleagues,[176] both of whom showed that dibutyryl cAMP had no effect on crop sac proliferation either by itself or when given in combination with prolactin. However, cCMP or its dibutyryl derivative enhanced responsiveness to locally injected cCMP in mediation of the effect of prolactin per se is open to question; evidence of cCMP involvement in cell proliferation in other systems is available.[177]

Prostaglandins have been implicated in the transduction of prolactin actions in mammary gland,[178] osmoregulatory surfaces,[108] liver of a teleost,[30,60] rat mesenteric vascular bed,[179] and rat liver.[180] We have tested the possibility that prostaglandins may be involved in the mediation of crop stimulation by determining if induction of $mRNA_{CP25}$ by prolactin could be mimicked by any of several prostaglandins. We made local injections of either prolactin or one of four different prostaglandins (PGE_1, PGE_2, $PGF_{1\alpha}$, and $PGF_{2\alpha}$) or arachidonic acid over hemicrops of two birds and then dissected out underlying crop sections, extracted mRNA, and translated the mRNA in reticulocyte lysate as described.[171] Analyses of the electrophoretic patterns of representative pairs of translation products are shown in Figure 4. PGE_2, and to a significantly lesser extent, $PGF_{2\alpha}$, induced expression of a mRNA coding for product comigrating with authentic CP25. Neither the other prostaglandins nor arachidonic acid caused any induction of putative $mRNA_{CP25}$, even though each of the treatments caused marked inflammatory hypertrophy near the injection site. The most obvious interpretation of these data is that prostaglandins act to mediate prolactin induction of $mRNA_{CP25}$. Alternative interpretations of these results include the possibility that PGE_2 (and $PGF_{2\alpha}$) amplified the effectiveness of the low titers of endogenously secreted prolactin present in the animals, or that the band observed is not authentic CP25. To specifically address the role of possible mediators, including prostaglandins, molecular probes for prolactin-induced gene products in crop would provide more unambiguous data.

We have begun to use molecular cloning to produce DNA complementary to prolactin-induced mRNAs from the crop. That work has thus far yielded clones for at least three separate induced gene products[187] which are being characterized. Because prolactin actions in tissues of all the vertebrate groups generally involve either differentiational or long-term metabolic changes, it is likely that a large fraction of the effects of prolactin are mediated through changes in gene expression. It is for this reason that use of molecular probes to analyze prolactin action in the crop holds promise of general importance. The only other system in which molecular probes of prolactin action are available is the mammary gland. While that system is very important, its endocrine and histological complexity limit its use in many ways. The columbid crop stands as an attractive alternative.

V. PROSPECTS AND PERSPECTIVES

While the wide variety of physiological functions attributable to prolactin has been impressive to endocrinologists for two decades, it has become clear that this is only the outward manifestation of a wider array of molecular changes brought about in response to the hormone. Many attempts have been made to "unify" the actions of prolactin into understandable, functional, or mechanistic categories.[34,181] Some have failed because they have not been generalizable (i.e., "osmoregulaton") and others because they include actions with widely different mechanistic bases (i.e., "parenting"). The temptation to look for generalizations is, however, irresistable, so I will willingly join. The exercise has been valuable, if only to focus attention on patterns and model systems amidst the constantly growing list of varied examples.

Today our attention is galvanized on the mechanisms of gene regulation in eukaryotes. That focus, and the different perspective that is necessary when one considers complex animals rather than simple prokaryotic systems, offers the prospect of comprehending a general prolactin mechanism while accommodating variety. The actions of agents which exert their effects through gene action are characteristically varied and

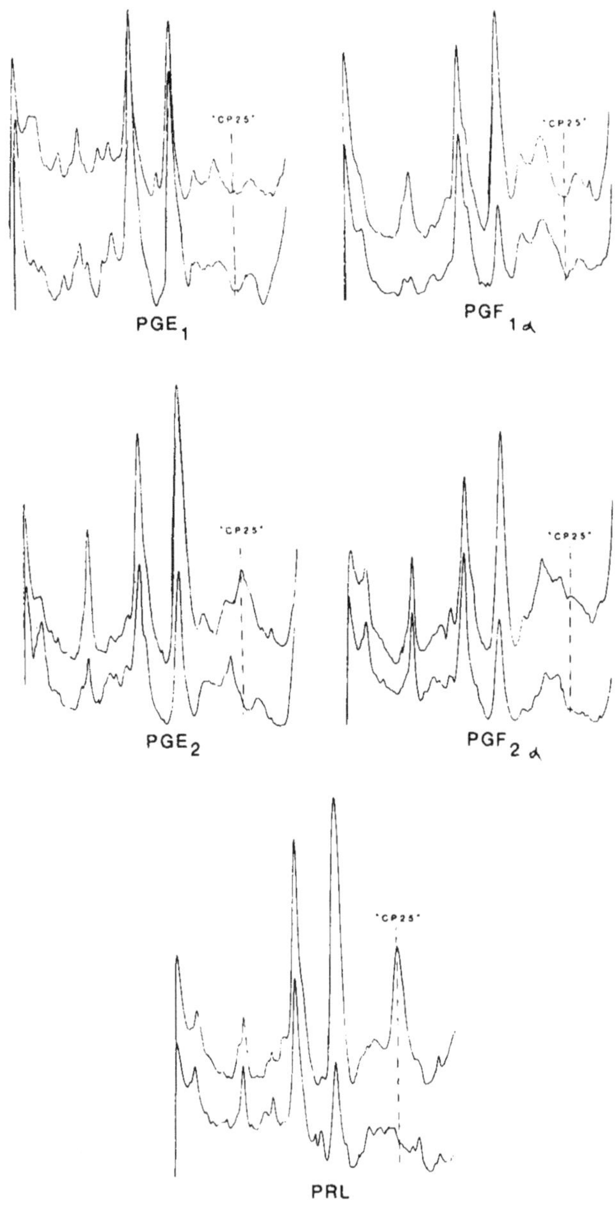

FIGURE 4. Analysis of translation products from mRNA of crops treated either with prolactin (PRL) or various prostaglandins. The figure includes densitometer tracings of five pairs of translation products such as those shown in Figure 3. The lower of each pair is the control hemicrop and the upper is the hemicrop treated with the agent. On each pair the migration position of authentic CP25 is marked. In the PRL pair, the large induction of CP25 is obvious from the major peak in the upper tracing. In neither PGE_1- or $PGF_{1\alpha}$-treated crops is there any evidence of CP25 induction. PGE_2 caused significant induction of product comigrating with CP25, and $PGF_{2\alpha}$ also caused minor induction of this band.

tissue specific. For example, estrogen induces vitellogenin in liver, uteroglobin in uterus, ovalbumin in oviduct, prolactin in pituitary gland, and other products in other tissues. If prolactin acts primarily (maybe exclusively) through alterations of gene expression, then we might understand the variety of its actions as expressions of the various differentiation states of its target tissues. This notion presumes a "two-step" process of gene product induction wherein some unknown "developmental factor(s)" yields different sets of potentially expressable genes in different tissues. The regulated products are thereafter expressed in response to a prolactin signal. This type of two-step model is widely applied in current concepts of gene regulation.[182,183] (In my present context, it is useful in that it asks us to look for a mechanism which is common to many target tissues by which prolactin induces specific gene expression.) The basis of the variation from one prolactin-regulated product to another (renal ATPase, mammary gland caseins, crop proteins, or hepatic enzymes) then becomes a "developmental" question rather than a question regarding prolactin action per se. Undoubtedly, not all of the actions of prolactin will be understood with this type of concept. Certainly some variations of prolactin action are attributable to multiple forms of the hormone (equivalent to placental lactogens in mammals) and the likely heterogeneity of prolactin receptors. However, the attractive possibility of discovering order and common mechanisms out of diversity is the challenge of understanding the actions of prolactin. The various nonmammalian models have given insight into this remarkable hormone and its related hormones and offer opportunities to develop a richer insight into the physiology and molecular biology of the actions of prolactin.

ACKNOWLEDGMENTS

I gratefully acknowledge the scientific and technical contributions of Lisa Nollin and Laurie Pukac, and thank Barbara DeNoyer for typing the manuscript. Some of the work described was supported by NSF grant PCM84-02368.

REFERENCES

1. Riddle, O. and Bates, R. W., Concerning anterior pituitary hormones, *Endocrinology*, 17, 1039, 1933.
2. Riddle, O., Prolactin in vertebrate function and organization, *J. Natl. Cancer Inst.*, 31, 1039, 1963.
3. Tata, J. R., What is so unique about hormone action?, *Mol. Cell. Endocrinol.*, 36, 17, 1984.
4. Liscia, D. S. and Vonderhaar, B. K., Purification of a prolactin receptor, *Proc. Natl. Acad. Sci. U.S.A.*, 79, 5930, 1982.
5. Posner, B. I., Kelly, P. A., and Friesen, H. G., Prolactin receptors in rat liver: possible induction by prolactin, *Science*, 188, 57, 1975.
6. Gout, P. W., Beer, C. T., and Noble, R. L., Prolactin-stimulated growth of cell cultures established from malignant Nb rat lymphomas, *Cancer Res.*, 40, 2433, 1980.
7. Tanaka, T., Shiu, R. P. C., Gout, P. W., Beer, C. T., Noble, R. L., and Friesen, H. G., A new and specific bioassay for lactogenic hormones: measurement of prolactin and growth hormone in human serum, *J. Clin. Endocrinol. Metab.*, 51, 1058, 1980.
8. Shiu, R. P. C., Elsholtz, H. P., Tanaka, T., Friesen, H. G., Gout, P. W., Beer, C. T., and Noble, R. L., Receptor-mediated mitogenic action of prolactin in a rat lymphoma cell line, *Endocrinology*, 113, 159, 1983.
9. Russell, D. H., Matrisian, L., Kibler, R., Larson, D., Magun, B. E., Poulos, B., and Copeland, J. G., Cyclosporin A inhibits the binding of prolactin to human peripheral blood lymphocytes, *Fed. Proc.*, 43, 818, 1984.
10. Russell, D. H. and Snyder, S. H., Amine synthesis in regenerating rat liver: effect of hypophysectomy and growth hormone on ornithine decarboxylase, *Endocrinology*, 84, 223, 1969.

11. Canellakis, E. S., Viceps-Madore, D., Kyriakidis, D. A., and Heller, J. S., The regulation and function of ornithine decarboxylase and of the polyamines, *Curr. Top. Cell. Regul.*, 15, 155, 1979.
12. Nicoll, C. S., Prolactin and growth hormone: specialists on one hand and mutual mimics on the other, *Perspect. Biol. Med.*, 25, 369, 1982.
13. Carr, F. E. and Jaffe, R. C., Prolactin and tadpole metamorphosis, evidence of prolactin receptors in premetamorphic *Rana catesbeiana* liver and tail fin, *Mol. Cell. Endocrinol.*, 17, 145, 1980.
14. Daughaday, W. H., Growth hormone and the somatomedins, in *Endocrine Control of Growth*, Daughaday, W. H., Ed., Elsevier, New York, 1981, 1.
15. Anderson, T. R., Rodriguez, J., Nicoll, C. S., and Spencer, E. M., The synlactin hypothesis: prolactin's mitogenic action may involve synergism with a somatomedin-like molecule, in *Insulin-Like Growth Factors/Somatomedins*, Spencer, E. M., Ed., Walter deGruyter, New York, 1983, 71.
16. Martial, J. and Cooke, N. E., Isolation and sequence analysis of the rat prolactin structural gene, in *Central Peripheral Regulation of Prolactin Function*, MacLeod, R. M. and Scapagnini, U., Eds., Raven Press, New York, 1980, 115.
17. Pestarino, M., Prolactinergic neurons in protochordate, *Cell Tissue Res.*, 233, 471, 1983.
18. Pestarino, M., Immunocytochemical demonstration of prolactin-like activity in the neural gland of the Ascidian *Styela plicata*, *Gen. Comp. Endocrinol.*, 54, 444, 1984.
19. Miller, W. L. and Eberhardt, N. L., Structure and evolution of the growth hormone gene family, *Endocrinol. Rev.*, 4, 97, 1983.
20. Specker, J. L., Ingleton, P. M., and Bern, H. A., Comparative physiology of the prolactin cell, in *Prolactin Secretion: A Multidisciplinary Approach*, Mena, F. and Velverde, C. M., Eds., Academic Press, New York, 1984, 17.
21. Rivas, R. J., Nishioka, R. S., Specker, J. L., and Bern, H. A., Control of secretion of two prolactins from the rostral pars distalis of tilapia, *Oreochromis (Sarotherodon)* mussambicus, *Am. Zool.*, 24, 117A, 1984.
22. Linzer, D. and Nathans, D., Growth-related changes in specific mRNA's of cultured mouse cells, *Proc. Natl. Acad. Sci. U.S.A.*, 80, 4271, 1983.
23. Meier, A. H., Chronoendocrinology of the vertebrates, in *Hormonal Correlates of Behavior*, Vol. 2, Eleftheriou, B. E. and Sprott, R. L., Eds., Plenum Press, New York, 1975, 469.
24. Meier, A. H., Trobec, T. N., Joseph, M. M., and John, T. M., Temporal synergism of prolactin and adrenal steroids in the regulation of fat stores, *Proc. Soc. Exp. Biol. Med.*, 137, 408, 1971.
25. Gala, R. R. and Haisenleder, D. J., Are both the nocturnal and diurnal prolactin surges necessary to maintain pseudopregnancy in the rat?, *Life Sci.*, 34, 179, 1984.
26. Marr, K. M. and Horseman, N. D., The role of a circadian rhythm of sensitivity to prolactin in pseudopregnancy, *Am. Zool.*, 24, 73A, 1984.
27. Meier, A. H. and Burns, J. T., Circadian hormone rhythms in lipid regulation, *Am. Zool.*, 16, 649, 1976.
28. Copinschi, G., DeLaef, M.-H., Brion, J. P., Leclercq, R., L'Hermite, M., Robyn, C., Virasoro, E., and VanCauter, E., Simultaneous study of cortisol, growth hormone and prolactin nyctohemeral variations in normal and obese subjects. Influence of prolonged fasting in obesity, *Clin. Endocrinol.*, 9, 15, 1978.
29. Cincotta, A. H. and Meier, A. H., Daily variations of lipogenesis and insulin receptor number in hepatocytes of the golden hamster: effects of prolactin and bromocryptine, *Am. Zool.*, 24, 119A, 1984.
30. Horseman, N. D. and Meier, A. H., Circadian phase-dependent prolactin mechanisms in hepatic lipogenesis of a teleost, *J. Endocrinol.*, 82, 367, 1979.
31. Burns, J. T. and Meier, A. H., Daily variations in pigeon crop sac responses to prolactin, *Experientia*, 27, 572, 1971.
32. Meier, A. H., Burns, J. T., Davis, K. B., and John, T. M., Circadian variations in sensitivity of the pigeon crop sac to prolactin, *J. Interdiscip. Cycle Res.*, 2, 161, 1971.
33. Horseman, N. D. and Nollin, L. J., The mitogenic, but not differentiative, response of pigeon crop to prolactin is circadian phase-dependent, *Endocrinology*, 116, 2085, 1985.
34. Nicoll, C. S., Physiological actions of prolactin, *1974 Handbook of Physiol.*, 4, 253, 1974.
35. Bailey, R. E., The incubation patch of passerine birds, *Condor*, 54, 121, 1952.
36. Jones, R. E., Hormonal control of incubation patch development in the California quail, *Lephertyx californicus*, *Gen. Comp. Endocrinol.*, 13, 1, 1969.
37. Egami, N. and Ishii, S., Hypophyseal control of reproductive functions in teleost fishes, *Gen. Comp. Endocrinol.*, Suppl. 1, 248, 1962.
38. Maurer, R. A., Estradiol regulates the transcription of the prolactin gene, *J. Biol. Chem.*, 257, 2133, 1982.
39. Liotta, A. S. and Krieger, D. T., Pro-opiomelanocortin-related and other pituitary hormones in the central nervous system, in *Brain Peptides*, Krieger, D. T., Brownstein, M. J., and Martin, J. B., Eds., John Wiley & Sons, New York, 1983, 613.

40. Buntin, J. D., personal communication, 1984.
41. Riddle, O., Bates, R. W., and Lahr, E. L., Prolactin induces broodiness in fowl, *Am. J. Physiol.*, 3, 352, 1935.
42. Ensor, D. M. and Ball, J. N., A bioassay for fish prolactin (paralactin), *Gen. Comp. Endocrinol.*, 11, 104, 1968.
43. Shiu, R. P. C. and Friesen, H. G., Mechanism of action of prolactin in the control of mammary gland function, *Annu. Rev. Physiol.*, 42, 83, 1980.
44. Donatsch, P. and Richardson, B., Localization of prolactin in rat kidney tissue using a double antibody technique, *J. Endocrinol.*, 66, 101, 1975.
45. Chomczyski, P. and Topper, Y. J., A direct effect of prolactin and placental lactogen on mammary epithelial nuclei, *Biochem. Biophys. Res. Commun.*, 60, 56, 1974.
46. Kledzik, G., Marshall, S., Gelato, M., Campbell, G., and Meites, J., Prolactin binding activity in the crop sacs of juvenile, mature, parent and prolactin-injected pigeons, *Endocrinol. Res. Commun.*, 2, 345, 1975.
47. White, B. A. and Nicoll, C. S., Prolactin receptors in *Rana catesbeiana* during development and metamorphosis, *Science*, 204, 851, 1979.
48. Foskett, J. K., Machen, T. E., and Bern, H. A., Chloride secretion and conductance of teleost opercular membrane: effects of prolactin, *Am. J. Physiol.*, 242, R380, 1982.
49. Giunta, C., Campantico, R., and Guardabassi, C., Biochemical response elicited in normal and prolactin treated *Xenopus laevis* (Daudin) specimens by environmental changes, *Comp. Biochem. Physiol.*, 47A, 323, 1973.
50. Anderson, T. R., Mayer, G. L., and Nicoll, C. S., Cyclic nucleotides and the control of epithelial cell proliferation: cyclic CMP may be a partial mediator of the response of the pigeon crop sac to prolactin, *J. Cyclic Nucl. Res.*, 7, 225, 1981.
51. Etkin, W. and Gona, A. G., Antagonism between prolactin and thyroid hormone in amphibian development, *J. Exp. Zool.*, 165, 457, 1967.
52. Doneen, B. A., Water and ion movements in the urinary bladder of the gobiid teleost *Gillichthys mirabilis* in response to prolactins and to cortisol, *Gen. Comp. Endocrinol.*, 28, 33, 1976.
53. Sage, M. and Bern, H. A., Assay of prolactin in vertebrate pituitaries by its dispersion of xanthopore pigment in the teleost *Gillichthys mirabilis*, *J. Exp. Zool.*, 180, 169, 172.
54. Spanos, E., Pike, J. W., Haussler, M. R., Colston, K. W., Evans, I. M. A., Goldner, A. M., McCain, T. A., and MacIntyre, J., Circulating 1α,25-dihydroxyvitamin D in the chicken: enhancement by injection of prolactin during egg-laying, *Life Sci.*, 19, 1751, 1976.
55. Spanos, E., Brown, D. J., Stevenson, J. C., and MacIntyre, I., Stimulation of 1,25-dihydroxycholecaliferol production by prolactin and related peptides in intact renal cell preparations *in vitro*, *Biochem. Biophys. Acta*, 672, 7, 1981.
56. Bikle, D. D., Spencer, E. M., Burke, W. H., and Rost, C. R., Prolactin but not growth hormone stimulates 1,25-dihydroxy vitamin D_3 production by chick renal preparations *in vitro*, *Endocrinology*, 107, 81, 1980.
57. Mayer, R. J., Hormonal factors in lipogenesis in mammary gland, *Vitam. Horm.*, 36, 101, 1978.
58. Goodridge, A. G. and Ball, E. G., The effect of prolactin on lipogenesis in the pigeon, *in vivo* studies, *Biochemistry*, 6, 1676, 1967.
59. Goodridge, A. G. and Ball, E. G., The effect of prolactin on lipogenesis in the pigeon, *in vitro* studies, *Biochemistry*, 6, 2335, 1967.
60. Horseman, N. D. and Meier, A. H., Circadian-dependent prolactin effects: hepatic RNA metabolism and prostaglandin mediation, *Gen. Comp. Endocrinol.*, 38, 269, 1979.
61. Burden, C. E., The failure of hypophysectomized *Fundulus heteroclitus* to survive in freshwater, *Biol. Bull.*, 110, 8, 1956.
62. Pickford, G. E. and Phillips, J. G., Prolactin, a factor in promoting survival of hypophysectomized killifish in fresh water, *Science*, 130, 454, 1959.
63. Pickford, G. E., Griffith, R. W., Torretti, J., Hendlez, E., and Epstein, F. H., Branchial reduction and renal stimulation of (Na^+, K^+)-ATPase by prolactin in hypophysectomized killifish in fresh water, *Nature (London)*, 228, 378, 1970.
64. Bern, H. A., Prolactin and osmoregulation, *Am. Zool.*, 15, 937, 1975.
65. Hirano, T., Prolactin and osmoregulation, in *Endocrinology 1980*, Cumming, I. A., Funder, J. W., and Mendelsohn, F. A. D., Eds., Australian Academy of Science, Canberra, 1980, 186.
66. Utida, S., Hirano, J., Oida, H., Ando, M., Toshusen, D. W., and Bern, H. A., Hormonal control of the intestine and urinary bladder in teleost osmoregulation, *Gen. Comp. Endocrinol.*, 3, 317, 1972.
67. Johnson, B. W., Endocrine control of hydromineral balance in teleosts, *Am. Zool.*, 13, 799, 1973.
68. Loretz, C. A. and Bern, H. A., Prolactin and osmoregulation in vertebrates, *Neuroendocrinology*, 35, 292, 1982.
69. Ensor, D. M., *Comparative Endocrinology of Prolactin*, Chapman and Hall, London, 1978.

70. Stanley, J. C. and Fleming, W. R., Effect of prolactin and ACTH on the serum and urine sodium levels of *Fundulus kansae, Comp. Biochem. Physiol.*, 20, 199, 1967.
71. Ball, J. N. and Ensor, D. M., Specific action of prolactin on plasma sodium levels in hypophysectomized *Poecilia latipinna* (Teleostei), *Gen. Comp. Endocrinol.*, 8, 432, 1967.
72. Dharmamba, M. and Maetz, J., Effects of hypophysectomy and prolactin on sodium balance of *Tilapia mossambica* in fresh water, *Gen. Comp. Endocrinol.*, 19, 175, 1972.
73. Utida, S., Matai, S., Hirano, T., and Kamemoto, T. I., Effect of prolactin on survival and plasma sodium levels in hypophysectomized medaka, *Oryzias latipes, Gen. Comp. Endocrinol.*, 16, 566, 1971.
74. Lam, T. J., The effect of prolactin on osmotic influx of water in isolated gills of the marine threespine stickleback, *Gasterosteus aculeatus L.*, form *Trachurus, Comp. Biochem. Physiol.*, 31, 909, 1969.
75. Ogawa, M., Yagasaki, M., and Yamazaki, F., The effect of prolactin on water influx in isolated gills of the goldfish, *Carassius auratus L., Comp. Biochem. Physiol.*, 44A, 1177, 1973.
76. Malarkey, W. B., George, J. M., and Baehler, R. W., Evidence for vasopressin contamination of ovine and bovine prolactin, *IRCS Med. Sci.*, 3, 391, 1975.
77. Bern, H. A., Bigbee, C. A., Collie, N. L., Foskett, J. K., Hughes, B., Loretz, C. A., and Marshall, W. S., Failure of ovine prolactin to elicit rapid responses by osmoregulatory surfaces, *Gen. Comp. Endocrinol.*, 44, 128, 1981.
78. Ensor, D. M. and Ball, J. N., Prolactin and freshwater sodium fluxes in *Poecilia latipinna* (Teleostei), *J. Endocrinol.*, 41, xvi, 1968.
79. Farmer, S. W., Papkoff, H., Bewley, T. A., Hayashida, T., Nishiota, R. S., Bern, H. A., and Li, C. H., Isolation and properties of teleost prolactin, *Gen. Comp. Endocrinol.*, 31, 60, 1977.
80. Maetz, J., Sawyer, W. H., Pickford, G. E., and Mayer, N., Evolution de la balance minérale du sodium chez *Fundulus heteroclitus* au cours du transfert d'eau de mer en eau douce: effets de l'hypophysectomie et de la prolactine, *Gen. Comp. Endocrinol.*, 8, 163, 1967.
81. Maetz, J., Mayer, N., and Chartier-Baraduc, M. M., La balance minérale du sodium chez *Anguilla anguilla* en eau de mer, en eau douce et au cours transfert d'un milieu à l'autre: effets de l'hypophysectomie et de la prolactine, *Gen. Comp. Endocrinol.*, 8, 177, 1967.
82. Schreibman, M. P. and Kallman, K. D., Endocrine control of freshwater tolerance in teleosts, *Gen. Comp Endocrinol.*, 6, 144, 1966.
83. Doneen, B. A. and Bern, H. A., *In vitro* effects of prolactin and cortisol on water permeability of the urinary bladder of the teleost *Gillichthys mirabilis, J. Exp. Zool.*, 187, 173, 1974.
84. Maetz, J., Fish gills, mechanisms of salt transfer in fresh water and sea water, *Phil. Trans. R. Soc. Lond.*, 262, 209, 1971.
85. Degnan, K. J., Karnaky, K. J., and Zadunaisky, J. A., Active chloride transport in the *in vitro* opercular skin of a teleost *(Fundulus heteroclitus)*, a gill-like epithelium rich in chloride cells, *J. Physiol.*, 271, 155, 1977.
86. Blanc-Livni, N. and Abraham, M., The influence of environmental salinity on the prolactin and gonadotrophins-secreting regions in the pituitary of *Mugil* (Teleostei), *Gen. Comp. Endocrinol.*, 14, 184, 1970.
87. Utida, S., Knaija, M., and Shirai, N., Relationship between the activity of Na^+, K^+ activated adenosine triphosphatase and the number of chloride cells in eel gills with special reference to sea water adaptation, *Comp. Biochem. Physiol.*, 38, 443, 1971.
88. Olivereau, M., Structure histologique du rein et electrolytes plastiques chez l'anguille a près autotransplantation de l'hypophyse, *Z. Vergl. Physiol.*, 71, 350, 1970.
89. Mizuhira, V., Amakawa, T., Yamashina, S., Shirai, N., and Utida, S., Electron microscopic studies on the localization of sodium ion and sodium-potassium-activated adenosine triphosphatase in chloride cells of eel gills, *Exp. Cell Res.*, 59, 346, 1970.
90. Foskett, J. K. and Scheffey, C., The chloride cell: definitive identification as the salt-secretory cell in teleosts, *Science*, 215, 164, 1982.
91. Kamiya, M., Hormonal effect on Na-K-ATPase activity in the gill of Japanese eel, *Anguilla japonica*, with special reference to sea water adaptation, *Endocrinol. Jpn.*, 19, 489, 1972.
92. Maetz, J., Mechanisms of salt and water transfer across membranes in teleosts in relation to the aquatic environment, *Mem. Soc. Endocrinol.*, 18, 3, 1970.
93. Gallis, J.-L., Lasserre, P., and Bellock, F., Freshwater adaptation in the euryhaline teleost, *Chelon labrusus*. I. Effects of adaptation, prolactin, cortisol and actinomycin D on plasma osmotic balance and (Na^+-K^+) ATPase in gill and kidney, *Gen. Comp. Endocrinol.*, 38, 1, 1979.
94. Leatherland, J. F. and Lam, T., Effect of prolactin on the density of mucous cells on the gill filaments of the marine form *(trachurus)* of the three-spined stickleback, *Gasterosteus aculeatus L., Can. J. Zool.*, 47, 787, 1969.
95. Mattheij, J. A. M. and Stroband, H. W. J., The effects of osmotic experiments and prolactin on the mucous cells in the skin and the ionocytes in the gills of the teleost *Cichlasoma biocellatum, Z. Zellforsch. Mikrosk. Anat.*, 121, 93, 1971.

96. Olivereau, M. and Lemoine, A. M., Action de la prolactine chez l'anguille intacte et hypophysectomisée, *Z. Vgl. Physiol.*, 73, 34, 1971.
97. Blüm, V. and Fiedler, K., Hormonal control of reproductive behavior in some cichlid fish, *Gen. Comp. Endocrinol.*, 5, 186, 1964.
98. Wendelarr Bonga, S. E. and Veenhuis, M., The effects of prolactin on the kidney cells of the stickleback, *Gasterosteus aculeatus*, after transfer from the sea to fresh water, studied by ultrathin sectioning and freeze-etching, *Gen. Comp. Endocrinol.*, 22, 382, 1974.
99. Gona, O., Localization of prolactin uptake by the kidney of the teleostean, *Colisa lalia*, *Gen. Comp. Endocrinol.*, 55, 289, 1984.
100. Leatherland, J. F. and Lam, T. J., Effect of prolactin on the density of mucous cells on the gill filaments of the marine form *(trachurus)* of the threespine stickleback, *Gasterosteus aculeatus L.*, *Can. J. Zool.*, 47, 787, 1969.
101. Johnson, D. W., Hirano, T., Sage, M., Foster, R. C., and Bern, H. A., Time course of response of storry flounder *(Platichthys stellatus)* urinary bladder to prolactin and to salinity transfer, *Gen. Comp. Endocrinol.*, 24, 373, 1974.
102. Doneen, B. A., Water and ion movements in the urinary bladder of the gobiid teleost *Gillichthys mirabilis* in response to prolactins and cortisol, *Gen. Comp. Endocrinol.*, 28, 33, 1976.
103. Hirano, T., Hayashi, S., and Utida, S., Stimulatory effect of prolactin on incorporation of (^{3}H) thymidine into the urinary bladder of the flounder *(Kareius bicoloratus)*, *J. Endocrinol.*, 56, 591, 1973.
104. Fryer, J. N., Prolactin-binding sites in tilapia *(Sarotherodon mossambicus)* kidney, *Gen. Comp. Endocrinol.*, 39, 397, 1979.
105. Edery, M., Young, G., Bern, H. A., and Steiny, S., Prolactin receptors in Tilapia *(Sarotherodon mossambicus)* tissues: binding studies using ^{125}I-labeled ovine prolactin, *Gen. Comp. Endocrinol.*, 56, 19, 1984.
106. Foskett, J. K., Chloride secretion by teleost opercular membrane: circuit analysis of the effects of prolactin, *Ann. N.Y. Acad. Sci.*, 372, 644, 1981.
107. Foskett, J. K., Machen, T. E., and Bern, H. A., Chloride secretion and conductance of teleost opercular membrane: effects of prolactin, *Am. J. Physiol.*, 242, 380, 1982.
108. Horseman, N. D. and Meier, A. H., Prostaglandin and the osmoregulatory role of prolactin in a teleost, *Life Sci.*, 2, 1485, 1978.
109. Berman, R., Bern, H. A., Nicoll, C. S., and Strohman, R. C., Growth-promoting effects of mammalian prolactin and growth hormone in tadpoles of *Rana catesbeiana*, *J. Exp. Zool.*, 156, 353, 1964.
110. Chadwick, C. S., Identity of prolactin with water drive factor in *Triturus viridescens*, *Proc. Soc. Exp. Biol. Med.*, 45, 335, 1940.
111. Dodd, M. I. H. and Dodd, J. M., The biology of metamorphosis, in *The Physiology of Amphibia*, Lofts, B., Ed., Academic Press, New York, 1976.
112. Etkin, W. and Gona, A. G., Antagonism between prolactin and thyroid hormone in amphibian development, *J. Exp. Zool.*, 165, 249, 1967.
113. Wright, M. L., Majerowski, M. A., Lukas, S. M., and Pike, P. A., Effect of prolactin on growth, development, and epidermal cell proliferation in the hindlimb of the *Rana pipiens* tadpole, *Gen. Comp. Endocrinol.*, 39, 53, 1979.
114. Yoshizato, K. and Yasumasu, I., Effect of prolactin on the tadpole tail fin. I. Stimulatory effect of prolactin on the collagen synthesis of the tadpole tail fin, *Devel. Growth Diff.*, 11, 305, 1970.
115. Yoshizato, K. and Yasumasu, I., Effect of prolactin on the tadpole tail fin. II. Stimulatory effect of prolactin on the synthesis of acid mucopolysaccharide of the tadpole tail fin, *Devel. Growth Diff.*, 12, 265, 1971.
116. Yoshizato, K. and Yasumasu, I., Effect of prolactin on the tadpole tail fin. IV. Effect of prolactin on the metabolic fate of hyaluronic acid, collagen and RNA with a special reference to catabolic process, *Devel. Growth Diff.*, 14, 119, 1972.
117. Yoshizato, K. and Yasumasu, I., Effect of prolactin on the tadpole tail fin. V. Stimulatory effect of prolactin on the incorporation of ^{3}H-thymidine into DNA of the tadpole tail fin, *Devel. Growth Diff.*, 14, 129, 1972.
118. Gona, A. G., Prolactin-induced stimulation of H^3-proline incorporation into the basement lamella of tadpole skin: light and electron microscope study, *Tissue Cell*, 3, 557, 1971.
119. Carr, F. E. and Jaffe, R. C., Solubilization and molecular weight estimation of prolactin receptors from *Rana catesbeiana* tadpole liver and tail fin, *Endocrinology*, 109, 943, 1981.
120. Jaffe, R. C. and Geschwind, I. I., Studies on prolactin inhibition of thyroxine-induced metamorphosis in *Rana catesbeiana* tadpoles, *Gen. Comp. Endocrinol.*, 22, 289, 1974.
121. Jaffe, R. C. and Geschwind, I. I., Influence of prolactin on thyroxine-induced changes in hepatic and tail enzymes and nitrogen metabolism in *Rana catesbeiana* tadpoles, *Proc. Soc. Exp. Biol. Med*, 146, 961, 1974.

122. Campantico, E., Giunta, C., Guardabassi, A., and Vietti, M., The stabilizing action of prolactin on lysosomes in tails from *Xenopus laevis* Daudin tadpoles, *Gen. Comp. Endocrinol.*, 18, 396, 1972.
123. Giunta, C., Campantico, E., Vietti, M., and Guastalla, A., Prolactin stabilizing action on lysosomes in the tail and gut from *Rana temporaria* tadpoles at prometamorphosis, *Gen. Comp. Endocrinol.*, 18, 568, 1972.
124. Gona, O., Uptake of ^{125}I-labeled prolactin by bullfrog kidney tubules: an autoradiographic study, *Endocrinology*, 93, 133, 1982.
125. Carr, F. E. and Jaffe, R. C., Characterization of solubilized prolactin receptors from *Rana catesbeiana* tadpole tissues, *Mol. Cell. Endocrinol.*, 25, 317, 1982.
126. Carr, F. E., Jacobs, P. J., and Jaffe, R. C., Changes in specific prolactin binding in *Rana catesbeiana* tadpole tissues during metamorphosis and following prolactin and thyroid-hormone treatment, *Mol. Cell. Endocrinol.*, 23, 65, 1981.
127. Chan, S. J., Noyes, B. E., Agarual, K. L., and Steiner, D. F., Construction and selection of recombinant plasmids containing full-length complementary DNAs corresponding to rat insulins I and II, *Proc. Natl. Acad. Sci. U.S.A.*, 76, 5036, 1979.
128. Chin, W. W., Kronenberg, H. M., Du, P. C., Maloof, F., and Habener, J. F., Nucleotide sequence of the mRNA encoding the pre-α-subunit of mouse thyrotropin, *Proc. Natl. Acad. Sci. U.S.A.*, 78, 5329, 1981.
129. Miller, W. L., Baxter, J. D., and Eberhardt, N. L., Peptide hormone genes: structure and evolution, in *Brain Peptides*, Krieger, D. T., Brownstein, M. J., and Martin, J. B., Eds., John Wiley & Sons, New York, 1983, 15.
130. Riddle, O. and Braucher, P. F., Studies on the physiology of reproduction in bird: control of the special secretion of the crop-gland in pigeons by an anterior pituitary hormone, *Am. J. Physiol.*, 97, 617, 1931.
131. Riddle, O., Bates, R. W., and Dykshorn, S. W., The preparation, identification and assay of prolactin — a hormone of the anterior pituitary, *Am. J. Physiol.*, 105, 191, 1933.
132. Reichert, L. E. and Wilhelmi, A. E., A summary of the biological activities of the various lots of pituitary hormones produced under the programs of the National Institute of Arthritis, Metabolism and Digestive Diseases, *Endocrinology*, 92, 1301, 1973.
133. Riddle, O. and Dykshorn, S. W., Secretion of crop-milk in the castrate male pigeon, *Proc. Soc. Exp. Biol. Med.*, 29, 1213, 1932.
134. Schooley, J. P. and Riddle, O., The morphological basis of pituitary function in pigeons, *Am. J. Anat.*, 62, 313, 1938.
135. Bates, R. W., Miller, R. A., and Garrison, M. M., Evidence in the hypophysectomized pigeon of a synergism among prolactin, growth hormone, thyroxine, and prednisone upon weight of the body, digestive tract, kidney and fat stores, *Endocrinology*, 71, 345, 1962.
136. Dumont, J. N., Prolactin-induced cytologic changes in the mucosa of the pigeon crop during crop-"milk" formation, *Z. Zellforsch.*, 68, 755, 1965.
137. Patel, M. D., The physiology of the formation of "pigeon's milk", *Physiol. Zool.*, 9, 129, 1936.
138. Goldsmith, A. R., Edwards, C., Kopucu, M., and Silver, R., Concentrations of prolactin and luteinizing hormone in plasma of doves in relation to incubation and development of the crop gland, *J. Endocrinol.*, 90, 437, 1981.
139. Cheng, M.-F. and Burke, W. H., Serum prolactin levels and crop-sac development in ring doves during a breeding cycle, *Horm. Behav.*, 17, 54, 1983.
140. Hansen, E. W., Squab-induced crop growth in ring dove foster parents, *J. Comp. Physiol. Psychol.*, 62, 120, 1966.
141. Buntin, J. D., Cheng, M.-F., and Hanen, E. W., Effect of parental feeding activity on squab-induced crop sac growth in ring doves *(Streptopelia risoria)*, *Horm. Behav.*, 8, 297, 1977.
142. Buntin, J. D. and Forsyth, I. A., Measurements of pituitary prolactin levels in breeding pigeons by crop sac radioreceptor assay, *Gen. Comp. Endocrinol.*, 37, 57, 1979.
143. Chadwick, A. and Jordan, B. J., The lipids of the crop epithelium of pigeons after injection with prolactin from the pituitaries of different vertebrate classes, *J. Endocrinol.*, 49, 51, 1971.
144. Horseman, N. D. and Will, C. L., Characterization of lipid storage associated proteins induced in crop tissue by prolactin, *J. Comp. Physiol. B*, 154, 237, 1984.
145. Cohen, S. and Taylor, J. M., Epidermal growth factor: chemical and biological characterization, *Rec. Prog. Horm. Res.*, 30, 533, 1974.
146. Nicoll, C. S., Bioassay of prolactin. Analysis of the pigeon crop-sac response to local prolactin injection by an objective and quantitative methods, *Endocrinology*, 80, 641, 1967.
147. Ben-David, M., A sensitive bioassay for prolactin based on H^3-methyl-thymidine uptake by the pigeon-crop mucous epithelium, *Proc. Soc. Exp. Biol. Med.*, 125, 705, 1967.
148. Jänne, J., Pösö, H., and Raina, A., Polyamines in rapid growth and cancer, *Biochem. Biophys. Acta*, 473, 241, 1978.

149. Russel, D. H., Byus, C. U., and Manen, C. A., Proposed model of major biochemical events of atrophic response, *Life Sci.*, 19, 1297, 1976.
150. Canellakis, Z. N. and Theoharides, T. C., Stimulation of ornithine decarboxylase synthesis and its control by polyamines in regenerating rat liver and cultured rat hepatoma cells, *J. Biol. Chem.*, 251, 4436, 1976.
151. Richards, J. R., Ornithine decarboxylase activity in tissues of prolactin-treated rats, *Biochem. Biophys. Res. Commun.*, 63, 292, 1975.
152. Rillema, J. A., Action of prolactin on ornithine decarboxylase activity in mammary gland explants of mice, *Endocrinol. Res. Commun.*, 3, 297, 1976.
153. Anderson, T. R., Pitts, D. S., and Nicoll, C. S., Prolactin's mitogenic action on the pigeon crop-sac mucosal epithelium involves direct and indirect mechanisms, *Gen. Comp. Endocrinol.*, 54, 236, 1984.
154. Hill, D. J., Francis, M. J. O., and Milner, R. D. G., Action of rat prolactin on plasma somatomedin levels in the rat and somatomedin release from perfused rat liver, *J. Endocrinol.*, 75, 137, 1977.
155. Francis, M. J. O. and Hill, D. J., Prolactin-stimulated production of somatomedin by rat liver, *Nature (London)*, 255, 167, 1975.
156. Holder, A. T. and Wallis, M., Regulation of serum somatomedin levels and growth in dwarf mice by growth hormone, prolactin and thyroxin, *J. Endocrinol.*, 71, 82P, 1976.
157. Hutchinson, R. E., Hinde, R. A., and Berdon, B. E., Oviduct development and its relation to other aspects of reproduction in domesticated canaries, *J. Zool. (London)*, 155, 87, 1967.
158. Vondehaar, B. K. and Topper, Y. J., A role of the cell cycle in hormone dependent differentiation, *J. Cell Biol.*, 63, 707, 1974.
159. Topper, Y. J. and Freeman, C. S., Multiple hormone interactions in the developmental biology of the mammary gland, *Physiol. Rev.*, 60, 1049, 1980.
160. Carr. F. E. and Jaffe, R. C., Solubilization and molecular weight estimation of prolactin receptors from *Rana catesbeiana* tadpole liver and tail fin, *Endocrinology*, 109, 943, 1981.
161. Buntin, J. D., Keskey, T. S., and Janik, D. S., Properties of hepatic binding sites for prolactin in ring doves *(Streptopelia risoria)*, *Gen. Comp. Endocrinol.*, 55, 418, 1984.
162. Posner, B. I., Kelly, P. A., and Friesen, H. G., Prolactin receptors in rat liver: possible induction by prolactin, *Science*, 188, 57, 1975.
163. Shani, J., Furr, B. J. A., and Forsyth, I. A., A re-examination of the antigonadal effect of prolactin in the male fowl, *Gallus domesticus*, *J. Reprod. Fertil.*, 34, 167, 1973.
164. Shani, J., Applebaum, J., and Goldhaber, G., Movement and localization of prolactin in the pigeon crop-sac: an autoradiographic study, *J. Endocrinol.*, 72, 397, 1977.
165. Kledzick, G., Marshall, S., Gelato, M., Campbell, G., and Meites, J., Prolactin binding activity in the crop sacs of juvenile, mature, parent and prolactin-injected pigeons, *Endocrinol. Res. Commun.*, 2, 345, 1975.
166. Forsyth, I. A., Buntin, J. D., and Nicoll, C. S., A pigeon crop sac radioreceptor assay for prolactin, *J. Endocrinol.*, 79, 349, 1978.
167. Shani, J., Barkey, R. J., and Amit, T., Endogenous prolactin maintains its own binding sites in the pigeon crop sac mucosa, *J. Rec. Res.*, 2, 407, 1982.
168. Sherry, W. E. and Nicoll, C. S., RNA and protein synthesis in the response of pigeon crop-sac to prolactin, *Proc. Soc. Exp. Biol. Med.*, 126, 824, 1967.
169. Frantz, W. L. and Rillema, J. A., Prolactin-stimulated uptake of amino acids-^{14}C and ^{3}H-OH in pigeon crop mucosa, *Am. J. Physiol.*, 215, 762, 1968.
170. Garrison, M. M. and Scow, R. O., Effect of prolactin on lipoprotein lipase in crop sac and adipose tissue of pigeons, *Am. J. Physiol.*, 228, 1542, 1975.
171. Pukac, L. A. and Horseman, N. D., Regulation of pigeon crop gene expression by prolactin, *Endocrinology*, 114, 718, 1984.
172. Walter, P. and Blobel, G., Translocation of proteins across the endoplasmic reticulum. II. Signal recognition protein (SRP) causes signal sequence-dependent and site-specific arrest of chain elongation that is released by microsomal membranes, *J. Cell Biol.*, 91, 557, 1981.
173. Shani, J., Sulimovici, S., Goldhaber, G., Givant, Y., Sulman, F. G., and Lunenfeld, B., c-AMP as a possible mediator in the proliferative effect of prolactin on the pigeon crop sac, *Gen. Comp. Endocrinol.*, 29, 253, 1976.
174. Shani, J., Sulimovici, S., Goldhaber, G., Givant, Y., Sulman, F. G., and Lunenfeld, B., Adenosine 3′,5′-cyclic monophosphate as a possible mediator in the proliferative effect of prolactin on the pigeon crop sac, *J. Endocrinol.*, 69, 169, 1976.
175. Sinha, Y. N. and Schmidt, G. H., Potentiation of the crop-sac stimulating action of prolactin by ATP, *Proc. Soc. Exp. Biol. Med.*, 133, 1279, 1970.
176. Anderson, T. R., Mayer, G. L., and Nicoll, C. S., Cyclic nucleotides and the control of epithelial cell proliferation: cyclic CMP may be a partial mediator of the response of the pigeon crop-sac to prolactin, *J. Cyclic Nucl. Res.*, 7, 225, 1981.

177. Anderson, T. R., Cyclic cytidine 3′-5′ monophosphate (cCMP) in cell regulation, *Mol. Cell. Endocrinol.*, 28, 373, 1982.
178. Rillema, J. A., Possible role of prostaglandin $F_{2\alpha}$ in mediating effect of prolactin on RNA synthesis in mammary gland explants of mice, *Nature (London)*, 253, 466, 1975.
179. Horrobin, D. F., Cellular basis of prolactin action: involvement of cyclic nucleotides, polyamines, prostaglandins, steroids, thyroid hormones, Na/K ATPases and calcium: relevance to breast cancer and the menstrual cycle, *Med. Hypotheses*, 5, 599, 1979.
180. Dave, J. R., Brown, N. U., and Knazek, R. A., Prolactin modifies the prostaglandin synthesis, prolactin binding and fluidity of mouse liver membranes, *Biochem. Biophys. Res. Commun.*, 108, 193, 1982.
181. Bern, H. A., On two possible primary activities of prolactins: osmoregulatory and developmental, *Verh. Dtsch. Zool. Ges.*, 1975, 40, 1975.
182. Weisbrod, S., Active chromatin, *Nature (London)*, 297, 289, 1982.
183. Darnell, J. E., Jr., Variety in the level of gene control in eukaryotic cells, *Nature (London)*, 297, 365, 1982.
184. Specker, J. L., personal communication.
185. Nicoll, C. S., personal communication.
186. Horseman, N. D., unpublished information.
187. Pukac, L. A. and Horseman, N. D., unpublished information.

Chapter 5

PROLACTIN REGULATION OF MEMBRANE FLUIDITY AND PROSTAGLANDIN FORMATION

Jitendra R. Dave

TABLE OF CONTENTS

I. Introduction ... 70

II. Prolactin Receptors and Their Regulation ... 70
  A. Mammary Gland Prolactin Receptors ... 70
  B. Prolactin Receptors in Other Major Target Tissues ... 71
    1. Ovarian Prolactin Receptors ... 71
    2. Kidney Prolactin Receptors ... 72
    3. Adrenal Gland Prolactin Receptors ... 72
    4. Seminal Vesicle and Testicular Prolactin Receptors ... 72
    5. Prostatic Prolactin Receptors ... 72
    6. Hepatic Prolactin Receptors ... 73

III. Membrane Lipid Microviscosity ... 74
  A. The Use of the Fluorescence Polarization Technique to Determine Membrane Lipid Microviscosity ... 74
  B. Factors Modulating Membrane Lipid Microviscosity ... 76

IV. Modification of Cellular Functions Mediated by Membrane Lipid Microviscosity Changes ... 78

V. Correlation Between Changes in Prolactin Receptors, Prostaglandin Synthesis, and Membrane Lipid Microviscosity ... 79
  A. Studies on Hepatic Membranes ... 79
  B. Studies on Prostatic Membranes ... 90

VI. Hypothesis That Prolactin Modifies Membrane Lipid Microviscosity and Prostaglandin Synthesis ... 96

VII. Conclusions ... 98

Acknowledgments ... 98

References ... 99

I. INTRODUCTION

Hormonal control of the growth and cellular function of both normal and neoplastic tissue first requires binding of the hormone to the target tissue cells. This binding, which determines the cellular responsiveness to the hormone, occurs at sites on or within the cells. Each hormone has high affinity, specific receptors, or binding sites to which it can bind to initiate its unique effect. Furthermore, the cellular effect of a hormone depends both on the form and levels of the circulating hormone and the responsiveness of the target tissue cell to that hormone. If a character of the receptor is altered, it is possible that its ability to interact with the hormone and, thus, the cellular function, may also be altered.

Our studies have been directed towards elucidating the mechanism(s) by which the functionality of such hormone receptors can be modulated. We have approached this by studying the receptor of prolactin (PRL), a phylogenetically ancient hormone that is known to affect lactation, growth, fertility, water/salt balance, adrenal function, immune system, and growth of certain tumors through an as yet poorly defined mechanism. In mammalian systems, PRL is considered to be a hormone primarily concerned with reproduction, including mammary gland growth and milk secretion. Other functions of PRL, such as its role in parental behavior and osmoregulation, which were essentially evolved in the lower vertebrates, still persist in mammals. In a review by Nicoll,[1] more than 80 actions of PRL have been listed.

II. PROLACTIN RECEPTORS AND THEIR REGULATION

The significance of determining PRL receptor levels becomes obvious when we consider the idea that this determination provides a quantitative index of the target tissue responsiveness to PRL. Furthermore, it is known that PRL, at least in rodents, plays a key role in mammary and prostatic tumorogenesis. This role of PRL would apparently involve its interaction with the membrane receptors. Determination of mammary or prostatic PRL receptors would thus provide a potential tool for assessing the cellular response to various therapeutic agents. The first report on the existence of specific PRL receptors was made by Turkington in mouse mammary gland using PRL bound to Sepharose beads[2] and ^{125}I-labeled PRL.[3] These findings were later confirmed by Birkinshaw and Falconer[4] in an autoradiographic study using radioiodinated PRL. Subsequently, specific binding sites for PRL were identified in the brain and a variety of other peripheral tissues, including the liver, ovary, uterus, pituitary, ventral and dorsolateral lobes of the prostate gland, seminal vesicles, adipose tissue, adrenal cortex, muscle, heart, Leydig cells of the testis, placenta, lung, etc.

A. Mammary Gland Prolactin Receptors

Since the mammary gland is thought to be the principle target tissue of PRL action, the mammary tissue from rat,[5-7] mouse,[8-13] or rabbit[14,15] was extensively used in earlier studies on PRL receptors. A variety of mammary tumors, including MTW9[16,17] and R3230AC,[5,18,19] transplantable mammary tumors in rat, 7,12-dimethylbenzanthracene (DMBA)-induced rat mammary tumors,[7,20-25] and human breast tumors[26-29] have also been studied to determine the existence of PRL receptors and their characteristics. Both immunohistochemical and autoradiographic studies have suggested the existence of cell membrane as well as intracellular binding sites for PRL. Nolin and Witorsch,[30] using immunohistochemical techniques, have reported intracellular PRL binding in alveolar cells of lactating mammary gland in the rat. These and other studies[31,32] have suggested that PRL can enter the cell, and its binding to intracellular organelles may produce physiological responses. In an autoradiographic study, Costlow and Mc-

Guire[20] reported that when DMBA-induced mammary tumor slices were incubated with ^{125}I-ovine PRL, tumor cells exhibited higher levels of specific PRL binding. However, nonspecific binding was confined to connective tissue and alveolar spaces. A number of studies have reported the presence of PRL receptor in human breast tumors.[27,28,33-38] An updated review on studies of PRL receptor in human tissue has been recently published.[39]

In characterizing PRL receptors, Shiu and Friesen[14,15,40,41] reported a time-dependent, temperature-dependent, saturable, reversible, and high-affinity binding of labeled PRL to mammary gland membranes. Furthermore, they also reported that PRL binding was influenced by the ionic strength and pH of the assay mixture and not by cyclic nucleotides, nucleotides, or steroids. The receptors for PRL from rabbit mammary gland have been purified and further characterized.[42,43]

Lactation is among the most studied factors known to modify mammary gland PRL receptors. The pattern of changes in mammary gland PRL-binding during pregnancy and lactation appears to be different in rabbit from that observed in rat and mouse. Some investigators have attributed these species differences in mammary gland PRL binding to the presence of high circulating placental lactogen during pregnancy in rat and mouse and the absence of the same in rabbit.[6,44,45] By competing with PRL for its receptors, placental lactogen at higher circulating levels may occupy a large proportion of available PRL receptors and thus present a lower value of the same.

PRL binding in the rat mammary gland is reported to remain low during pregnancy, followed by a rise in the same after parturition, and it remains high during the entire lactation period.[6,46] However, in rabbit mammary gland, PRL receptors first rise during mid-pregnancy, then decline, and again rise from late pregnancy through parturition and lactation.[47,48] Sakai et al.[49] reported a higher number, and not the affinity, of PRL-binding sites on collagenase-dispersed mammary epithelial cells obtained from virgin or lactating mice than that obtained from pregnant mice. These investigators suggested that progesterone may be responsible for this decrease in PRL binding during pregnancy. A similar antagonistic effect of progesterone on PRL binding in rabbit mammary gland has been reported.[47] Specific binding sites for PRL has been identified on dispersed epithelial cells of rabbit mammary gland.[50] The authors reported approximately a doubling of the number of PRL receptors per cell between mid-gestation and early lactation.

Various other factors, such as male and female sex hormones, PRL, insulin, polyamines, etc., have also been reported to modify PRL binding in normal and neoplastic mammary gland.[7,19,21,51,52] Detailed accounts of these and other studies have been provided in a number of reviews.[53-56]

B. Prolactin Receptors in Other Major Target Tissues

1. Ovarian Prolactin Receptors

Along with gonadotropins and estrogens, PRL is known to play a key role in the maintenance and formation of a functional corpora lutea. Furthermore, PRL is also reported to exert a variety of actions on ovarian function. The presence of PRL in the ovary of rats, cow, pig, human, etc. has been demonstrated.[57-64] In a classical study, Bohnet et al.[65] demonstrated that treatment of female rats with antiprolactin antibody resulted in an increase in the number of old corpora lutea. Using autoradiographic, immunohistochemical, and radioreceptor techniques, a number of studies subsequently characterized and confirmed the existence of PRL receptors in theca, luteal, and granulosa cells.[58,60,66-70] In a recent review by Bonifacino and Dufau,[71] the function, chracterization, and factors modulating ovarian PRL receptors are discussed in great detail.

2. Kidney Prolactin Receptors

A decade ago, in a number of studies, the existence of specific PRL-binding sites in kidney cells of male and female rats was reported.[31,72,73] In an autoradiographic study, Costlow and McGuire[74] reported that kidney tubules in the cortex display specific localization of ^{125}I-labeled PRL, whereas the medulla was devoid of PRL-binding activity. Marshall et al.[75,76] investigated the effects of glucocorticoids and thyroid status on PRL binding to kidney membranes. These investigators reported that while thyroidectomy reduced PRL binding in the kidney membranes, adrenalectomy increased the same. Administration of thyroxine to thyroidectomized animals and glucocorticoids to adrenalectomized animals reversed the effects of thyroidectomy and adrenalectomy on PRL binding in kidney membranes.

3. Adrenal Gland Prolactin Receptors

The existence of specific PRL-binding sites in adrenal gland was reported in earlier studies by Marshall et al.[72] This group reported that unilateral nephrectomy and water deprivation increase serum PRL levels, and that a similar treatment significantly increases PRL binding in adrenals. Costlow and McGuire,[74] using an autoradiographic technique, confirmed the existence of adrenal PRL-binding sites and reported that adrenal medulla is almost devoid of PRL-binding activity; however, the zona reticularis and, to a lesser extent, the zona fasciculata, exhibit PRL-binding activity. Marshall et al.[75] also studied the effects of glucocorticoids on PRL binding to adrenal and reported that low doses of dexamethasone produced a consistent reduction in PRL binding. The binding in adrenal glands of male rats was reported to be higher than that in female rats.[76] Furthermore, PRL binding was reported by this group to remain unchanged following thyroidectomy and thyroxine treatment.[76]

4. Seminal Vesicle and Testicular Prolactin Receptors

The evidence that PRL plays a significant role in the regulation of male reproductive function has led to a number of investigations on PRL receptors in male reproductive organs. In 1973, Turkington et al.[8] reported PRL-binding activity in a number of tissues, including the testes and seminal vesicles. Later, many investigators confirmed the existence of specific PRL-binding sites in these tissues of rats and mice.[77,78] Charreau et al.[79] defined the nature of PRL-binding sites in the rat testis. These binding sites could bind either PRL or human growth hormone, and were localized exclusively in the interstitial tissue and not in the seminiferous tubules. Aragona et al.[80] also reported that binding sites for PRL are primarily located on the Leydig cells, and that a maximal testicular PRL binding is observed at 45 days of age in the rat. This group also reported that bromocriptine treatment of male rats from days 20 to 40 decreased PRL secretion and blocked the binding of luteinizing hormone (LH) to Leydig cells without affecting PRL and follicle-stimulating hormone (FSH) binding, indicating that PRL may regulate testicular LH receptors. These and other studies suggested that the known PRL/LH interplay may involve a direct effect on the Leydig cells.

5. Prostatic Prolactin Receptors

Next to mammary gland and liver, a large number of investigations have used the prostate gland, especially the ventral prostate lobe in the rodents, to understand the mechanism of PRL action. The direct action of some nongonadotropic pituitary factor on the prostate was predicted by early observations that prostatic atrophy was more marked following hypophysectomy than following castration, and that androgen stimulation of prostatic growth was impaired after hypophysectomy.[84,85] Subsequently, using other functional parameters, Yamanaka et al.[86] reported that PRL is the hypophyseal prostatotrophic factor. The existence of lactogenic hormone receptors in the prostate was suggested by Sonenberg and Money[87] in 1955. They reported that prostate gland retained radioactivity after administration of ^{131}I-PRL to male rats. Relatively recently, many laboratories have reported specific, high-affinity, reversible, and saturable ^{125}I-PRL binding to prostate gland, which is primarily modulated by androgens.[77,88] Using immunohistochemical techniques, intracellular PRL binding in normal and abnormal rat and human prostate gland has been identified.[89-92] More recently, Witorsch surveyed the literature and presented his immunohistochemical studies on PRL-binding sites in prostate tissue in a review article.[93]

6. Hepatic Prolactin Receptors

A large number of studies has used hepatic tissue for detailed characterization of PRL receptors (including demonstration of intracellular PRL receptors and PRL receptor internalization). In fact, in a series of publications surveying PRL research literature, Horrobin[56] has indicated that "liver has been and is a favorite organ for conducting PRL-binding studies."

The hepatic tissue obtained from male animals has been reported to be virtually devoid of PRL receptors and administration of estrogens or castration is known to induce the same, suggesting an antagonistic action of androgens on hepatic PRL receptors. Most studies on hepatic PRL receptors have been carried out using liver from female animals. PRL receptor in the liver of female mouse, rat, and rabbit is absent, or is very low, during postnatal through prepubertal periods and reaches the adult level at or around puberty.[94,95] In a number of physiological and pathophysiological conditions, the pattern of changes in hepatic PRL receptors is often opposite to that of mammary gland PRL receptors. For example, in the mammary gland of mouse and rat, PRL receptors are low during pregnancy and high during lactation, whereas hepatic PRL receptors are high during pregnancy and are low during lactation.[94,96-98] Smith et al.[7] have reported changes in hepatic and mammary gland PRL binding in the rat during pregnancy and lactation. The hepatic PRL binding was found to reach a maximal level during pregnancy and then decline during lactation, whereas the binding in mammary gland increased with the advancement of pregnancy, reached a peak on day 11 of lactation, and then declined.[7] In a recent study, hepatic PRL binding in mouse during pregnancy and lactation was found to follow the pattern of changes similar to that during pregnancy and lactation in the rat.[98]

Unlike DMBA-induced mammary tumors, which have been reported to exhibit only a slight decrease in PRL receptors following hypophysectomy,[99] PRL receptors in rat liver are rapidly decreased after the removal of the pituitary.[99-105] Bohnet et al.[105] reported that a combined treatment of PRL, growth hormone, and adrenocorticotropic hormone is most effective in maintaining PRL receptors in the liver of hypophysectomized rats.

The Snell dwarf is a recessive mutant strain of mice, and the pituitary gland of the dwarf mouse is unable to synthesize and secrete TSH, growth hormone, and PRL. In a few earlier studies, it was reported that PRL binding is absent in the liver of dwarf mice, and that administration of PRL or pituitary grafting can induce the receptor.[78,106-108] It was thus suggested that PRL may modulate its own receptors in the liver. In an earlier study, when we injected either ovine PRL or purified bovine growth hormone (100 μg) to dwarf mice every 8 hr for a period of 2 to 4 days, there was a >20-fold increase in hepatic PRL binding.[108,180] We were able to achieve the induction in hepatic PRL binding by growth hormone or PRL administration to dwarf mice pretreated with cycloheximide for 4 days (Table 1). These results suggested that induction of PRL receptor after bovine growth hormone or ovine PRL was not entirely due to

Table 1
EFFECTS OF IN VIVO ADMINISTRATION OF OVINE PROLACTIN AND PURIFIED BOVINE GROWTH HORMONE ON SPECIFIC ^{125}I-OVINE PROLACTIN BINDING TO HEPATIC MEMBRANES IN THE SNELL DWARF MICE

Treatment	K_a ($\times 10^9\ M^{-1}$)	Q (fmol/mg protein)	Specific binding (cpm/mg protein)
	Experiment I[180]		
Tris-saline controls	0	0	1,500 ± 450
oPRL for 2 days[a]	8.3 ± 2.3	83 ± 14	33,000 ± 3,500
oPRL for 2 days + CHX for 4 days[b]	7.8 ± 1.2	89 ± 18	35,500 ± 4,200
	Experiment II[108]		
Tris-saline controls	0	0	1,300 ± 300
bGH for 2 days	8.5 ± 1.0	59 ± 7	22,600 ± 1,400
bGH for 2 days + CHX for 4 days	8.0 ± 2.5	77 ± 38	24,000 ± 5,400

[a] A dose of 100 μ of ovine PRL or purified bovine growth hormone was intraperitoneally injected every 8 hr. Cycloheximide treatment began 2 days prior to oPRL or bGH treatment and was injected subcutaneously in a 140 μg/day dose once each day. Values are mean ± S.D. of three to seven observations.

[b] CHX, cycloheximide.

the synthesis of new receptor protein, and that this treatment in dwarf mice might have uncovered pre-existing PRL receptors which were otherwise inaccessible. This induction of PRL receptor in the liver of dwarf mice could also be a result of some biophysical alterations in the membrane lipid bilayer. Biophysical alterations in membrane lipid bilayer can be monitored by apparent lipid-microviscosity measurements. Thus, to determine if changes in PRL binding and membrane lipid microviscosity are correlated, studies on membrane lipid microviscosity were undertaken.

III. MEMBRANE LIPID MICROVISCOSITY

A. The Use of the Fluorescence Polarization Technique to Determine Membrane Lipid Microviscosity

The microviscosity (inversely related to fluidity) of biological membranes appears to represent the mobility of their lipid molecules. This, in turn, affects the mobility of the mosaic of protein molecules within the membrane bilayer. Several approaches are available for the determination of membrane microviscosity, including nuclear magnetic resonance (NMR), electron spin resonance (ESR), and fluorescence polarization. Among these techniques, fluorescence polarization with an appropriate lipid probe is a practical method for measuring lipid microviscosity in the biomembranes.[109] A large number of fluorescent dyes have been used as probes in membrane studies.[110,111] For lipid microviscosity measurements by the fluorescence polarization technique, the choice of probe should possess a series of well-defined spectral and physical characteristics.[112] These probes can be divided into two major classes. First, there are the fluorescent dyes which are incorporated spontaneously into a well-defined lipid region of biomembranes; the fluorescence polarization characteristics of such probes directly relates to the lipid microviscosity of the labeled region. In the second class of probes,

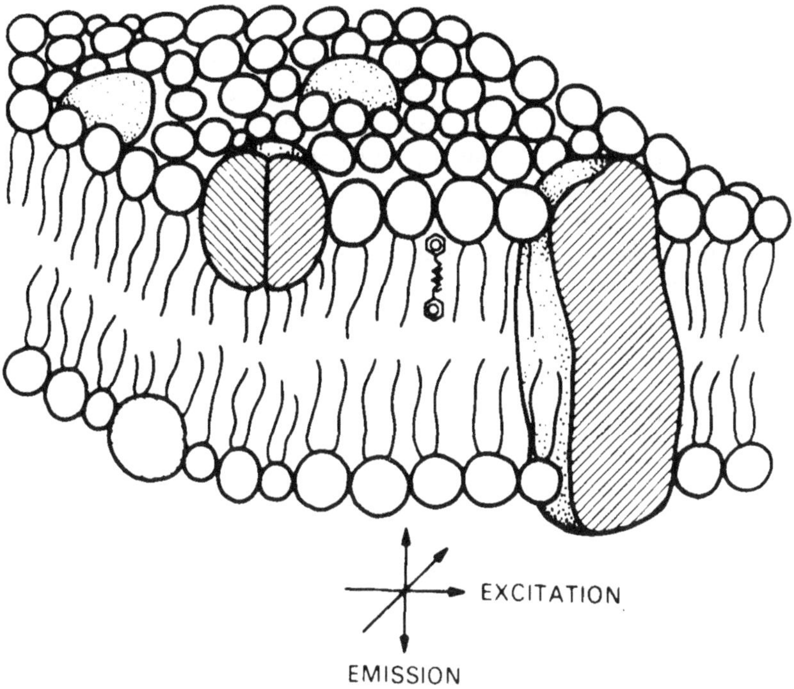

FIGURE 1. 1,6-Diphenylhexatriene (DPH) is one of the most efficient fluorescent probes used for determining the fluidity of membrane lipids by fluorescence polarization technique. The approximate orientation of a DPH molecule in membrane bilayer is schematically shown in the figure.

the fluorescent dye is covalently attached to phospholipids or fatty acids at different loci. Because some rotational modes of the fluorescent moiety in such compounds are markedly hindered, additional considerations are needed to make a correlation between microviscosity and fluorescence polarization. 1,6-Diphenylhexatriene (DPH) belongs to the first class, and is one of the most efficient fluorescent probes for the hydrocarbon regions of lipids.[113] Furthermore, DPH possesses a series of excellent spectral properties for a fluidity probe and can be dispersed in aqueous media to form microaggregates which are almost void of fluorescence.[113] When this dispersion is mixed with a lipid-containing system, the molecule of DPH is very efficiently incorporated into the hydrophobic region and displays a sharp increase in fluorescence signal.[113] Furthermore, DPH partitions equally into solid and fluid lipid domains.[114] Thus, in a heterogenous system such as biological membranes, DPH would be evenly distributed in the lipid layer near the polar head as schematically shown in Figure 1, and the derived value of microviscosity is therefore an average of all lipid domains. Involvement of any light-scattering depolarization effect on the measurement of fluorescence polarization could be eliminated by selecting various ratios of DPH to total membrane lipid and using the most appropriate ratio for routine measurements. In a review by Shinitzky and Barenholz,[109] these and other aspects of microviscosity determination are discussed. The determination of membrane microviscosity by the fluorescence polarization method using DPH involves the steps outlined below.

1. Incubation of membranes with the probe: DPH can be dissolved in a suitable organic solvent (e.g., tetrahydrofuran or ether) to make a stock solution (it is

advisable to protect the stock solution from light). An aliquot of the stock solution is dispersed in an appropriate assay buffer containing the membrane preparation to provide a desired ratio of DPH to total membrane lipids. This mixture is incubated at a constant temperature (22 to 25°C) for 30 min to 1 hr to allow the partitioning of the probe into membranes.

2. Determination of fluorescence polarization: membrane preparations, after removal of unincorporated DPH, are subjected to fluorescence polarization analysis in a fluorometer equipped with polarizers at a constant temperature. To determine the changes in membrane lipid phase transition, these measurements are made at different temperatures ranging from 0 to 40°C. From the observed values of fluorescence intensities, the value of steady-state polarization, P (directly proportional to membrane lipid microviscosity and inversely related to membrane fluidity), is computed using the following equation:

$$P = I_{\parallel} - I_{\perp}/I_{\parallel} + I_{\perp}$$

Where, $I_{\parallel}$ and $I_{\perp}$ are the fluorescence intensities polarized, parallel, and perpendicular to the polarization of the excitation beam. This value of polarization constant, P, can be used to determine the fluorescence anisotropy (r) and the apparent lipid microviscosity ($\bar{\eta}$) as described by Shinitzky and Barenholz.[109] These authors have discussed the theoretical aspects involved in derivation of the above-mentioned equation.

B. Factors Modulating Membrane Lipid Microviscosity

The organization of lipid in biomembranes is complex and heterogeneous. The fluid mosaic model of Singer and Nicolson[115] describes the lipids forming a fluid pool in which glycoproteins and proteins are considered to float. In membranes with a higher protein to lipid ratio, protein molecules may form a loosely bound network in the fluid pool of lipids. Depending on the physiological state of the biomembrane, the protein molecules can change their position and undergo reorganization. However, the redistribution in the lipid compartment is much more prominent than that in the proteins. The lipid reorganization or mobility is common and frequent within a monolayer; however, redistribution of lipids between the monolayers is extremely slow and rare.[116] There is also a considerable degree of coupling between the two monolayers of most biomembranes. This coupling is commonly achieved mainly by long chain fatty acyl moieties of phospholipids, and partly by certain special proteins which are embedded in bilayer at axes perpendicular to the plane of the membrane. According to Shinitzky and Henkart,[117] "By analogy to macroscopic liquids, fluidity, the reciprocal of viscosity, is proportional to the molecular free volume and is inversely proportional to the packing density of the fluid molecules." Thus, from the lipid composition of a membrane, its relative fluidity could be predicted. For example, an unsaturated fatty acyl chain will occupy a relatively larger volume than a fully saturated acyl chain; hence, an unsaturated fatty acyl chain may be classified as a "fluidizer". On the other hand, cholesterol has the rigid planer structure which will occupy a relatively small volume and is considered as a "rigidizer".

A number of studies using model membranes has described a variety of factors which modulate the lipid microviscosity. Changes in the membrane lipid microviscosity could be induced by local modifications of phospholipid to cholesterol molar ratio,[118,119] degree of unsaturation and length of the fatty acid moieties within the phospholipids,[114,120,121] concentrations of specific phosphatides, or the ratio of lipid to protein.[113,119,122] In most mammalian membranes, the presence of cholesterol, due to its rigid planer structure, will decrease "membrane fluidity" and increase the lipid microviscosity. Shinitzky and Inbar[119] reported that in liposomes with phospholipid composition identical to that in human erythrocytes, increasing the molar ratio of cholesterol to phospholipid (C:P ratio) from 0 to 1.4 increased the value of apparent lipid microviscosity from 0.5 to 6 *P*, suggesting a crucial role of cholesterol in the modulation of lipid microviscosity in biological systems.

Even though membrane cholesterol is constantly exchanged with the serum cholesterol pool, the C:P ratio of biomembranes remains fairly constant under normal physiological conditions. This is due to a steady equilibrium between these two cholesterol pools. Furthermore, this ratio becomes very crucial for cells such as erythrocytes and platelets,[118,123,124] which are constantly bathed in a cholesterol-rich environment of serum and also are devoid of any metabolic ability to process it. In contrast to this, cells which are metabolically active can maintain a constant ratio of C:P under normal in vivo condition by internally regulated processes such as intracellular synthesis (to increase) or esterification (to decrease) of cholesterol, receptor-mediated uptake of cholesterol-rich lipoproteins, or by passive translocation.

Chapman and Wallach[125] reported that the degree of unsaturation in fatty acid moieties of phospholipids influences the microviscosity of membranes. This is due to the fact that the presence of naturally occurring unsaturated fatty acids of phospholipids in *cis* configuration occupy greater specific volume than saturated fatty acids. The intracellular metabolic processes may mediate changes in the degree of unsaturation in fatty acyl chains. This phenomenon appears to be a major mechanism of membrane microviscosity adaptation in nutritional or metabolic disorders.[126] In a recent study, rats maintained on a regimen of intermittent starvation followed by refeeding a fat-free diet were reported to have a higher content of unsaturated fatty acyl chains, a lower ratio of C:P, and increased lipid fluidity of hepatocyte plasma membranes than that in controls.[127] Furthermore, Lentz et al.[114] reported that this effect of degree of fatty acyl chain unsaturation on microviscosity is dependent on the chain length. The effect is greater when the fatty acyl chain is shorter and vice versa.

The third major determinant of membrane microviscosity is the concentration of individual phospholipids in the membrane. Shinitzky and Henkart[117] have listed the "microviscosity index" of various phospholipid lyposomes relative to those constituted with a human erythrocyte phospholipid (HEP) mixture. Taking HEP as 1, the relative microviscosity index of phosphatidylcholine, phosphatidylserine, phosphatidylglycerol, and cardiolipin was found to be <1, whereas that of phosphatidylethanolamine, phosphatidylinositol, glangliosides, and sphingomyelin was found to be >1. The authors designated phospholipids with a relative microviscosity index of <1 as "fluidizers" and >1 as "rigidizers".[117] Among various phospholipids, sphingomyelin exhibited the highest relative microviscosity index of 3.4 (at 37°C) which, however, is far below the index of 12.1 for cholesterol.[117] Furthermore, among lecithins, phosphatidylcholine is a "fluidizer" and phosphatidylethanolamine is a "rigidizer". This would then suggest that methylation of phosphatidylethanolamine to phosphatidylcholine, a common biochemical reaction, could alter membrane microviscosity.

In the last major determinants of membrane microviscosity, i.e., the ratio of protein to lipid, the qualitative effects of proteins are similar to that of cholesterol. Thus, membranes with a higher ratio of protein to lipid would exhibit a higher value of microviscosity and vice versa. This effect of proteins on membrane microviscosity may be related to the specific volume in the bilayer.

There are a number of concepts illustrating the interaction of protein and lipid in biomembranes. According to the "vertical displacement of membrane proteins" concept of Shinitzky,[128,129] the bulk proteins become more exposed to the aqueous medium in membranes with higher lipid microviscosity (a phenomenon similar to the "squeezing out" effect) whereas the same would tend to sink into the membrane interior in more fluid membranes. One can look at this concept from an entirely different angle and hypothesize that the reverse of this phenomenon may also be occurring. In other words, higher lipid microviscosity may push the proteins into the interior of the bilayer and restrict their exposure to the surface whereas lower microviscosity may increase the mobility of the proteins and allow them to be more exposed to the aqueous medium. It is, however, difficult to hypothesize a single type of membrane protein displacement in biomembrane, and perhaps depending on the tertiary structure and size of the protein, many types of displacement may be occurring in vivo. This biophysical process has great physiological significance since mobility of proteins in the bilayer provides a mechanism for modulation of a variety of membrane-associated phenomena.

IV. MODIFICATION OF CELLULAR FUNCTIONS MEDIATED BY MEMBRANE LIPID MICROVISCOSITY CHANGES

Among the available approaches, studying the effect of lipid microviscosity on cellular function is one of the best approaches for understanding the mechanisms of membrane processes. Changes in lipid fluidity may modulate the expression of plasma membrane-mediated cellular functions by altering the degree of exposure and rotational and lateral mobility of membrane proteins. A variety of membrane-associated phenomena, including (1) membrane receptors or binding proteins for lipid, protein, and peptide hormones, and other functional lipids, proteins, peptides, and vitamins, (2) antigens, and (3) enzymes, etc. are likely to be partially affected by membrane lipid microviscosity changes. In the past few years, a number of studies have demonstrated the effect of changes in membrane lipid microviscosity or lipid environment on membrane-associated parameters. Some of the key observations made by other investigators on this aspect are briefly discussed in this section.

Cooper and co-workers have carried out extensive studies on the effect of fluidity on human platelet and red blood cell function.[123,124,130-135] The fluidity of red blood cells is decreased in spur-cell anemia due to an abnormal C:P ratio, and in abetalipoproteinemia due to an abnormal sphingomyelin to lecithin ratio. This decrease in fluidity is associated with a folded and scalloped red cell contour in both disorders. As a result of these abnormalities, the red blood cells have reduced ability to flow and short survival in vivo. Increasing the C:P ratio of red blood cell membranes has been known to decrease the cell membrane permeability to a variety of substances that enter either by diffusion or are transported through a carrier.[126,136] In vitro enrichment of platelets with cholesterol has been reported to decrease membrane fluidity and increase epinephrine-induced platelet sensitivity to aggregation, a specific functional characteristic of platelets.[124,136] This functional abnormality and increased C:P ratio has been reported to occur in platelets of Type IIa hyperlipoproteinemia patients.[134,137] The activity of platelet membrane-associated adenylate cyclase has also been reported to be modulated by membrane lipid fluidity changes. A decrease in platelet membrane fluidity was found to increase the basal level of adenylate cyclase activity and blocked the prostaglandin E_1- or sodium fluoride-induced stimulation of the same.[138]

Table 2 presents a summary of other cellular functions modified by changes in membrane lipid composition or lipid microviscosity.

Table 2
EFFECT OF CHANGES IN LIPID ENVIRONMENT OR MICROVISCOSITY ON CELL MEMBRANE-ASSOCIATED PHENOMENA

Sources of membranes	Observed or presumed changes in microviscosity[a]	Functional parameter studied	Effect of microviscosity on the parameter[a]	Ref.
1	2	3	4	5
Effect on enzymes				
Rat hepatic plasma membrane	−	Adenylate cyclase	+	139†
	−	5′-Nucleotidase	+	139†
	−	Na^+,K^+-ATPase	+	140†
	−	Na^+,K^+-ATPase	+	127
Rat pituitary membrane	+	Adenylate cyclase	−	141
Effect on uptake and transport				
Rat adipocyte membrane	−	Basal glucose uptake	−	142†
	−	Insulin-stimulated glucose uptake	−	142†
	−	Basal and insulin-stimulated 2-deoxy D-glucose uptake	−	142†
Human erythrocyte	−	Glucose uptake	−	143 144
Novikoff hepatoma cell	−	2-Deoxy D-glucose uptake	−	145
Mouse 3T3 fibroblasts	+	Uptake of 2-deoxy D-glucose and α-aminoisobutyric acid	+	146†
Macrophage	+	Phagocytosis	−	126†
Effect on receptor				
Rabbit neutrophils	−	Chemotactic peptide receptors	+	147
Human polymorphonuclear leukocytes	−	Chemotactic peptide receptors	+	146†
	−	Chemotactic peptide receptors	+	148
Rat brain membrane	−	Alpha-adrenergic receptors	+	149
Mouse bone marrow cell	−	Transferrin receptors	+	150
	+	Transferrin receptors	−	150
Mouse brain: crude and synaptosomal membrane	+	Serotnin receptors	+	151†
	−	Serotonin receptors	−	151†
Rhesus monkey corpus luteum membrane	−	Human luteinizing hormone receptors	+	152
Rat testicular membrane	+	Human luteinizing hormone receptors	−	153†

Note: The studies marked with † have reported changes in microviscosity and other studies have suggested the influence of microviscosity on the functional parameter.

[a] The plus sign indicates an increase in microviscosity or activity of the parameter; the minus sign indicates a decrease in the same.

V. CORRELATION BETWEEN CHANGES IN PROLACTIN RECEPTORS, PROSTAGLANDIN SYNTHESIS, AND MEMBRANE LIPID MICROVISCOSITY

A. Studies on Hepatic Membranes

As described in Sections III and IV, the plasma membrane appears to be a dynamic matrix in which proteins and glycoproteins can be considered to float. It was then

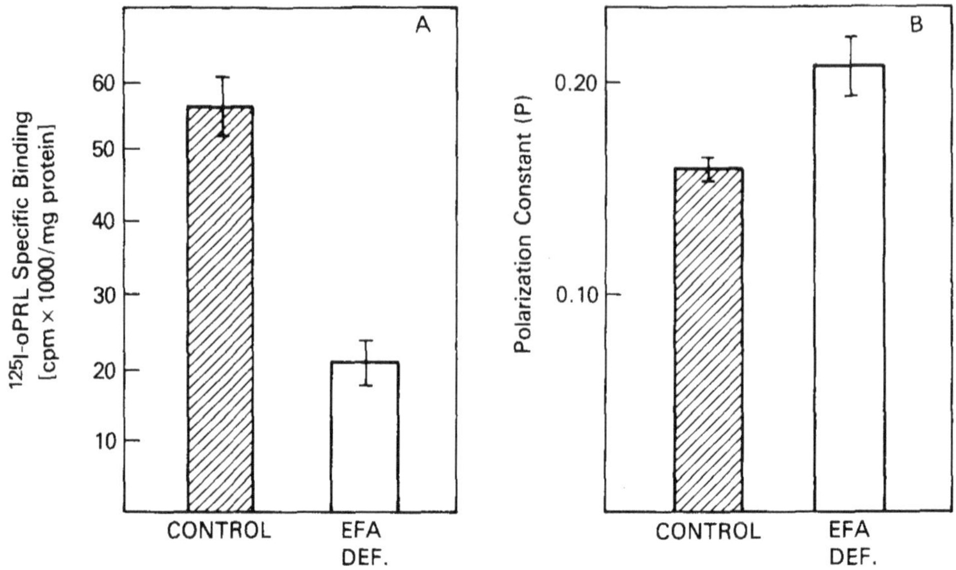

FIGURE 2. The dietary deficiency of essential fatty acids (EFA) decreases hepatic PRL binding (A) and membrane lipid microviscosity (B). Female C_3H mice were fed with a diet deficient in EFA for 32 weeks.[155,156]

thought that the induction of PRL receptors in dwarf mouse liver following bovine growth hormone or ovine PRL treatment in the absence of protein synthesis might be due to physical alterations of the membrane. Such alteration of the bilayer may permit the receptor protein to change its position or conformation, thereby exposing an active pre-existing binding site. If this thesis is true, then changes in the availability of PRL receptor might be reflected on membrane lipid microviscosity changes and vice versa. In other words, that decrease in microviscosity might allow the receptors to move within the bilayer more easily, and thereby increase the accessibility of receptor for the ligand. This argument was supported by an earlier observation that relaxation of membrane bilayer accomplished by exposing hepatocytes in vitro to phospholipase A_2 resulted in a 60% increase in the number of PRL receptors.[154] A series of investigations, reported earlier in part and summarized in this section, were undertaken to observe if changes in PRL binding correlated with membrane lipid microviscosity changes.

Direct alteration of the hepatic plasma membrane lipids may be achieved in vivo by dietary means. As discussed in an earlier section, the proportion of unsaturated fatty acids in membrane is one of the determinants of lipid microviscosity. Arachidonic acid, a polyunsaturated fatty acid, is an important constituent of most mammalian cells, and is synthesized in animals from linoleic acid. The latter, a dienoic acid, is not synthesized in animal tissue and, hence, is dietary "essential". By feeding a diet deficient in essential fatty acids (EFA) for a long period, one may induce dramatic alterations in lipid composition of tissue membranes. In an earlier study, female C_3H mice maintained on an EFA-deficient diet showed a time-related loss of hepatic PRL binding which was less than 50% after 32 weeks of treatment.[155,156] Furthermore, treatment of EFA-deficient mice with exogenous PRL did not reverse this loss in hepatic PRL binding.[156] As anticipated, the lipid microviscosity of hepatic membranes obtained from EFA-deficient mice was greater than control values (Figure 2). These results suggested that changes in membrane lipid microviscosity are inversely related to the number of accessible PRL binding sites, and that EFA are required for the maintenance of PRL receptors and physiological values of lipid microviscosity.

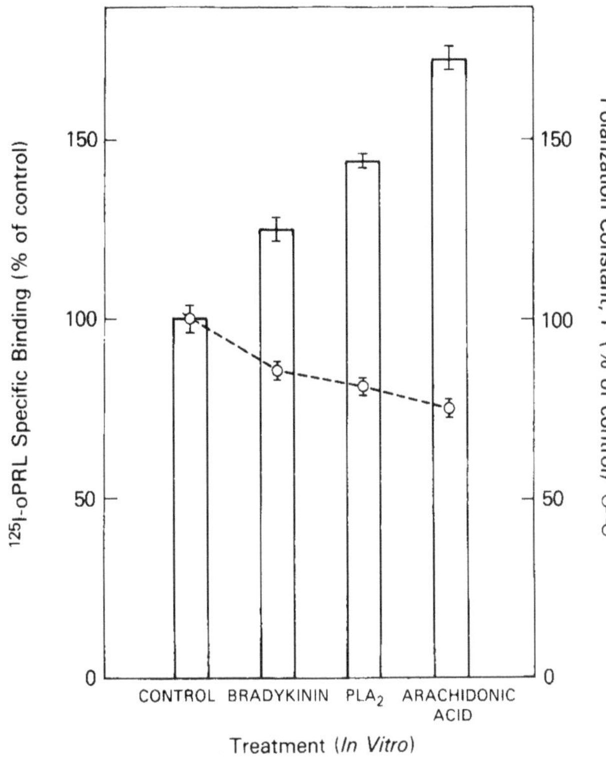

FIGURE 3. Induction of specific ^{125}I-ovine PRL binding to microsomal (× 15,000 to 100,000 g pellet) membrane fraction of C_3H female mice after the liver homogenate was incubated with bradykinin (5 μg/mℓ), phospholipase A_2 (0.026 units/mℓ) and arachidonic acid (4 μg/mℓ) for 30 min at 28°C. Values are mean of six observations ± S.E. (From Dave, J. R., Knazek, R. A., and Liu, S. C., *Biochem. Biophys. Res. Commun.*, 103, 727, 1981. With permission.)

Such a role of EFA was confirmed in one of the in vitro studies in which female mouse liver homogenate was treated for 30 min at 28°C with graded amounts of arachidonic acid. The membrane preparation isolated after this treatment exhibitied a dose-dependent increase and then a decline in specific binding of ^{125}I-ovine PRL. A maximal increase of approximately 75% in PRL binding was observed at a 4 μg/mℓ concentration of arachidonic acid (Figure 3).[154]

The enzyme, phospholipase A_2, metabolizes phospholipids into arachidonic acid and lysophospholipids. In order to determine if exogenously added phospholipase A_2 would also modify PRL binding by acting on membrane phospholipids and releasing endogenous arachidonic acid, in another study, female mouse liver homogenate was treated with graded amounts of phospholipase A_2 at 28°C for various time intervals. The membrane preparations isolated after this in vitro treatment exhibited a dose-related increase and then decline in specific ^{125}I-oPRL binding. At all the time intervals studied, a maximal increase of 30 to 60% in PRL binding was observed at 0.026 units of phospholipase A_2 per milliliter (Figure 3).[154]

Certain activators of phospholipase A_2, such as bradykinin, a decapeptide, are known to stimulate membrane-associated phospholipase A_2 activity. To determine if in vitro addition of bradykinin would stimulate endogenous phospholipase A_2 and mimic the effects produced by exogenously added phospholipase A_2 on PRL binding,

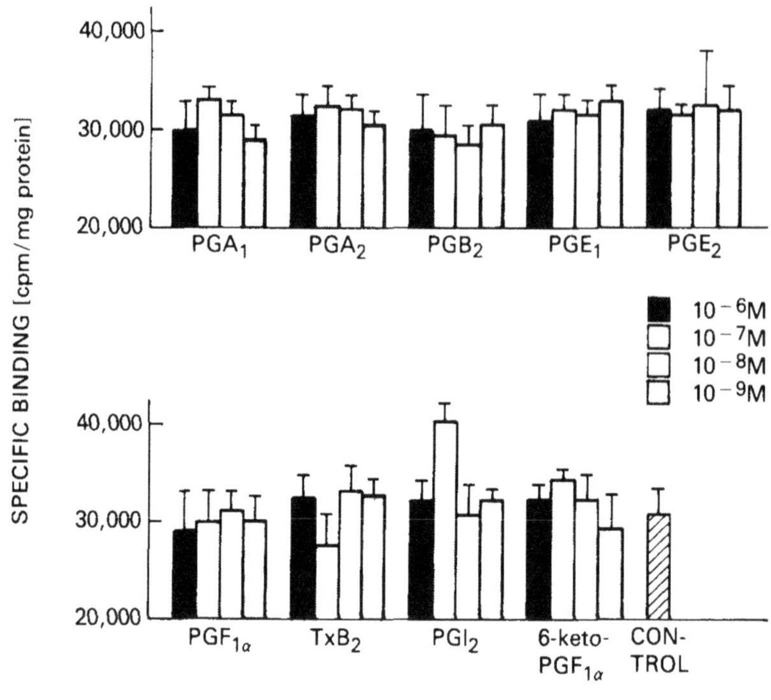

FIGURE 4. Effect of treatment of mouse liver homogenates with various prostaglandins on PRL-binding capacities in liver membranes. Liver homogenates were incubated with 1 to 1000 n*M* prostaglandins for 30 min at 29°C. These suspensions were centrifuged, and micrsomal membrane fractions were subsequently obtained and used for PRL binding. Values are mean ± S.D. (From Dave, J. R. and Knazek, R. A., *Proc. Natl. Acad. Sci., U.S.A.*, 77, 6597, 1980. With permission.)

female mouse liver homogenate was treated with various amounts of bradykinin at 28°C for various time intervals. This treatment stimulated specific ^{125}I-oPRL binding in a dose-related fashion, with a maximal increase of approximately 25% observed at 5 μg of bradykinin per milliliter within 30 min (Figure 3).[154]

This induction of hepatic PRL binding after in vitro treatment with arachidonic acid, phospholipase A_2, and bradykinin, occurring within 30 min of treatment period, was found to increase the number, rather than the affinity, of PRL receptor sites. Under the same conditions, the optimal stimulatory concentrations of arachidonic acid, phospholipase A_2, and bradykinin produced approximately 25, 20, and 15% decreases in membrane microviscosity, respectively. These results suggested that changes in PRL binding and membrane lipid microviscosity are interrelated and that arachidonic acid may have a direct or indirect role in the modulation of these parameters in vivo.

In the biosynthesis of prostaglandins, arachidonic acid plays one of its better studied and physiologically important roles. This fatty acid is a precursor for the "2" series of prostaglandins. As a logical extension of the in vitro studies, the effects of various prostaglandins on hepatic PRL binding in female mice was studied next. In this study, liver homogenate of adult C_3H female mice was treated with various concentrations of prostaglandins at 29°C for 30 min. Membrane preparations isolated after this treatment were assayed for specific ^{125}I-labeled ovine PRL binding. Only membrane preparations treated with prostaglandin I_2 (prostacyclin) showed an increase in specific PRL binding (Figure 4).[157] The maximal stimulation of 40 to 50% was observed at 100 n*M* prostaglandin I_2 and was due to an increase in the number of receptor sites. Under

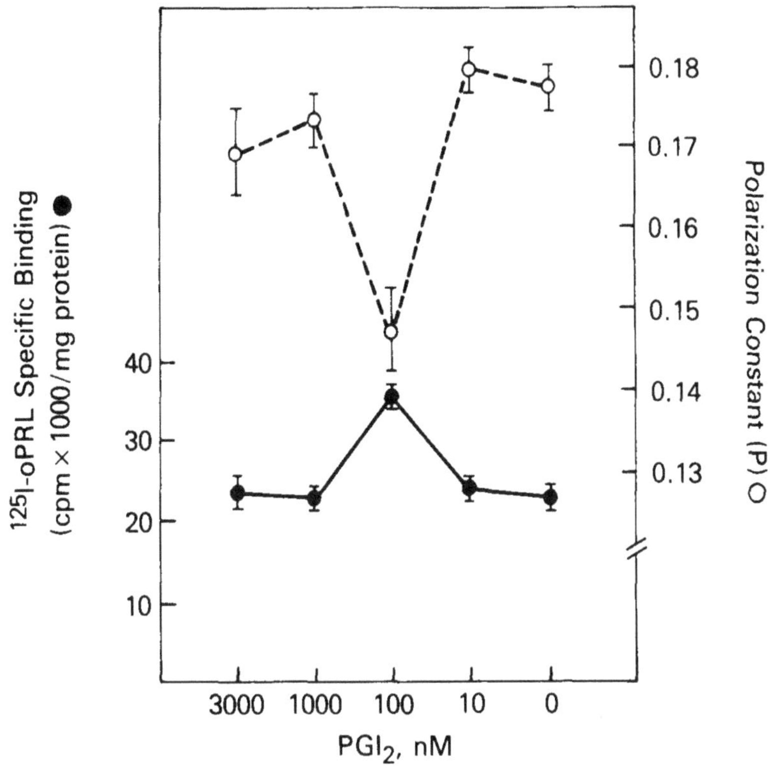

FIGURE 4A. Effects of various concentrations of prostaglandin I_2 on induction of specific ^{125}-I-ovine PRL binding and membrane lipid microviscosity. 1,6-Diphenylhexatriene (1 μmol) was used to measure lipid microviscosity by the fluorescence polarization technique. Microviscosity measurement of liver microsomal (× 15,000 to 100,000 g pellet) membrane preparations of C_3H female mice were carried out at 24°C after the liver homogenate was incubated with PGI_2 for 30 min at 29°C. Values are mean ± S.D. (From Dave, J. R. and Knazek, R. A., *Proc. Natl. Acad. Sci. U.S.A.*, 77, 6597, 1980. With permission.)

the same conditions, prostaglandin I_2 produced a 17% decrease in membrane microviscosity (Figure 4A). These data suggested that specific prostaglandins may modulate the fluidity of the lipid bilayer and the subsequent ease with which receptors can assume active configurations within the matrix. Furthermore, the loss of PRL binding in EFA-deficient animals and the stimulation of PRL binding after in vitro arachidonic acid, phospholipase A_2, and bradykinin treatments may be mediated via the prostaglandin cascade.

To confirm these observations and to support the idea that the prostaglandin cascade may be involved in the modification of hepatic PRL receptors and membrane microviscosity, studies were undertaken to determine the effects of indomethacin on these parameters. This nonsteroidal antiinflammatory drug inhibits the cascade and blocks the synthesis of prostaglandins. Indomethacin treatment suppressed the number of detectable hepatic PRL receptors in both male and female mice. This occurred in a dose-dependent fashion with 7.5 μg of indomethacin per gram of body weight injected every 4 hr, exerting a maximal effect within 20 hr. Administration of 50 μg of PRL every 4 hr stimulated hepatic PRL binding in controls but not in the indomethacin-treated mice (Figure 5A).[158] Indomethacin treatment increased and exogenous PRL treatment decreased the lipid microviscosity of mouse hepatic membranes (Figure 5B).

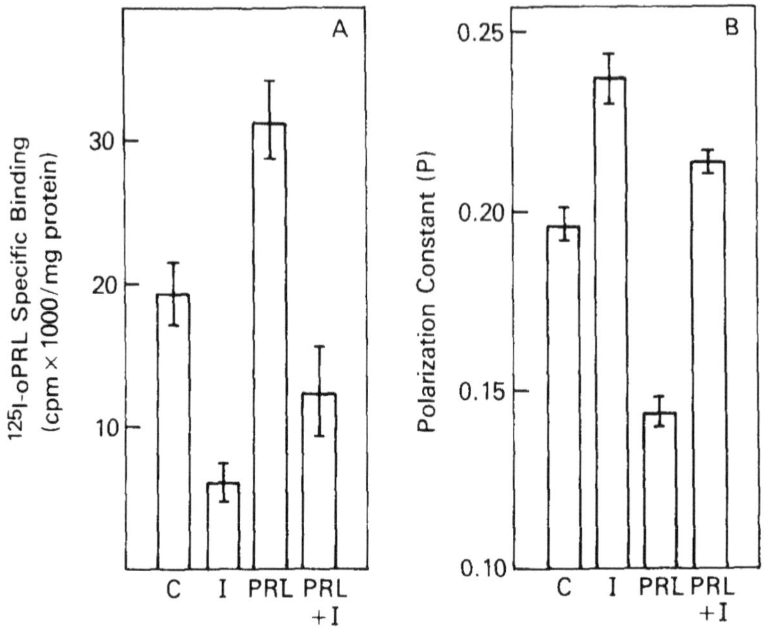

FIGURE 5. Specific PRL binding (A) and lipid microviscosity (B) of male C_3H mice treated with 7.5 μg of indomethacin (I) per gram of body weight and/or 50 μg of ovine PRL every 4 hr for a period of 20 hr. Each bar represents mean of three animals, the livers of which were assayed individually in triplicate. Vertical lines represent 1 S.E.M. (From Knazek, R. A., Liu, S C., Dave, J. R., Christy, R. J., and Keller, J. A., *Prostaglandins Med.*, 6, 403, 1981. With permission.)

Administration of PRL to indomethacin-treated animals, however, could not completely reverse the effect of the drug on lipid microviscosity. Determination of the molar ratio of phospholipids to cholesterol in hepatic membranes obtained from mice in various groups revealed an inverse relationship to the polarization constant (Figure 6).[158] This observation suggested that the effect of indomethacin on hepatic membrane lipid microviscosity was primarily mediated via an alteration in cholesterol and/or phospholipid metabolism. These data (1) support the proposed involvement of prostaglandins in the modulation of PRL binding, (2) support the existence of a correlation between hepatic PRL binding and membrane lipid microviscosity, and (3) suggest that PRL-induced up regulation of its hepatic receptors may involve modifications of prostaglandin synthesis, phospholipid to cholesterol ratios, and membrane lipid microviscosity.

The preceding studies also raised a question regarding the ability of PRL to alter the synthesis of prostaglandins. In another study, to determine if PRL indeed modified the rate of prostaglandin synthesis, male mice were injected subcutaneously with 100 μg of ovine PRL every 4 hr for 0, 24, or 48 hr and were sacrificed 8 hr after administration of the last PRL injection. The number of PRL binding sites in hepatic membranes increased with the duration of PRL injections. Increases of 79 and 144% above the control value were observed after 24 and 48 hr of injections, respectively (Figure 7).[159] This treatment also lowered the hepatic membrane lipid microviscosity in a time-dependent fashion (Figure 7).

In order to determine the effect of PRL administration on prostaglandin synthesis, hepatic membranes obtained from PRL-treated and untreated animals were incubated in vitro at 37°C for 2 hr in the presence or absence of 0.5 mM of indomethacin. The

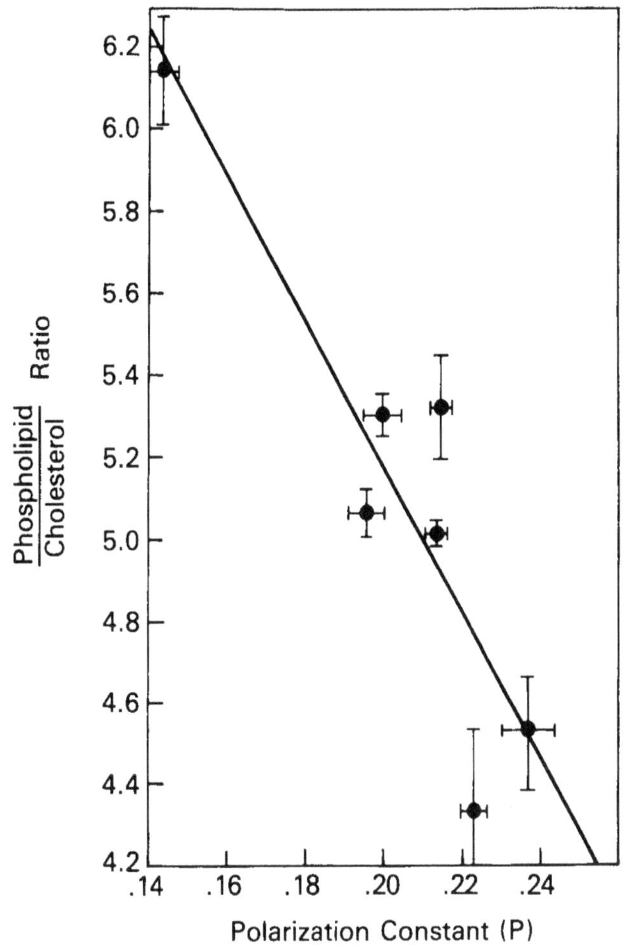

FIGURE 6. The ratio of phospholipid to cholesterol was inversely proportional to the polarization constant, P, an index of membrane lipid microviscosity. Each parameter was measured independently with vertical and horizontal bars representing 1 S.E.M. The line was drawn by the least squares method. (From Knazek, R. A., Liu, S. C., Dave, J. R., Christy, R. J., and Keller, J. A., *Prostaglandins Med.*, 6, 403, 1981. With permission.)

prostaglandins were partially purified by silicic acid column chromatography and the amounts of prostaglandin E and $F_{2\alpha}$ were determined by specific radioimmunoassays. The values obtained by radioimmunoassay were corrected for the recovery, and values obtained in the presence of indomethacin were subtracted from the values obtained in the absence of indomethacin (total prostaglandins − endogenously present prostaglandins = newly synthesized prostaglandins). Each value was normalized for the amount of membrane protein present.

Administration of PRL produced a time-related increase in synthesis of both prostaglandin E and $F_{2\alpha}$ by hepatic membranes (Figure 8).[159] These results are thus compatible with the hypothesis that prostaglandin synthesis is involved either directly or indirectly in the induction and maintenance of the PRL receptors, at least in liver, and that modification of the physical characteristics of the surrounding lipid matrix may modulate the functionality of this receptor. In other words, these results imply that the PRL receptors may be controlled directly or indirectly through changes in the mem-

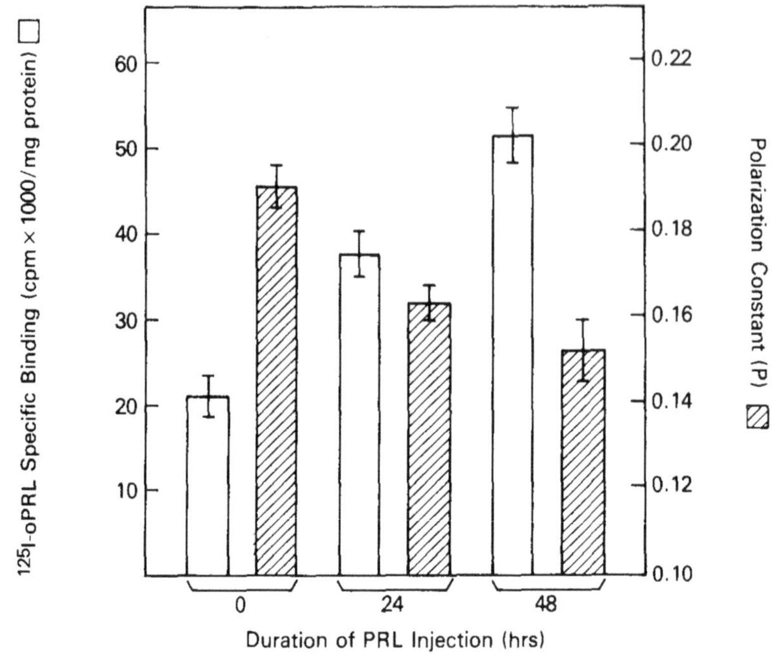

FIGURE 7. Specific binding of PRL (open bars) to mouse heptic membranes increased with duration of in vivo PRL treatment. Fluorescence polarization measurements (cross hatched bars), an index of membrane microviscosity, decreased concomitantly with duration of PRL treatment. Male mice were injected subcutaneously with 100 μg of ovine PRL every 4 hr for 0, 24, or 48 hr and were sacrificed 8 hr after receiving the last injection. Values are mean ± S.E.M. of 4 to 5 animals and each sample was assayed individually in triplicate. (From Dave, J. R., Brown, N. V., and Knazek, R. A., *Biochem. Biophys. Res. Commun.*, 108, 193, 1982. With permission.)

brane microviscosity and that changes in local prostaglandin levels and synthesis may influence the lipid microviscosity of these membranes.

In a logical extension of the preceding studies, to determine if this thesis held true under in vivo conditions involving alterations of plasma PRL level, the same parameters were measured at various stages of gestation and lactation in female mice. The values for PRL-binding capacity of hepatic membranes from C_3H female mice at different stages of gestation and lactation are shown in Figure 9.[98] Their PRL-binding capacity increased as pregnancy advanced. A significant increase was observed by day 8 of gestation which subsequently reached a maximum of more than threefold greater than the nonpregnant, nonlactating (NPNL) levels, at 16 days of gestation. A dramatic fall to NPNL levels in this activity was noted over the next 2 days with the values remaining near this level throughout the remainder of gestation and the entire period of lactation. The values of PRL binding were 128, 145, 289, 325, 93, and 95% of those observed for NPNL females for 4, 8, 12, 16, 18, and 20 days of gestation, respectively, and 115, 112, 126, 123, 108, and 104% of NPNL values for 0, 4, 8, 12, 16, or 20 days of lactation, respectively.

As also shown in Figure 9, the values of membrane microviscosity were significantly decreased until 20 days of gestation when a return to near NPNL levels was observed, which remained unchanged thereafter. These values of fluorescence polarization were 80, 75, 75, 92, 101, 96, 101, and 102% of the NPNL values for 4, 8, 16, or 20 days of gestation and 4, 8, 16, or 20 days of lactation, respectively. The phospholipid to cholesterol ratio measured in each of the groups was inversely related to the fluorescence polarization constants observed during gestation and lactation.

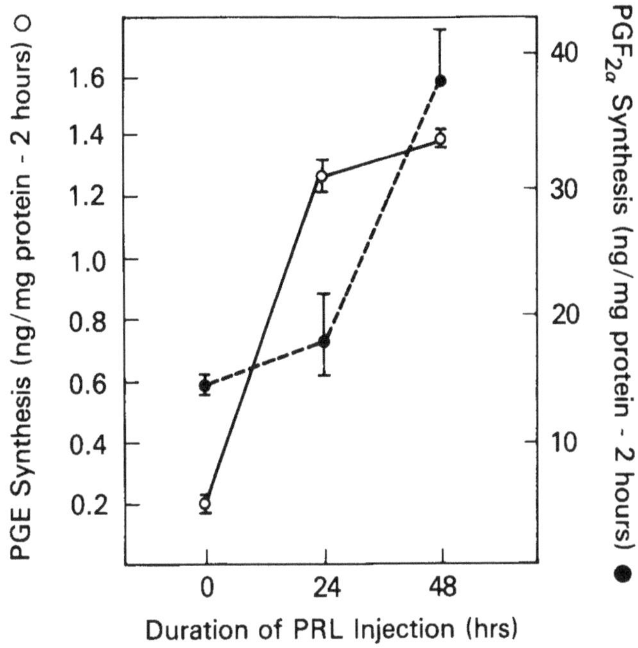

FIGURE 8. Demonstration that in vivo PRL administration increased synthesis of prostaglandin (PG) E and $F_{2\alpha}$ by mouse hepatic membranes in vitro. The hepatic membrane preparations from 3 to 4 animals in each group were pooled. Each point represents 3 incubations with or without indomethacin, assayed in duplicate. Vertical bars = 1 S.E.M. (From Dave, J. R., Brown, N. V., and Knazek, R. A., *Biochem. Biophys. Res. Commun.*, 108, 193, 1982. With permission.)

The amount of prostaglandin synthesized in different groups, after incubating the membranes for 2 hr at 37°C in the presence and absence of indomethacin, is shown in Figure 10. The amount of prostaglandin E synthesized was highest in mid-pregnancy, whereas the increase in prostaglandin $F_{2\alpha}$ synthesis was more rapid and sustained. The amount of prostaglandin E synthesized increased from NPNL values to peak (0.60 ng/mg protein) at 12 days of gestation. Thereafter, a steady decline in the synthesis was observed, with NPNL levels being reached by 18 days of gestation and a further decrease to 0.07 ng/mg protein by the day of parturition. Recovery to NPNL levels occurred by day 20 of lactation. Synthesis of prostaglandin $F_{2\alpha}$ remained at maximal levels from days 12 to 18 of gestation followed by a striking fall on the day of parturition to 7.8 ng/mg protein. The values remained at this NPNL level throughout the entire period of lactation.

These results suggest that PRL-binding capacity, membrane fluidity, and prostaglandin-synthesizing capacity of mouse hepatic membranes are increased during early to mid-pregnancy, probably in response to raised circulating PRL. There is a dramatic decrease in all of these activities at term and during lactation. These results are also compatible with the thesis proposed earlier, but need additional justifications in view of the fact that pregnancy and lactation also involve the synthesis of new PRL receptors in response to higher levels of circulating PRL. Mice in early pregnancy have relatively low levels of serum PRL which increase with the advancement of pregnancy and are associated with a surge[160] near term. Serum PRL levels increase further during lactation (tenfold) to reach a peak at day 10 of lactation and return to normal levels

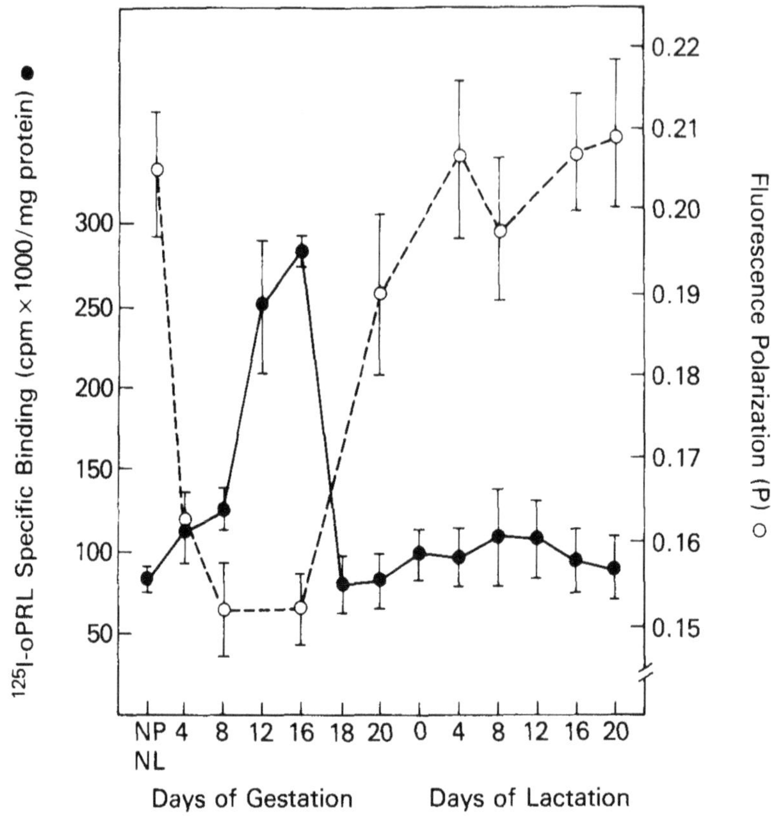

FIGURE 9. Changes in PRL binding and lipid microviscosity of hepatic microsomal (× 15,000 to 100,000 g pellet) membranes from C_3H female mice during pregnancy and lactation. NPNL signifies a nonpregnant, nonlactating mouse. Each solid circle represents PRL binding data (± S.E.M.) from livers obtained from four or five animals, assayed individually in triplicate. Each open circle represents fluorescence polarization constant of pooled membrane preparations from 3 to 4 mice in each group assayed in triplicate. (From Dave, J. R., Richardson, L. L., and Knazek, R. A., *J. Endocrinol.*, 99, 99, 1983. With permission.)

near day 20 of lactation. Serum levels of placental lactogen begin to increase at midgestation, reaching a plateau during the week before parturition. Other investigators have also shown the existence of a surge of another lactogen peaking on day 10 of gestation. The increase of PRL-binding capacity in early gestation seen in the preceding study may be due to increased fluidity of the membranes. However, the further increase from days 8 to 12 of gestation probably reflects the synthesis of new receptors at this stage, in response to raised lactogen levels. The changes in fluidity observed at this stage are an apparent result of a change in the ratio of phospholipids to cholesterol and may play a role in maximally exposing the already present and newly synthesized receptors during late pregnancy. These changes in microviscosity may be brought about by increases in the ability of the membranes to synthesize prostaglandins. However, the very high circulating PRL levels seen at parturition and during lactation are probably responsible for the lower PRL receptor levels and increased microviscosity. Similar changes in PRL receptors and fluidity were observed in other studies in which hypophysectomized female rats were injected with high doses of PRL every hour.

In this series of investigations, hypophysectomized, immature, female rats having subcutaneous diethyl stillbesterol implants were treated with pregnant mare serum go-

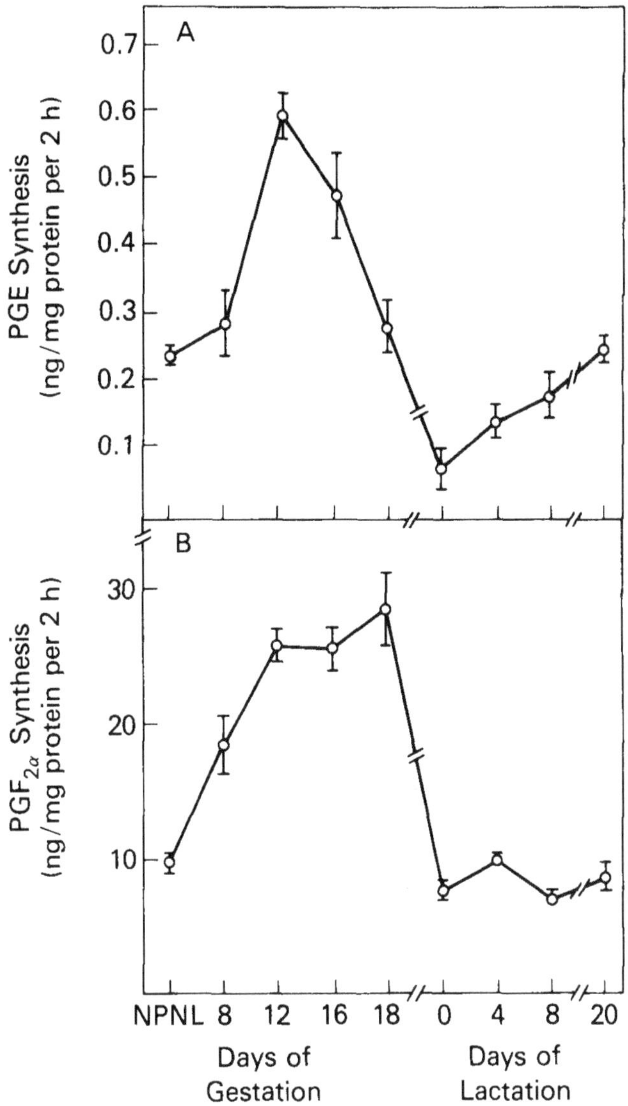

FIGURE 10. Prostaglandin (PG) E (A) and $PGF_{2\alpha}$ (B) synthesis by mouse liver microsomal membrane preparations during gestation and lactation. NPNL signifies a nonpregnant, nonlactating mouse. The membrane preparations from three to four animals in each group were pooled. Each point represents the mean of three incubations (± S.E.M.) with or without indomethacin, assayed in duplicate. (From Dave, J. R., Richardson, L. L., and Knazek, R. A., *J. Endocrinol.*, 99, 99, 1983. With permission.)

nadotropin and human chorionic gonadotropin. In addition, graded doses of 0, 0.25, 0.5, 1.0, 2.5, or 50 μg of ovine PRL were injected hourly for the 48 hr prior to sacrifice at 28 days of age. This treatment produced a biphasic response in PRL binding, lipid fluidity, and phospholipid to cholesterol ratio of hepatic membranes. Administration of physiological amounts of PRL increased all three parameters, whereas higher or lower amounts of PRL produced minimal or no effects.[161] PRL binding was increased three- or fivefold in hepatic membranes obtained from 0.5 or 1.0 μg of ovine PRL-injected groups, respectively. The ratio of phospholipid to cholesterol in hepatic membranes was increased, and lipid microviscosity was decreased in these two groups.

These data suggest that administration of small amounts of PRL to immature, hypophysectomized, female rats decreases the lipid microviscosity of hepatic membranes whereas the same increases when hypo- or hyperprolactinemic states are achieved.

The studies described in this section strongly suggested an involvement of prostaglandin synthesis in the modulation of hepatic PRL receptors which are parallelled by changes in membrane fluidity. As an obvious extension of these studies, and to determine if the interrelation we observed between changes in PRL binding, membrane microviscosity, and prostaglandin synthesis in hepatic tissue held true in other "PRL target tissue", studies were undertaken on prostate gland.

B. Studies on Prostatic Membranes

Except for the effects of androgen, little is known about the regulation of PRL receptor levels in prostatic tissue. The studies described in this section were undertaken to determine if their response was similar to hepatic membranes.

1. Indomethacin treatment could modify microviscosity of prostatic membranes and in turn influence the level of prostatic PRL receptors.
2. Administration of PRL could modify these parameters in prostatic membranes.
3. Agents known to modify membrane lipid microviscosity in vitro could produce changes in PRL binding.

To determine the effects of indomethacin, adult male rats were treated with vehicle or either of two doses of indomethacin (3.8 or 7.5 μg/g of body weight) every 8 hr for 1, 2, and 3 days. Similar to its effects on hepatic membranes, indomethacin decreased PRL binding in rat ventral and dorsolateral lobes of prostate gland in a time- and dose-dependent fashion (Figure 11).[162] In ventral prostate membranes, the higher dose of indomethacin decreased PRL binding by approximately 80% after 3 days of treatment. PRL binding to membrane preparations from dorsolateral prostate gland of control animals was 10% of that observed for ventral prostate and decreased to undetectable levels within 24 to 48 hr after the start of indomethacin treatment. The effects of indomethacin on membrane lipid microviscosity were inversely related to that of PRL binding. Both ventral and dorsolateral prostatic membrane lipid microviscosity increased after indomethacin treatment in a time- and dose-dependent manner. These effects of indomethacin on prostatic PRL binding and lipid microviscosity suggest that as observed in liver, these phenomena are interrelated and that prostaglandins are involved in the maintenance of prostatic PRL in vivo.

The effects of PRL administration on prostatic PRL binding and membrane lipid microviscosity were determined recently.[163] In this study, adult male rats were injected with 0, 100, or 400 μg of ovine PRL every day subcutaneously for a period of 5 days. PRL administration upregulated its receptors in ventral and dorsolateral lobes of the prostate gland in a dose-related manner. This increase in PRL binding was greater in magnitude in dorso-lateral lobe than that in ventral lobe. This treatment also produced a dose-related decrease in lipid microviscosity of membranes obtained from ventral and dorsolateral lobes and the effect was greater in dorsolateral lobe than that in ventral lobe. These data supported the earlier thesis and suggested that PRL administration may produce generalized effects on lipid microviscosity.

In the studies described so far, membrane lipid microviscosity measurements were made under the conditions producing modifications in PRL binding. We then decided to find out if factors known to alter lipid microviscosity in vitro would also produce changes in PRL binding. Studies were, therefore, undertaken to determine (1) if the age-related changes in ^{125}I-labeled ovine PRL-specific binding of rat ventral prostate were correlated with changes in membrane lipid microviscosity and (2) if membrane

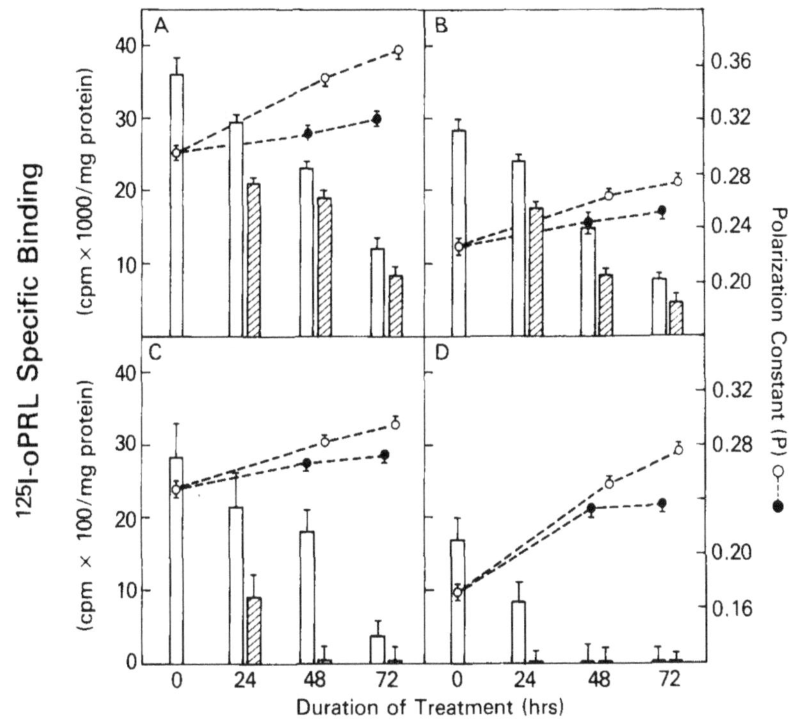

FIGURE 11. Demonstration that indomethacin treatment decreased specific PRL binding and lipid fluidity (inversely related to polarization constant) of rat prostatic membranes. (A) PRL binding to a × 15,000 g (× 125 to 15,000 g pellet) membrane fraction of ventral prostate in rats treated with the lower (3.8 μg/g of body weight every 8 hr; open bars) or higher (7.5 μg/g of body weight every 8 hr; crosshatched bars) doses of indomethacin for 1, 2, and 3 days; (B) PRL binding to a × 100,000 g (× 15,000 to 100,000 g pellet) membrane fraction of ventral prostate in rats treated with the lower or higher doses of indomethacin; (C) PRL binding to a × 15,000 g membrane fraction of dorsolateral prostate in rats treated with the lower or higher doses of indomethacin; and (D) PRL binding to a × 100,000 g membrane fraction of dorsolateral prostate in rats treated with the lower or higher doses of indomethacin. Values of polarization constant are presented as broken lines and solid (higher dose of indomethacin) or open (lower dose of indomethacin) circles. Each point represents data from a pool of three to four ventral prostates and five to six dorsolateral prostates assayed in triplicate. Vertical bars = 1 S.E.M. (From Dave, J. R. and Witorsch, R. J., *Prostate*, 4, 119, 1983. With permission.

fluidizers (aliphatic alcohols) in vitro produced age-dependent effects on PRL binding of prostatic membranes. Membrane preparations of ventral prostate glands obtained from immature (24 to 25 days old), young adult (80 to 90 days old), and aged (550 to 610 days old) male rats were used for these studies. Figure 12 shows an age-dependent inverse relationship in prostatic PRL-binding capacity and membrane lipid microviscosity. Relative to immature rats, prostatic PRL binding decreased approximately 50% in young adult rats and 75% in aged rats.[164] Membrane lipid microviscosity, relative to immature rats, was increased 72% in young adult rats and 140% in aged rats. The in vitro exposure of membrane preparations to alcohols produced an effect which was dependent on the lipid microviscosity of membranes. Prostatic membrane preparations obtained from immature animals having a relatively lower lipid microviscosity value exhibited no significant effects of this in vitro treatment on PRL binding whereas those obtained from aged animals having higher lipid microviscosity exhibited maximal increase in PRL binding. As shown in Figure 13A, ethanol at concentrations ranging from 1 to 16.7% had no significant effects on specific PRL binding to prostatic mem-

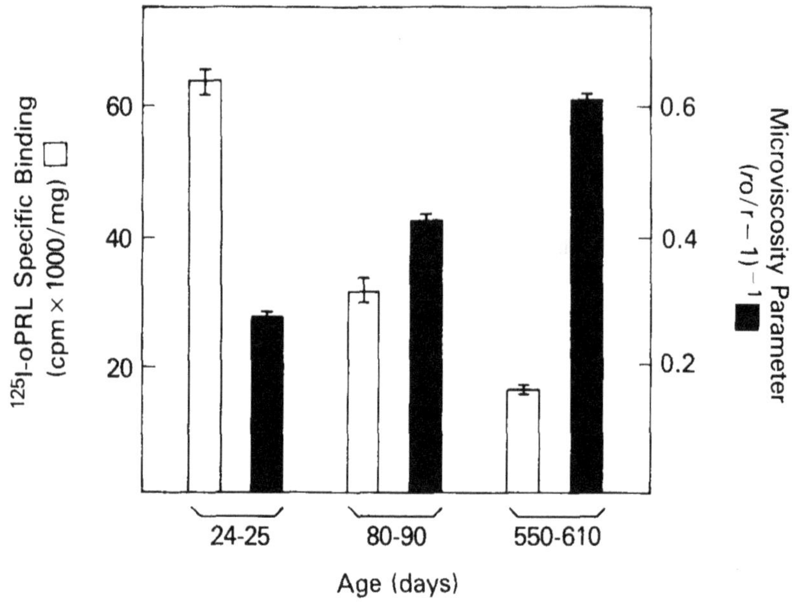

FIGURE 12. Demonstration that specific binding of PRL (open bars) to prostatic membranes decreased and membrane lipid microviscosity (solid bars) increased as a function of age. Each value represents the mean ± S.D. of three different experiments, each assayed individually in triplicate. (From Dave, J. R. and Witorsch, R. J., *Biochim. Biophys. Acta*, 772, 321, 1984. With permission.)

branes obtained from immature rats. However, it produced a dose-related increase and decline in PRL binding of prostatic membranes from older rats. A maximal increase in PRL binding of approximately 60% in young adult rats and 170% in aged rats was observed at 4.8% ethanol. Propanol at concentrations ranging from 1 to 16.7% and butanol from 1 to 9.1% concentrations produced a progressive decline in PRL binding in immature rats and a dose-related increase and decline in the same in older rats (Figure 13B and 13C). A maximal increase in PRL binding of approximately 50 and 40% in young adult rats and 140 and 110% in aged rats was observed at 3.8% propanol and 1.0% butanol, respectively.

The changes produced by exposure to alcohols on prostatic membranes lipid microviscosity expressed both as the measured values of steady-state polarization and calculated values of the microviscosity paramter $[(r_o/r) - 1]^{-1}$ are shown in Table 3. Exposure to alcohols decreased the value of the prostatic membrane microviscosity parameter in young adult and aged rats. However, no significant changes in the same were observed in immature rat prostatic membranes. The relative degree of fluidization by the alcohols was greater in magnitude in membranes obtained from aged rats than in young adult rats. These data strongly support the thesis, and suggest that modulation of membrane lipid microviscosity could alter the availability of PRL receptors. Furthermore, changes in membrane lipid microviscosity with aging may be of prime importance in modulating the accessibility of PRL receptors of prostate gland.

Most studies in the literature, including ours, on hepatic and prostatic PRL receptors have been carried out using partially purified membrane preparations exhibiting maximal density of the PRL receptors. In liver tissue, this happens to be a "crude microsomal" membrane fraction (× 15,000 to 100,000 g pellet in 0.3 *M* sucrose) which consists largely of broken plasma membranes, Golgi, and microsomal membranes. In the prostate gland, due to the presence of excessive secretory material, these subcellular organelles are pelleted in 15,000 g (× 125 to 15,000 g pellet in 0.3 *M* sucrose) fraction.[181]

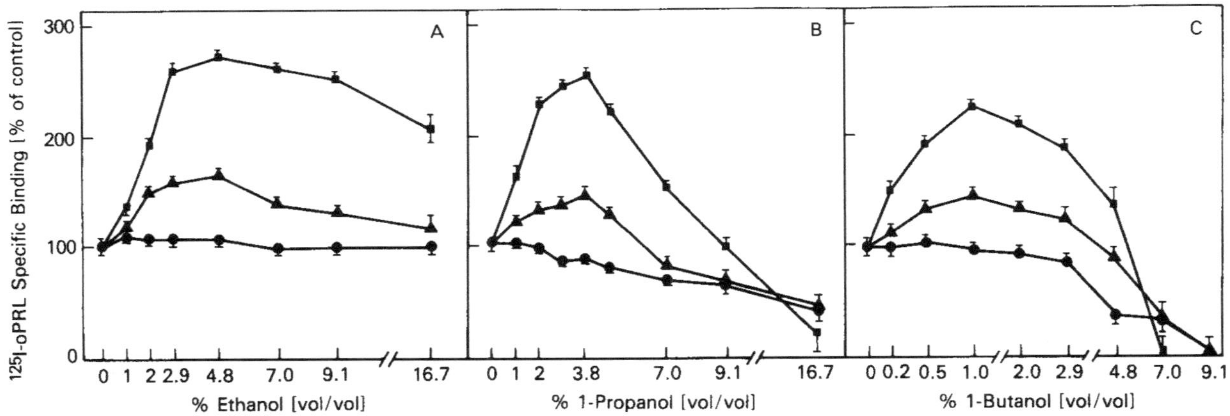

FIGURE 13. A representative experiment showing that membrane fluidizers in vitro produced age-dependent effects on specific PRL binding. PRL binding of ventral prostatic membranes co-incubated with ^{125}I-labeled ovine PRL and various concentrations of ethanol (A), or 1-propanol (B), or 1-butanol (C). Alcohols were co-incubated overnight with membranes and ^{125}-I-labeled ovine PRL during the binding assay. Each point represents means from prostatic membranes obtained from animals at 24 to 25 days (circles), 80 to 90 days (triangles), and 550 to 610 days (squares) of age assayed in triplicate. Vertical bars = 1 S.E.M. (From Dave, J. R. and Witorsch, R. J., *Biochim. Biophys. Acta,* 772, 321, 1984. With permission.)

Table 3
EFFECTS OF IN VITRO EXPOSURE OF ALIPHATIC ALCOHOLS ON STEADY-STATE POLARIZATION AND MICROVISCOSITY PARAMETER OF PROSTATIC MEMBRANES OBTAINED FROM ANIMALS AT DIFFERENT AGES

Age (days)	Treatment (in vitro)	Polarization constant, P	Microviscosity parameter[a] $(r_0/r - 1)^{-1}$
24 to 25	0	0.112 ± 0.002	0.272 ± 0.004
	Ethanol 4.8%	0.109 ± 0.002	0.263 ± 0.004 (96)
	Propanol 3.8%	0.118 ± 0.003	0.292 ± 0.006 (107)
	Butanol 1.0%	0.113 ± 0.001	0.276 ± 0.002 (101)
80 to 90	0	0.154 ± 0.001	0.426 ± 0.002
	Ethanol 4.8%	0.132 ± 0.001	0.341 ± 0.002 (80)
	Propanol 3.8%	0.142 ± 0.001	0.376 ± 0.001 (88)
	Butanol 1.0%	0.145 ± 0.001	0.390 ± 0.001 (92)
550 to 610	0	0.192 ± 0.002	0.607 ± 0.004
	Ethanol 4.8%	0.145 ± 0.002	0.390 ± 0.003 (64)
	Propanol 3.8%	0.153 ± 0.001	0.422 ± 0.001 (70)
	Butanol 1.0%	0.156 ± 0.001	0.435 ± 0.001 (72)

[a] Microviscosity parameter was calculated as described by Shinitzky and Barenholz.[109] Each value represents the mean from two experiments assayed individually in triplicate ± S.D. Values in parentheses are precent of control values.

From Dave, J. R. and Witorsch, R. J., *Biochim. Biophys. Acta,* 772, 321, 1984. With permission.

Among various subcellular fractions, high density of PRL-binding sites exists mainly in the Golgi and plasma membrane fractions. The use of partially purified membrane fractions enabled us to define if the correlation between changes in lipid microviscosity and PRL receptors exists in all the subcellular fractions, or whether it is exhibited in only a particular fraction. An answer to this question was sought by recent studies in which we determined PRL binding and membrane lipid microviscosity in purified plasma membrane and Golgi fractions from female rat liver and male rat prostate gland.

Figure 14A shows that, compared to Golgi fraction, PRL binding in plasma membranes was approximately five times lower in female rat liver and male rat prostate gland. This has also been reported by other investigators. In order to determine if PRL binding in these fractions is dependent on membrane lipid microviscosity changes, the Golgi and plasma membrane fractions were exposed in vitro to the optimal concentration of ethanol. As shown in Figure 14B, this in vitro treatment increased PRL binding in the plasma membrane fraction by approximately 60% and that in the Golgi fraction by approximately 12%. These data suggested that PRL binding in plasma membrane is modulated to a greater extent by membrane lipid microviscosity changes than that in the Golgi fraction. These observations were further confirmed in an in vivo manipulation in which prostatic plasma membrane and Golgi fractions from animals treated with bromocriptine or vehicle were used to determine these parameters. The administration of bromocriptine to adult male rats for 48 hr resulted in a 70% decrease in PRL binding to prostatic plasma membrane fractions and only a 15% decrease in the same in the Golgi fraction. The in vitro treatment of these membrane fractions to an optimal concentration of ethanol increased PRL binding to prostatic plasma membranes in

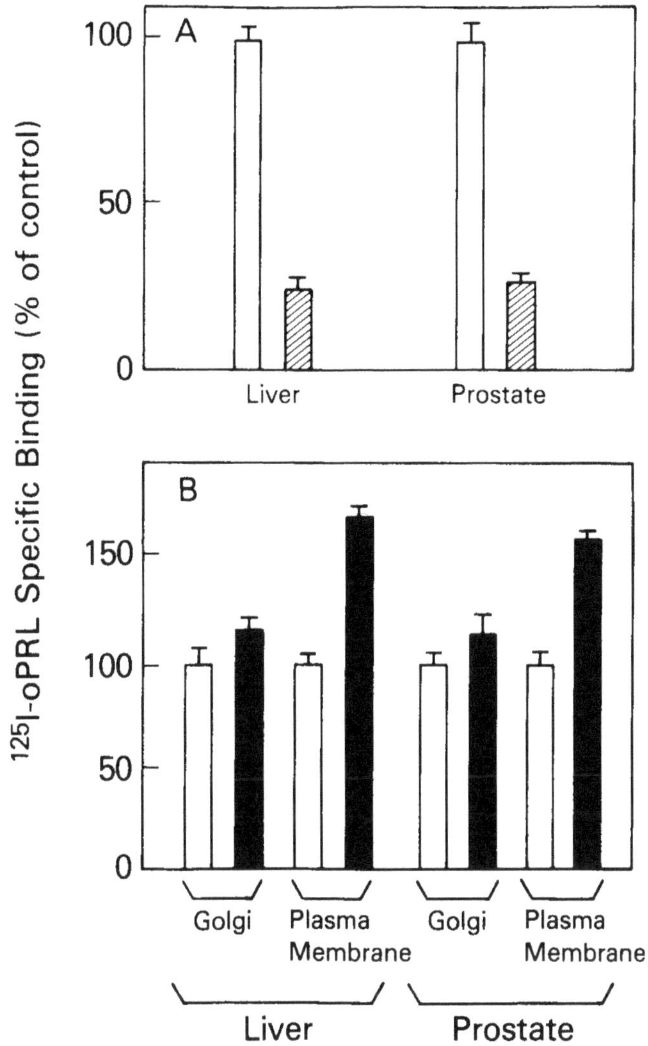

FIGURE 14. (A) Specific ^{125}I-labeled PRL binding to purified Golgi fraction (open bars) and plasma membranes (crosshatched bars) of female rat liver and male rat ventral prostate gland; (B) demonstration that specific binding of PRL to female rat hepatic and male prostatic plasma membrane is increased after exposure of the membranes to ethanol (solid bars) in virto. Prostatic membrane fractions were incubated with ^{125}I-labeled ovine PRL at 4.8% ethanol concentration, and hepatic membranes were incubated with ^{125}I prolactin at 1.0% ethanol concentration. Each bar represents mean ± S.D. of three different pools of membrane fractions, each assayed individually in triplicate.

control animals by approximately 55% and in bromocriptine-treated animals by approximately 120%. In contrast to this, the in vitro exposure of the prostatic Golgi fractions obtained from controls and bromocriptine-treated animals showed 10 to 15% increases in PRL binding. This in vivo bromocriptine treatment produced an approximately 20% increase in prostatic plasma membrane lipid microviscosity; however, no significant changes in lipid microviscosity of the Golgi membrane were evident.[181] These data supported the hypothesis and suggested that PRL receptors associated with the plasma membrane are modulated to a greater extent by changes in the serum PRL levels than those associated with the Golgi fraction. Furthermore, PRL receptors associated with the plasma membrane fraction appear to be dependent on membrane lipid microviscosity changes. Since plasma membrane "lipid pool" of a variety of tissues is in constant exchange with plasma "lipid pool", these data also raised a question whether these changes in prostatic plasma membrane lipid microviscosity following bromocriptine treatment are a localized effect occurring only in "PRL-target" tissues or are a result of PRL-mediated generalized effect of overall lipid metabolism? A generalized effect of bromocriptine on lipids would certainly involve changes in serum and erythrocytes. Thus, in another study, we determined serum lipid microviscosity and erythrocyte membrane microviscosity following bromocriptine treatment for 24 or 48 hr. This treatment produced a time-dependent increase in both serum and erythrocyte membrane lipid microviscosity which, after 48 hr of treatment, was approximately 20 to 30% above the vehicle-treated control values. To avoid the possibility of the drug-induced changes in this parameter, physiological amounts of rat PRL and bromocriptine were simultaneously administered to another group of rats. This treatment prevented bromocriptine-induced changes in both serum and erythrocyte membrane lipid microviscosity.[181]

The observations described in this section prompted us to hypothesize that the mechanism by which PRL may be regulating its own receptors involves alterations in membrane lipid microviscosity and prostaglandin formation. These ideas are shared in the following section.

VI. HYPOTHESIS THAT PROLACTIN MODIFIES MEMBRANE LIPID MICROVISCOSITY AND PROSTAGLANDIN SYNTHESIS

The results presented in earlier sections suggest that there are at least two types of control mechanisms involved in the modulation of PRL receptor levels. One appears to be a short-term or localized control involving the target tissue membranes under study. As elucidated in the proposed model (Figure 15), the prostaglandin synthesis cascade may be involved in this short-term control mechanism. By a mechanism yet unknown, binding of a PRL molecule to its receptor triggers the activity of membrane-associated phospholipase(s) (possibly phospholipase A_2), which acts on membrane phospholipids to release arachidonic acid and lysophospholipids. According to this hypothesis, arachidonic acid is then acted upon by another enzyme, prostaglandin synthetase (activity of which is also believed to be dependent on prolactin), to form various prostaglandins. Appropriate concentrations of prostaglandin(s) are then thought to modify the membrane lipid microviscosity and, in turn, increase the availability of cryptic or unavailable PRL receptors. Compatible with this hypothesis are studies published by Rillema and Wild.[165] They reported that PRL (NIH-PS-13) had a direct action on phospholipase A_2 in microsomal preparations from mouse mammary gland. However, in subsequent studies with other PRL preparations, including NIH-PS-14, no effects of PRL on phospholipase A_2 activity in mammary tissue were observed.[166] We, however, recently observed that using purified plasma membrane fraction, and not the crude microsomal fraction, physiological concentrations of PRL stimulated phospholipase A_2 activity in female rat liver and male rat prostate gland.[181] Furthermore, the observations that phospholipase C, phospholipase A_2, and the prostaglandins enhance the rate of RNA synthesis in cultured mammary tissues, and these factors, in combination with polyamines, stimulate casein biosynthesis,[167-169] are in support of the proposed hypothesis.

The second type of control mechanism, on the other hand, appears to be relatively long term in which circulating PRL is thought to modify (possibly via prostaglandin cascade) the serum and membrane lipid composition. These modifications are presumably brought about by changes in cholesterol metabolism (and thus changing phos-

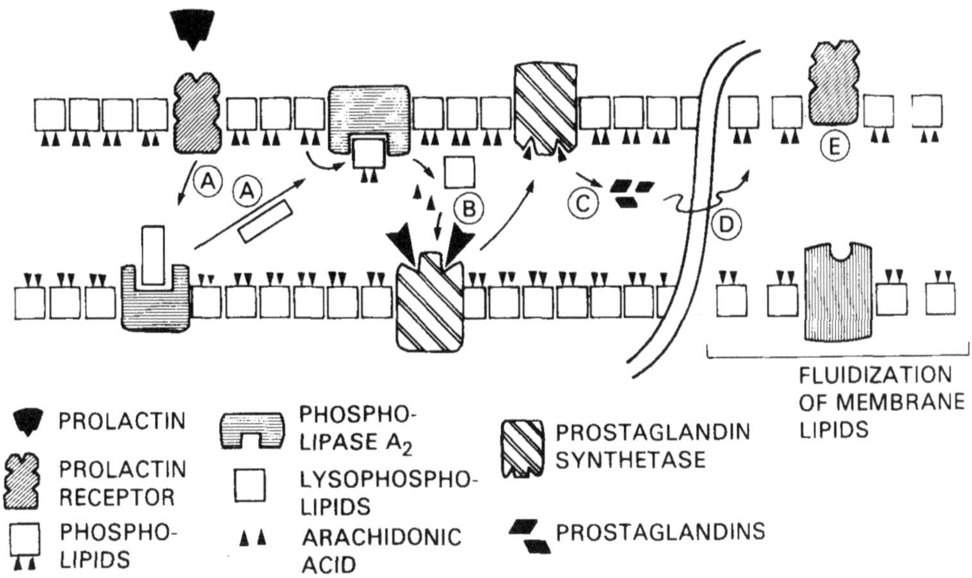

FIGURE 15. A schematic presentation of the hypothetical model for modulation of PRL receptors. According to this model, binding of PRL to its receptor triggers the cascade and activities membrane-associated phospholipase A_2 (A) by an as yet unknown mechanism. Activation of phospholipase A_2, by releasing arachidonic acid and lysophospholipids from membrane phospholipids, would then stimulate prostaglandin synthesis (B); Stimulation of prostaglandin-synthetase complex would generate various prostaglandins from arachidonic acid (C); prostaglandin(s) at a narrow concentration range may modify, dircetly or indirectly, membrane lipid microviscoity (D). This modification of membrane lipid microviscosity, under optimal conditions, may alter the availability of cryptic PRL receptors by allowing the receptor protein to increase its mobility within the bilayer (E).

pholipid to cholesterol molar ratio) which, in turn, would modify membrane lipid microviscosity and hence the availability of PRL receptors. The major difference between these two hypothetical phenomena is the mechanism by which changes in lipid microviscosity are brought about. In the short-term or in vitro situation, there may be local factor(s) (e.g., methylation of phospholipids) which are responsible for altering the lipid microviscosity whereas in the long-term or in vivo effect, probably an alteration of overall tissue/body lipid metabolism could have involved.

Though this hypothetical model has many pitfalls and has generated many more questions, there is equally valid evidence in its favor. For example, the sequence of the proposed cascade is uncertain and the direct in vitro effects of PRL on certain events in the proposed model are questionable. In preliminary studies using highly purified plasma membranes of rat ventral prostate gland, it was observed that addition of physiological quantities of purified rat prolactin stimulated phospholipase A_2 activity and also produced changes in membrane microviscosity. However, additional studies are needed to exhibit the time-dependent changes occurring in these parameters to determine the actual sequential events.

The second type of control mechanism involved in the modulation of PRL receptors is postulated to consist mainly of certain key biochemical event(s) that are highly sensitive to the circulating levels of PRL. When the hormone levels are altered, the entire biochemical assembly is thought to be modified. This modification may involve lipid/lipoprotein metabolism which in turn may alter serum lipid/lipoprotein composition and subsequent translocation of lipids between cellular membranes and serum. As a result of these alterations, the dynamics and composition of membrane lipids may be modified. There are considerable direct and indirect evidences available to support the existence of such effects of PRL and a few of them are listed below.

1. We recently observed that bromocriptine or perphenezine treatment in vivo alters the serum lipid fluidity, erythrocyte membrane fluidity, and prostatic membrane fluidity.
2. Simultaneous administration of bromocriptine and rat PRL prevented the changes in these parameters.
3. In earlier studies, hypophysectomized female rats treated with graded doses of PRL exhibited dose-dependent changes in hepatic membrane lipid fluidity and cholesterol to phospholipid molar ratios.[161]
4. PRL is reported to modify cholesterol metabolism in the corpus luteum, favoring its storage in the absence of other hypophyseal factors.[170,171]
5. PRL is reported to alter cholesterol metabolism and the relative distribution of free and esterified cholesterol in testis[172,173] and steroidogenesis and cholesterol metabolism in the ovary.[170,174-177]
6. Significant alterations in phospholipid content of a variety of tissues (e.g., brain, kidney, liver, muscle, testis, prostate, etc.) in adult male rats[178] and in metabolism and serum levels of free fatty acids in dogs[179] are reported after in vivo PRL administration.

These alterations brought about by the action of PRL may not be strictly to modulate its own receptors in target tissues, but may imply that PRL might be producing orchestral effects on a variety of PRL-dependent phenomena which may be secondary to its general primary action.

VII. CONCLUSIONS

The studies described herein suggest the following:

1. It appears that one of the important functions of PRL in the body is to modify serum and membrane lipid microviscosity. These alterations also result in the modulation of PRL receptor levels in "PRL-target" tissues.
2. The prostaglandin cascade is probably involved in these modifications.
3. Circulating PRL levels appear to have an important role in the modulation of membrane and serum lipid microviscosity since hyper- and hypoprolactinemic states increase membrane lipid microviscosity.
4. PRL receptors associated with "target tissue" plasma membranes exhibit greater sensitivity or dependence to microviscosity changes, and thus might provide a control mechanism for the entry of PRL into the cell.

The observations that PRL modifies the lipid microviscosity of hepatic and prostatic plasma membranes, erythrocyte membranes, and serum may explain, in part, how a hormone primarily associated with lactation may have a variety of other roles in both male and female systems. Since many of the membrane-bound enzymes, immune responses, and a variety of metabolic processes are dependent on membrane lipid microviscosity, it is postulated that PRL may be exerting its wide range of actions by altering the membrane lipid microviscosity. Establishing such a role for PRL may open up an entirely new understanding of this hormone and its physiological significance.

ACKNOWLEDGMENTS

The author wishes to thank Dr. Richard A. Knazek, Laboratory of Pathophysiology, National Cancer Institute, Bethesda, Md., and Dr. Raphael J. Witorsch, Department of Physiology, Medical College of Virginia, Richmond, for their generous help, critical comments, outstanding guidance, and thoughtful suggestions in the course of these investigations. The data presented in this chapter on hepatic membranes were generated in the laboratory of Dr. Knazek, and those on prostatic membranes were generated in the laboratory of Dr. Witorsch. A part of the investigations reported herein was funded by NIH grant 5R26CA 23653 awarded to Dr. Witorsch. Sincere thanks are also due to Dr. P. M. Gullino, Chief, Laboratory of Pathophysiology, NCI, Bethesda, Md., for his constant encouragement and deep interest in these studies and Dr. I. J. Dave for critical reading of the typescript. The author wishes to thank Sue Liu, Laura Richardson, Nicole Brown, Robert Christy, and Robert Vick for their valuable help at different stages of these studies.

REFERENCES

1. Nicoll, C. S., Physiological actions of prolactin, in *Handbook of Physiology, Section 7,* Vol. 4, Knobil, E. and Sawyer, W. H., Eds., American Physiological Society, Washington, D. C., 1974, 253.
2. Turkington, R. W., Stimulation of RNA synthesis in isolated mammary cells by insulin and prolactin bound to Sepharose, *Biochem. Biophys. Res. Commun.*, 41, 1362, 1970.
3. Turkington, R. W., Effector-receptor relations in the action of prolactin, in *Human Prolactin,* Pasteels, J. L. and Robyn, C., Eds., Excerpta Medica, New York, 1973, 24.
4. Birkinshaw, M. and Falconer, I. R., The localization of prolactin labelled with radioactive iodine in rabbit mammary tissue, *J. Endocrinol.*, 55, 323, 1972.
5. Costlow, M. W., Buschow, R. A., and McGuire, W. L., Prolactin stimulation of prolactin receptors in rat liver, *Science,* 184, 85, 1974.
6. Holcomb, H. H., Costlow, M. E., Buschow, R. A., and McGuire, W. L., Prolactin binding in rat mammary gland during pregnancy and lactation, *Biochim. Biophys. Acta,* 428, 104, 1976.
7. Smith, R. D., Hilf, R., and Senior, A. E., Prolactin binding to 7,12-demethylbenz(a)-anthracene-induced mammary tumors and liver in diabetic rats, *Cancer Res.*, 36, 3726, 1976.
8. Turkington, R. W., Frantz, W. L., and Majumder, G. C., Effector-receptor relations in the action of prolactin, in *Human Prolactin,* Pasteels, J. L. and Robyn, C., Eds., Excerpta Medica, New York, 1973, 24.
9. Frantz, W. L. and Turkington, R. W., Formation of biologically active ^{125}I-prolactin by enzymatic iodination, *Endocrinology,* 91, 1545, 1972.
10. Frantz, W. L., McIndoe, J. H., and Turkington, R. W., Prolactin receptors: characteristics of the particulate fraction binding activity, *J. Endocrinol.*, 60, 485, 1974.
11. Nagasawa, H. and Yanai, R., Mammary prolactin receptor and pituitary prolactin secretion in lactating mice with different lactational performance, *Acta Endocrinol.*, 88, 94, 1978.
12. Sakai, S., Kohmoto, K., and Johke, T., A receptor site for prolactin in lactating mouse mammary tissues, *Endocrinol. Jpn.*, 22, 379, 1975.
13. Sheth, N. A., Ranadive, K. J., and Sheth, A. R., *In vitro* binding of radioiodinated human placental lactogen to murine mammary gland, *Eur. J. Cancer,* 10, 653, 1974.
14. Shiu, R. P. C. and Friesen, H. G., Properties of a prolactin receptor from the rabbit mammary gland, *Biochem. J.*, 140, 301, 1974.
15. Shiu, R. P. C., Kelly, P. A., and Friesen, H. G., Radioreceptor assay for prolactin and other lactogenic hormones, *Science,* 180, 968, 1973.
16. Costlow, M. W., Buschow, R. A., Richiert, N. J., and McGuire, W. L., Prolactin and estrogen binding in transplantable hormone-dependent and autonomous rat mammary carcinoma, *Cancer Res.*, 35, 970, 1975.
17. Powell, B. L., Diamond, E. J., Kopaek, S., and Hollander, V. P., Prolactin binding in ovariectomy-nonresponsive rat mammary carcinoma, *Cancer Res.*, 37, 1328, 1977.
18. Turkington, R. W., Prolactin receptors in mammary carcinoma cells, *Cancer Res.*, 34, 758, 1974.
19. Smith, R. D., Hilf, R., and Senior, A. E., Prolactin binding to R3230AC mammary carcinoma and livers in hormone-treated and diabetic rats, *Cancer Res.*, 37, 595, 1977.
20. Costlow, M. E. and McGuire, W. L., Autoradiographic localization of prolactin receptors in 7,12-dimethylbenzathracene-induced rat mammary carcinoma, *J. Natl. Cancer Inst.*, 58, 1173, 1977.
21. Costlow, M. E., Buschow, R. A., and McGuire, W. L., Prolactin receptors and androgen-induced regression of 7,12-dimethylbenz(a)anthracene-induced mammary carcinoma, *Cancer Res.*, 36, 3324, 1976.

22. DeSombre, E. R., Kledzik, G., Marshall, S., and Meites, J., Estrogen and prolactin receptor concentrations in rat mammary tumors and response to endocrine ablation, *Cancer Res.*, 36, 354, 1976.
23. Holdaway, I. M. and Friesen, H. G., Correlation between hormone binding and growth response of rat mammary tumor, *Cancer Res.*, 36, 1562, 1976.
24. Kelly, P. A., Bradley, C., Shiu, R. P. C., Meites, J., and Friesen, H. G., Prolactin binding to rat mammary tumor tissue, *Proc. Soc. Exp. Biol. Med.*, 146, 816, 1974.
25. Kledzik, G. S., Bradley, C. J., Marshall, S., Campbell, G. A., and Meites, J., Effects of high doses of estrogen on prolactin-binding activity and growth of carcinogen-induced mammary cancers in rats, *Cancer Res.*, 36, 3265, 1976.
26. Friesen, H. G., The role of prolactin in breast cancer, in *Recent Results in Cancer Research*, St. Arneault, G., Bond, P., and Israel, L., Eds., Springer-Verlag, Berlin, 1976, 143.
27. Holdaway, J. M. and Friesen, H. G., Hormone binding by human mammary carcinoma, *Cancer Res.*, 37, 1946, 1977.
28. Morgan, L., Raggatt, P. R., DeSouza, I., Salih, H., and Hobbs, J. R., Prolactin receptors in human breast tumors, *J. Endocrinol.*, 73, 17P, 1977.
29. DeSouza, I., Hobbs, J. R., Morgan, L., and Salih, H., Localization of prolactin in human breast tumors, *J. Endocrinol.*, 73, 17P, 1977.
30. Nolin, J. M. and Witorsch, R. J., Determination of endogenous immunoreactive prolactin in rat mammary epithelial cells during lactation, *Endocrinology*, 99, 949, 1976.
31. Donatsch, P. and Richardson, B., Localization of prolactin in rat kidney tissue using a double antibody technique, *J. Endocrinol.*, 66, 101, 1975.
32. Chomczynski, P. and Topper, Y. J., A direct effect of prolactin and placental lactogen on mammary epithelial nuclei, *Biochem. Biophys. Res. Commun.*, 60, 56, 1974.
33. McGuire, W. L., Chamness, G. C., Horwitz, K. B., and Zava, D. T., Prolactin and estrogen receptors in breast cancer, *Patholbiol. Annu.*, 7, 191, 1977.
34. Pearson, O. H., Manni, A., Chambers, M., Brodkey, J., and Marshall, J. S., Role of pituitary hormones in the growth of human breast cancer, *Cancer Res.*, 38, 4323, 1978.
35. DiCarlo, R. and Muccioli, G., Prolactin receptor in human mammary carcinoma, *Tumor*, 65, 695, 1979.
36. Partridge, R. K. and Hahnel, R., Prolactin receptors in human breast carcinoma, *Cancer Res.*, 43, 643, 1979.
37. Bohnet, H. G., Determination and properties of proteohormone receptors in malignant gynecological tumors with special reference to lactogen receptors in human breast cancer, *Arch. Gynecol.*, 229, 333, 1980.
38. DiCarlo, R., Muccioli, G., and DiCarlo, F., Hormone receptors in benign breast tumors, in *Endocrinology of Cystic Breast Disease*, Angeli, A., Bradlow, H. L., and Dogliotti, L., Eds., Raven Press, New York, 1983, 211.
39. Hermite-Baleriaux, M. L. and Hermite, M. L., Prolactin receptors in human tissues, in *Hormone Research in Growth and Reproduction*, Saxena, B. B., Catt, K. J., Birnbaumer, L., and Martini, L., Eds., Raven Press, New York, 1984, 291.
40. Shiu, R. P. C. and Friesen, H. G., Solubilization and purification of a prolactin receptor from the rabbit mammary gland, *J. Biol. Chem.*, 249, 7902, 1974.
41. Shiu, R. P. C. and Friesen, H. G., Prolactin receptors, in *Methods in Receptor Research*, Blecher, M., Ed., Marcel Dekker, Basel, 1976, 565.
42. Shiu, R. P. C. and Friesen, H. G., Interaction of cell-membrane prolactin receptor with its antibody, *Biochem. J.*, 157, 619, 1976.
43. Haeuptle, M. T., Aubert, M. L., Djiane, J., and Kraehenbuhl, J.-P., Binding sites for lactogenic and somatogenic hormones from rabbit mammary gland and liver, *J. Biol. Chem.*, 258, 305, 1983.
44. Kelly, P. A., Tsushima, T., Shiu, R. P. C., and Friesen, H. G., Lactogenic and growth hormone-like activities in pregnancy determined by radioreceptor assays, *Endocrinology*, 99, 765, 1976.
45. McNeilly, A. S. and Friesen, H. G., Binding of prolactin to the rabbit mammary gland during pregnancy, *J. Endocrinol.*, 74, 507, 1977.
46. Bohnet, H. G., Gomez, F., and Friesen, H. G., Prolactin and estrogen binding sites in the mammary gland of lactating and non-lactating rat, *Endocrinology*, 101, 1111, 1977.
47. Djiane, J., Durand, P., and Kelly, P. A., Evaluation of prolactin receptors in rabbit mammary gland during pregnancy and lactation, *Endocrinology*, 100, 1348, 1977.
48. Djiane, J. and Durand, P., Prolactin-progesterone antagonism in self regulation of prolactin receptors in the mammary gland, *Nature (London)*, 266, 641, 1977.
49. Sakai, S., Enami, J., Nandi, S., and Banerjee, M. R., Prolactin receptors on dissociated mammary epithelial cells at different stages of development, *Mol. Cell Endocrinol.*, 12, 285, 1978.
50. Aubert, M. L., Suard, Y., Sizonenko, P. C., and Kraehenbuhl, J.-P., Receptors of lactogenic hormones: study with dispersed cells from rabbit mammary gland, in *Progress in Prolactin Physiology and Pathology*, Robyn, C. and Harter, M., Eds., Elsevier/North-Holland, New York, 1978, 45.

51. Sheth, N. A., Tikekar, S. S., Ranadive, K. J., and Sheth, A. R., Enhancement of *in vitro* binding of rat prolactin to murine mammary gland by polyamines, *IRCS Med. Sci.*, 4, 323, 1976.
52. Gandilhon, P., Melancon, R., Gandilhon, F., Djiane, J., and Kelly, P. A., Prolactin receptors in N-nitroso-N-methylurea-induced rat mammary tumors: relationship to tumor age and down-regulation in short-term explant culture, *Anticancer Res.*, 3, 203, 1983.
53. Nagasawa, H., Sakai, S., and Banerjee, M. R., Prolactin receptors, *Life Sci.*, 24, 193, 1979.
54. Nagasawa, H., Prolactin: its role in the development of mammary tumors, *Med. Hypotheses*, 5, 1117, 1979.
55. Nagasawa, H., Prolactin and human breast cancer: a review, *Eur. J. Cancer*, 15, 267, 1979.
56. Horrobin, D., in *Prolactin*, Eden Press, Montreal, 1975, 1.
57. Saito, T. and Saxena, B. B., Specific receptors for prolactin in the ovary, *Acta Endocrinol.*, 80, 126, 1975.
58. Rolland, R. and Hammond, J. M., Demonstration of a specific receptor for prolactin in porcine granulosa cells, *Endocrinol., Res. Commun.*, 2, 281, 1975.
59. Rolland, R., Gunsalus, G. L., and Hammond, J. M., Demonstration of specific binding of prolactin by porcine corpora lutea, *Endocrinology*, 98, 1083, 1976.
60. Richards, J. S. and Williams, J. J., Luteal cell receptor content for prolactin (PRL) and luteinizing hormone (LH): regulation by LH and PRL, *Endocrinology*, 99, 1571, 1976.
61. Cheng, K. W., Changes in rat ovaries of specific binding for LH, FSH, and prolactin during the oestrous cycle and pregnancy, *J. Reprod. Fertil.*, 48, 129, 1976.
62. Poindexter, A. N., Buttram, V. C., Besch, P. K., and Smith, R. G., Prolactin receptors in the ovary, *Fertil. Steril.*, 31, 273, 1979.
63. Davies, T. F., Katikineni, M., Chan, V., Harwood, J. P., Dufau, M. L., and Catt, K. J., Lactogenic receptor regulation in hormone-stimulated steroidogenic cells, *Nature (London)*, 283, 863, 1980.
64. Koppelman, M. C. S. and Dufau, M. L., Prolactin receptors in luteinized rat ovaries: unmasking of specific binding sites with detergent treatment, *Endocrinology*, 111, 1350, 1982.
65. Bohnet, H. G., Shiu, R. P. C., Grinwich, D., and Friesen, H. G., *In vivo* effects of antisera to prolactin receptors in female rats, *Endocrinology*, 102, 1657, 1978.
66. Wang, C., Hsueh, A. J. W., and Erickson, G. F., Induction of functional prolactin receptors by follicle-stimulating hormone in rat granulosa cells *in vivo* and *in vitro*, *J. Biol. Chem.*, 254, 11330, 1979.
67. Nolin, J. M., Incorporation of endogenous prolactin by granulosa cells and dictyate oocytes in the postpartum rat: effects of estrogen, *Biol. Reprod.*, 22, 417, 1980.
68. Dunaif, A. E., Zimmerman, E. A., Friesen, H. G., and Frantz, A. G., Intracellular localization of prolactin receptor and prolactin in the rat ovary by immunocytochemistry, *Endocrinology*, 110, 1465, 1982.
69. Magoffin, D. A. and Erickson, G. F., Prolactin inhibition of luteinizing hormone-stimulated androgen synthesis in ovarian interstitial cells cultured in defined medium: mechanism of action, *Endocrinology*, 111, 2001, 1982.
70. Navickis, R. J., Jones, P. B. C., and Hsueh, A. J. W., Modification of prolactin receptors in cultured rat granulosa cells by FSH, LH and GnRH, *Mol. Cell. Endocrinol.*, 27, 77, 1982.
71. Bonifacino, J. S. and Dufau, M. L., Prolactin receptors in the ovary, in *Hormone Receptors in Growth and Reproduction*, Saxena, B. B., Catt, K. J., Birnbaumer, L., and Martini, L., Eds., Raven Press, New York, 1984, 149.
72. Marshall, S., Gelato, M., and Meites, J., Serum prolactin levels and prolactin binding activity in adrenals and kidneys of male rats after dehydration, salt loading and unilateral nephrectomy, *Proc. Soc. Exp. Biol. Med.*, 149, 185, 1975.
73. Reddy, S. and Watkins, W. B., Uptake of ^{125}I-labeled human placental lactogen and human placental lactogen by the tissues of normal and lactating rats, *J. Endocrinol.*, 65, 183, 1975.
74. Costlow, M. E. and McGuire, W. L., Autoradiographic localization of the binding of ^{125}I-labeled prolactin to rat tissues *in vitro*, *J. Endocrinol.*, 75, 221, 1977.
75. Marshall, S., Huang, H. H., Kledzik, G. S., Campbell, G. A., and Meites, J., Glucocorticoid regulation of prolactin receptors in kidneys and adrenals of male rats, *Endocrinology*, 102, 869, 1978.
76. Marshall, S., Bruni, J. F., and Meites, J., Effects of hypophysectomy, thyroidectomy, and thyroxine on specific prolactin receptor sites in kidneys and adrenals of male rats, *Endocrinology*, 104, 390, 1979.
77. Aragona, C. and Friesen, H. G., Specific prolactin binding sites in the prostate and testis of rats, *Endocrinology*, 97, 677, 1975.
78. Bohnet, H. G. and Friesen, H. G., Effect of prolactin and growth hormone on prolactin and LH receptors in the dwarf mouse, *J. Reprod. Fertil.*, 48, 307, 1976.
79. Charreau, E. H., Attramadal, A., Torjesen, P. A., Purvis, K., Calandra, R., and Hansson, V., Prolactin binding in rat testis: specific receptors in interstitial cells, *Mol. Cell. Endocrinol.*, 6, 303, 1977.

80. Aragona, C., Bohnet, H. G., and Friesen, H. G., Localization of prolactin binding in prostate and testis: the role of serum prolactin concentration on the testicular LH receptors, *Acta Endocrinol. (kbh)*, 84, 402, 1977.
81. Klemcke, H. G., Bartke, A., and Borer, K. T., Testicular prolactin receptors and serum growth horomone in golden hamsters: effects of photoperiod and time of day, *Biol. Reprod.*, 29, 605, 1983.
82. Klemcke, H. G., Bartke, A., and Borer, K. T., Regulation of testicular prolactin and luteinizing hormone receptors in golden hamsters, *Endocrinology*, 114, 594, 1984.
83. Bartke, A., Klemcke, H. G., Amador, A. G., and Talamantes, F., Differential effects of cryptorchidism and short photoperiod on testicular prolactin and LH/human chorionic gonadotrophin receptors in the golden hamster, *J. Endocrinol.*, 103, 227, 1984.
84. Grayhack, J. T., Bunce, P. L., Kearns, J. W., and Scott, W. W., Influence of the pituitary on prostate response to androgen in the rat, *Johns Hopkins Hosp. Bull.*, 96, 154, 1955.
85. Huggins, C. and Russell, P. S., Quantitative effects of hypophysectomy on testis and prostate of dogs, *Endocrinology*, 39, 1, 1946.
86. Yamanaka, H., Kirdani, R. Y., Saroff, J., Murphy, G. P., and Sandberg, A. A., Effects of testosterone and prolactin on rat prostatic weight, 5 alpha-reductase and arginase, *Am. J. Physiol.*, 229, 1102, 1975.
87. Sonenberg, M. and Money, W., The fate and metabolism of anterior pituitary hormones, *Rec. Prog. Horm. Res.*, 11, 43, 1955.
88. Hanlin, N. L. and Yount, A. P., Prolactin binding in the rat ventral prostate, *Endocrinol. Res. Commun.*, 2, 489, 1975.
89. Witorsch, R. J. and Smith, J. P., Evidence for androgen-dependent intracellular binding of prolactin in rat ventral prostate gland, *Endocrinology*, 101, 929, 1977.
90. Witorsch, R. J., Immunohistochemical studies of prolactin-binding in sex accessory organs of the male rats, *J. Histochem. Cytochem.*, 26, 565, 1978.
91. Witorsch, R. J., Immunohistochemical localization of prolactin-binding sites in R3327 rat prostatic cancer cells, *Hormone Res.*, 10, 268, 1979.
92. Witorsch, R. J., The application of immunoperoxidase methodology for the visualization of prolactin binding sites in human prostate tissue, *Human Path.*, 10, 521, 1979.
93. Witorsch, R. J., Visualization of prolactin binding sites in prostate tissue, in *The Prostatic Cell: Structure and Function*, Murphy, G. P., Sandberg, A. A., and Karr, J. P., Eds., Alan R., Liss, New York, 1981, 89.
94. Kelly, P. A., Posner, B. I., Tsushima, T., and Friesen, H. G., Studies of insulin, growth hormone and prolactin binding: ontogenesis, effects of sex and pregnancy, *Endocrinology*, 95, 532, 1974.
95. Dave, J. R. and Knazek, R. A., Changes in the prolactin-binding capacity of mouse hepatic membranes with development and aging, *Mech. Ageing Dev.*, 23, 235, 1983.
96. Posner, B. I., Characterization and modulation of growth hormone and prolactin binding in mouse liver, *Endocrinology*, 98, 645, 1976.
97. Hayden, T. J., Bonney, R. C., and Forsyth, I. A., Ontogeny and control of prolactin receptors in the mammary gland and liver of virgin, pregnant and lactating rats, *J. Endocrinol.*, 80, 259, 1979.
98. Dave, J. R., Richardson, L. L., and Knazek, R. A., Prolactin-binding capacity, prostaglandin synthesis and fluidity of murine hepatic membranes are modified during pregnancy and lactation, *J. Endocrinol.*, 99, 99, 1983.
99. Costlow, M. E., Buschow, R. A., and McGuire, W. L., Prolactin receptors in 7,12-dimethylbenzanthracene-induced mammary tumors following endocrine ablation, *Cancer Res.*, 36, 3941, 1976.
100. Posner, B. I., Kelly, P. A., and Friesen, H. G., Induction of a lactogenic receptor in rat liver: influence of estrogen and the pituitary, *Proc. Natl. Acad. Sci. U.S.A.*, 71, 2407, 1974.
101. Kelly, P. A., Posner, B. I., and Friesen, H. G., Effects of hypophysectomy, ovariectomy and cycloheximide on specific binding sites for lactogenic hormones in rat liver, *Endocrinology*, 97, 1408, 1975.
102. Posner, B. I., Kelly, P. A., and Friesen, H. G., Prolactin receptors in rat liver: possible induction by prolactin, *Science*, 188, 57, 1975.
103. Costlow, M. E., Buschow, R. A., and McGuire, W. L., Prolactin stimulation of prolactin receptors in rat liver, *Life Sci.*, 17, 1457, 1975.
104. Posner, B. I., Regulation of lactogen specific binding sites in rat liver: studies on the role of lactogens and estrogen, *Endocrinolgoy*, 99, 1168, 1976.
105. Bohnet, H. G., Aragona, C., and Friesen, H. G., Induction of lactogenic receptors. I. In the liver of hypophysectomized female rats, *Endocrinol. Res. Commun.*, 3, 187, 1976.
106. Frantz, W. L., Mann, L. C., and Welsch, C. W., Induction of specific binding sites for prolactin in the liver of DW/J dwarf mice by injected prolactin, *IRCS Med. Sci.*, 5, 32, 1977.
107. Knazek, R. A., Liu, S. C., and Gullino, P. M., Induction of lactogenic binding sites in the liver of the Snell dwarf mouse, *Endocrinology*, 101, 50, 1977.

108. Knazek, R. A., Liu, S. C., Graeter, R. L., Wright, P. C., Mayer, J. R., Lewis, R. H., Gould, E. B., and Keller, J. A., Growth hormone causes rapid induction of lactogenic receptor activity in the Snell dwarf mouse liver, *Endocrinology*, 103, 1590, 1978.
109. Shinitzky, M. and Barenholz, Y., Fluidity parameters of lipid regions determined by fluorescence polarization, *Biochim. Biophys. Acta*, 515, 367, 1978.
110. Azzi, A., The application of fluorescent probes in membrane studies, *Q. Rev. Biophys.*, 8, 237, 1975.
111. Bashford, C. L., Morgan, C. G., and Radda, G. K., Measurement and interpretation of fluorescence polarizations in phospholipid dispersions, *Biochim. Biophys. Acta*, 426, 157, 1976.
112. Shinitzky, M., Dianonx, A. C., Gitler, C., and Weber, G., Microviscosity and order in the hydrocarbon region of micelles and membranes determined with fluorescent probes. I. Synthetic micelles, *Biochemistry*, 10, 2106, 1971.
113. Shinitzky, M. and Barenholz, Y., Dynamics of the hydrocarbon layer in lyposomes of lecithin and sphingomyelin containing dicetylphosphate, *J. Biol. Chem.*, 249, 2652, 1974.
114. Lentz, R. B., Barenholz, Y., and Thompson, T. E., Fluorescence depolarization studies of phase transitions and fluidity in phospholipid bilayers. I. Single component phosphatidylcholine liposomes, *Biochemsitry*, 15, 4521, 1976.
115. Singer, S. J. and Nicolson, G. L., The fluid mosaic model of the structure of cell membranes, *Science*, 175, 720, 1972.
116. Rousselet, A., Guthmann, C., Matricon, J., Bienvenue, A., and Deuaux, P. F., Study of the transverse diffusion of spin labelled phospholipids in biological membranes. I. Human red blood cells, *Biochim. Biophys. Acta*, 426, 357, 1976.
117. Shinitzky, M. and Henkart, P., Fluidity of cell membranes — current concepts and trends, *Int. Rev. Cytol.*, 60, 121, 1979.
118. Vanderkooi, J., Fischkoft, S., Chance, B., and Cooper, A., Fluorescent probe analysis of the lipid architecture of natural and experimental cholesterol-rich membranes, *Biochemistry*, 13, 1589, 1974.
119. Shinitzky, M. and Inbar, M., Microviscosity parameter and protein mobility in biological membranes, *Biochim. Biophys. Acta*, 433, 133, 1976.
120. Cogan, U., Shinitzky, M., Weber, G., and Nishida, T., Microviscosity and order in the hydrocarbon region of phospholipid and phospholipid-cholesterol dispersions determined with fluorescence probes, *Biochemistry*, 12, 521, 1973.
121. Cossins, A. R., Adaptation of biological membranes to temperature. The effect of temperature acclimation of goldfish upon the viscosity of synaptosomal membranes, *Biochim. Biophys. Acta*, 470, 395, 1977.
122. Schmidt, C. F., Barenholz, Y., and Thompson, T. E., A nuclear magnetic resonance study of sphingomyelin in bilayer systems, *Biochemistry*, 16, 2649, 1977.
123. Cooper, R. A., Arner, E. C., Wiley, J. S., and Shattil, S. J., Modification of red cell membrane structure by cholesterol-rich lipid dispersions: a model for the primary spur cell defect, *J. Clin. Invest.*, 55, 115, 1975.
124. Shattil, S. J., Anaya-Galindo, R., Bennett, J., Colman, R. W., and Cooper, R. A., Platelet hypersensitivity induced by cholesterol incorporation, *J. Clin. Invest.*, 55, 636, 1975.
125. Chapman, D. and Wallach, D. F. H., Recent physical studies of phospholipids and natural membranes, in *Biological Membranes*, Chapman, D., Ed., Academic Press, New York, 1968, 125.
126. Cooper. R. A., Abnormalities or cell-membrane fluidity in the pathogenesis of disease, *N. Engl. J. Med.*, 297, 371, 1977.
127. Storch, J. and Schachter, D., Dietary induction of acyl chain desaturases alters the lipid composition and fluidity of rat hepatocyte plasma membranes, *Biochemistry*, 23, 1165, 1984.
128. Borochov, H. and Shinitzky, M., Vertical displacement of membrane proteins mediated by changes in microviscosity, *Proc. Natl. Acad. Sci. U.S.A.*, 73, 4526, 1976.
129. Shinitzky, M. and Ravnay, B., Degree of exposure of membrane proteins determined by fluorescence quenching, *Biochemistry*, 16, 982, 1977.
130. Cooper, R. A., Anemia with spur cells: a red cell defect acquired in serum and modified in the circulation, *J. Clin. Invest.*, 48, 1820, 1969.
131. Cooper, R. A., Kimball, D. B., and Durocher, J. R., Role of the spleen in membrane conditioning and hemolysis of spur cells in liver disease, *N. Engl. J. Med.*, 290, 1279, 1974.
132. Shattil, S. J. and Cooper, R. A., Membrane microviscosity and human platelet function, *Biochemistry*, 15, 4832, 1976.
133. Cooper, R. A., Durocher, J. R., and Leslie, M. H., Increased microviscosity of red cell membrane lipids in abetalipoproteinemia, *J. Clin. Invest.*, 60, 115, 1972.
134. Shattil, S. J., Bennett, J. S., Colman, R. W., and Cooper, R. A., Abnormalities of cholesterol-phospholipid composition in platelets and low-density lipoproteins of human hyperbetalipoproteinemia, *J. Clin. Invest.*, 89, 341, 1977.

135. Cooper, R. A., Deloy-Puray, M., Lando, P., and Greenberg, M. S., An analyses of lipoproteins, bile acids and red cell membrane associated with target cells and spur cells in patients with liver disease, *J. Clin. Invest.*, 51, 3182, 1972.
136. Wiley, J. S. and Cooper, R. A., Inhibition of cation cotransport by cholesterol enrichment of human red cell membranes, *Biochim. Biophys. Acta*, 413, 425, 1975.
137. Carvalho, A. C. A., Colman, R. W., and Lees, R. S., Platelet function in hyperlipoproteinemia, *N. Engl. J. Med.*, 290, 434, 1974.
138. Sinha, A. K., Shattil, S. J., and Colman, R. W., Cyclic AMP metabolism in cholesterol-rich platelets, *J. Biol. Chem.*, 252, 3310, 1977.
139. Dipple, I. and Houslay, M. D., The activity of glucagon-stimulated adenylate cyclase from rat liver plasma membranes is modulated by the fluidity of its lipid environment, *Biochem. J.*, 174, 179, 1978.
140. Keeffe, E. B., Scharschmidt, B. F., Blankenship, N. M., and Ockner, R. K., Studies of relationships among bile flow, liver plasma membrane Na K-ATPase, and membrane microviscosity in the rat, *J. Clin. Invest.*, 64, 1590, 1979.
141. Dave, J. R., Eiden, L. E., Karanian, J. W., and Eskay, R. L., Ethanol exposure decreases pituitary corticotropin-releasing factor binding, adenylate cyclase activity, pro-opiomelanocortin biosynthesis and plasma beta-endorphin levels in the rat, *Endocrinology*, 118, 280, 1986.
142. Sauerheber, R. D., Esgate, J. A., and Kuhn, C. E., Alcohols inhibit adipocyte basal and insulin-stimulated glucose uptake and increase the membrane lipid fluidity, *Biochim. Biophys. Acta*, 691, 115, 1982.
143. Lacko, K., Wittke, B., and Geck, P., Interaction of alcohols with the transport system of glucose in human erythrocytes, *J. Cell. Physiol.*, 83, 267, 1973.
144. Lacko, L., Wittke, B., and Lacko, I., Inhibition of glucose transport in human erythrocytes by benzylalcohol, *J. Cell. Physiol.*, 96, 199, 1978.
145. Plagemann, P. G. W. and Erbe, J., Temperature-dependent changes in activation energies of the transport systems for nucleosides, choline and deoxyglucose of cultured Novikoff rat hepatoma cells and effects of cytochalasin B and lipid solvents, *J. Membr. Biol.*, 25, 381, 1975.
146. Yuli, I., Tomonaga, A., and Snyderman, R., Chemoattractant receptor functions in human polymorphonuclear leukocytes are divergently altered by membrane fluidizers, *Proc. Natl. Acad. Sci. U.S.A.*, 79, 5906, 1982.
147. Liao, C. S. and Freer, R. J., Cryptic receptor for chemotactic peptides in rabbit neutrophils, *Biochem. Biophys. Res. Commun.*, 93, 566, 1980.
148. Fletcher, M. P., Seligmann, B. E., and Gallin, J. I., Correlation of human neutrophil secretion, chemoattractant receptor mobilization, and enhanced functional capacity, *J. Immunol.*, 128, 941, 1982.
149. Ciofalo, F. R., Ethanol and alpha-adrenergic receptors, *Proc. West. Pharmacol. Soc.*, 21, 267, 1978.
150. Muller, C. and Shinitzky, M., Modulation of transferrin receptors in bone marrow cells by changes in lipid fluidity, *Br. J. Hematol.*, 42, 355, 1979.
151. Heron, D. S., Shinitzky, M., Hershkowitz, M., and Samuel, D., Lipid fluidity markedly modulates the binding of serotonin to mouse brain membranes, *Proc. Natl. Acad. Sci. U.S.A.*, 77, 7463, 1980.
152. Cameron, J. C. and Stouffer, R. L., *In vitro* effects of ethanol on primate luteal membranes: exposure of additional gonadotropin receptor sites, *Endocrinology*, 110, 1451, 1982.
153. Dave, J. R. and Witorsch, R. J., Indomethacin decreases both luteinizing hormone binding and fluidity of testicular microsomal membranes in rat, *Prostaglandins Leukotriene Med.*, 12, 371, 1983.
154. Dave, J. R., Knazek, R. A., and Liu, S. C., Arachidonic acid, bradykinin and phospholipase A_2 modify both prolactin binding capacity and fluidity of mouse hepatic membranes, *Biochem. Biophys. Res. Commun.*, 103, 727, 1981.
155. Knazek, R. A., Liu, S. C., St. Amand, L. M., Dave, J. R., and Christy, R. J., Cholesterol accentuates the effect of unsaturated fatty acid deficiency on mammary gland development, *Proc. Am. Assoc. Cancer Res.*, 22, 28, 1981.
156. Knazek, R. A. and Liu, S. C., Dietary essential fatty acids are required for maintenance and induction of prolactin receptors, *Proc. Soc. Exp. Biol. Med.*, 162, 346, 1979.
157. Dave, J. R. and Knazek, R. A., Prostaglandin I_2 modifies both prolactin binding capacity and fluidity of mouse liver membranes, *Proc. Natl. Acad. Sci. U.S.A.*, 77, 6597, 1980.
158. Knazek, R. A., Liu, S. C., Dave, J. R., Christy, R. J., and Keller, J. A., Indomethacin causes a simultaneous decrease of both prolactin binding and fluidity of mouse liver membranes, *Prostaglandins Med.*, 6, 403, 1981.
159. Dave, J. R., Brown, N. V., and Knazek, R. A., Prolactin modifies the prostaglandin synthesis, prolactin binding and fluidity of mouse liver membranes, *Biochem. Biophys. Res. Commun.*, 108, 193, 1982.
160. Sinha, Y. N., Selby, F. W., and VanderLaan, W. P., Relationship of prolactin and growth hormone to mammary function during pregnancy and lactation in the C_3H/ST mouse, *J. Endocrinol.*, 61, 219, 1974.

161. Dave, J. R., Knazek, R. A., and Liu, S. C., Prolactin modifies the fluidity of rat liver membranes, *Biochem. Biophys. Res. Commun.*, 100, 45, 1981.
162. Dave, J. R. and Witorsch, R. J., Indomethacin decreases both prolactin binding and membrane fluidity of ventral and dorso-lateral lobes of rat prostate gland, *Prostate*, 4, 119, 1983.
163. Dave, J. R. and Witorsch, R. J., Prolactin increases serum lipid fluidity and prolactin binding and lipid fluidity of rat prostatic membranes, *Am. J. Physiol.*, 248, E687, 1985.
164. Dave, J. R. and Witorsch, R. J., Modulation of prolactin binding sites *in vitro* by membrane fluidizers. II. Age-dependent effects on rat ventral prostatic membranes, *Biochim. Biophys. Acta*, 772, 321, 1984.
165. Rillema, J. A. and Wild, B. E., Prolactin activation of phospholipase A_2 activity in membrane preparations from mammary gland, *Endocrinology*, 100, 1219, 1977.
166. Rillema, J. A., Mode of action of prolactin on normal and neoplastic mammary tissues, in *Hormonal Regulation of Experimental Mammary Tumors*, Vol. 2, Eden Press, Montreal, 1982, 77.
167. Rillema, J. A., Linebaugh, B. E., and Mulder, J. A., Regulation of casein synthesis by polyamines in mammary gland explants from mice, *Endocrinology*, 100, 529, 1977.
168. Rillema, J. A., Mechanism of prolactin action, *Fed. Proc.*, 39, 2593, 1980.
169. Rillema, J. A., Prolactin-like actions of phospholipase C on RNA and casein synthesis in mouse mammary gland explants, *Horm. Metabol. Res.*, 16, 532, 1984.
170. Armstrong, D. T., Miller, L. S., and Knudsen, K. A., Regulation of lipid metabolism and progesterone production in rat corpora lutea and ovarian interstitial elements by prolactin and luteinizing hormone, *Endocrinology*, 85, 393, 1969.
171. Armstrong, D. T., Knudsen, K. A., and Miller, L. S., Effect of prolactin upon cholesterol metabolism and progesterone biosynthesis in corpora lutea of rats hypophysectomized during pseudopregnancy, *Endocrinolgoy*, 86, 634, 1970.
172. Bartke, A., Prolactin changes cholesterol stores in the mouse testis, *Nature (London)*, 224, 700, 1969.
173. Bartke, A., Effects of prolactin and luteinizing hormone on the cholesterol stores in the mouse testis, *J. Endocrinol.*, 49, 317, 1971.
174. Bartosik, D., Romanoff, E. B., Watson, D. J., and Scricco, E., Luteotropic effects of prolactin in the bovine ovary, *Endocrinology*, 81, 186, 1967.
175. Hillard, J., Spies, H. G., Lucas, L., and Sawyer, C. H., Effect of prolactin on progestin release and cholesterol storage by rabbit ovarian interstitium, *Endocrinology*, 82, 122, 1968.
176. Zarrow, M. X., Clark, J. H., and Denenberg, V. H., The onset of the diurnal rhythm in ovarian cholesterol levels in the rat, in relation to the onset of puberty and ovarian and uterine weight changes, *Neuroendocrinol.*, 4, 270, 1969.
177. Hashimoto, I. and Wiest, H. G., Leuteotrophic and leuteolytic mechanisms in rat corpora lutea, *Endocrinolgoy*, 84, 886, 1969.
178. Rao, M. R. C., Gajalakshmi, V., Murthy, M. S. R., Indira, K., and Govindappa, S., Effect of prolactin adminstration on phospholipid content in selected reproductive and non-reproductive tissues of albino rats, *Ind. J. Exp. Biol.*, 16, 254, 1978.
179. Winkler, B., Rathgeb, I., and Altszuler, N., Effect of ovine PRL administration on free fatty acid metabolism in the dog, *Endocrinology*, 88, 1349, 1971.
180. Dave, J. R. and Knazek, R. A., unpublished observations.
181. Dave, J. R. and Witorsch, R. J., unpublished observations.

Chapter 6

ROLE OF CALCIUM IONS AND THE PHOSPHOLIPIDS IN THE PROLACTIN REGULATION OF ITS TARGET CELLS

James A. Rillema

TABLE OF CONTENTS

I. Introduction 108

II. Phospholipids and Prolactin 108
- A. Phospholipid Changes During Lactation 108
- B. General Role of Phospholipids in Cellular Regulation 109
- C. Effects of Phospholipase Inhibitors on Prolactin Responses 111
- D. Effects of Phospholipases on Metabolic Processes in Mammary Cells 112
  - 1. Phospholipase A_2 112
  - 2. Phospholipase C 113
- E. Possible Mechanism by Which Phospholipase C Affects Mammary Cells 113
- F. Possible Relationship of Phospholipase A_2 to Phospholipase C 115

III. Calcium and Prolactin 115
- A. Calcium Concentrations and Prolactin Responses 115
- B. Studies with A23187, a Calcium Ionophore 116
- C. Calmodulin 117

IV. Summary 117

References 117

I. INTRODUCTION

It now appears that most, if not all, cellular responses to different stimuli involve an altered concentration of calcium ions within cells. These changes can be caused by a number of mechanisms, including altered calcium ion fluxes either across the plasma membrane and/or across the membranes of intracellular organelles. One of the primary events responsible for altered rates of calcium ion fluxes appears to be perturbations in the phospholipids contained in the lipid portions of cellular membranes. These perturbations can be initiated by altered activities of the phospholipase enzymes associated with membrane structures. For further details about these mechanisms, the reader is referred to several recent review articles in which the roles of the phospholipids and calcium ions in regulating metabolic processes in a number of different cell types are discussed.[1-13] Recent studies with prolactin-responsive tissues suggest that prolactin actions on both mitogenic and certain differentiative processes may also be triggered, at least in part, by an initial effect on the phospholipase enzymes and a consequential altered calcium ion distribution in its target cells.

Although prolactin is known to have a variety of actions in a multiple number of target cells,[14] the discussion in this chapter will focus on the prolactin effects in mammary tissues, obviously because this is the tissue in which most investigative work concerning the mechanism of action of prolactin has been carried out. Studies concerning the actions of prolactin in target cells other than mammary cells will be discussed, however, where appropriate.

II. PHOSPHOLIPIDS AND PROLACTIN

In mammary and other tissues, prolactin has several actions on phospholipid metabolism which are related to both secretory processes and possibly, the primary action of this hormone. Milk product formation in the mammary gland is known to involve the synthesis and secretion of massive amounts of phospholipids; prolactin is known to stimulate these processes. In addition, the primary action of prolactin on its target cells may involve perturbations in membrane phospholipids. The first indication that this might be true was derived from experiments published from our laboratory several years ago. Phospholipase A_2, when added exogenously to cultured mouse mammary tissues, was found to stimulate several metabolic processes in a fashion similar to that of prolactin.[15] Prior to publication of these studies, however, several laboratories had reported marked changes in phospholipid metabolism in the mammary gland during pregnancy and lactation. These studies were elegantly summarized in a review chapter by Kinsella and Infante in 1978.[16] Relevant portions of that chapter along with more recently published information will be included in this discussion.

A. Phospholipid Changes During Lactation

Most, if not all, the phospholipids found in milk are secreted as components of the milk fat globule membrane.[16] The fats, mostly triglycerides, found in milk are synthesized within the mammary alveolar cells and are secreted via exocytosis (reverse pinocytosis) into the lumen of the alveoli. The milk fat globule membrane, which encompasses the secreted triglycerides, is therefore derived from the apical plasma membrane of the alveolar cells. During lactation, the amount of phospholipids secreted into bovine milk is estimated to be about 6% per day of the total tissue content of phospholipids.[16] In view of the role of prolactin in stimulating lactational processes, it was therefore not surprising to find that prolactin has a profound effect[17] on stimulating the rate of $[^{32}P]O_4$ incorporation into phospholipids in cultured mouse mammary gland

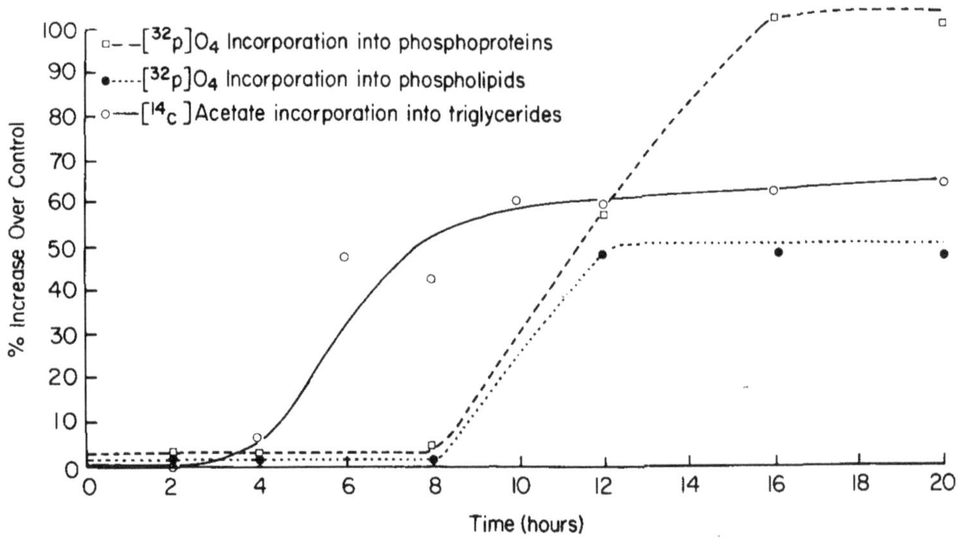

FIGURE 1. Time course of prolactin stimulation of triglyceride, phosphoprotein, and phospholipid synthesis in cultured mouse mammary tissues. Mammary tissues from 12 to 14-day pregnant mice were cultured for 24 to 36 hr with 1 μg/mℓ of insulin and 10^{-7} *M* of cortisol. Prolactin (1 μg/mℓ) was then added to the cultures and incubations were continued for the times indicated. When the rate of $[^{32}P]O_4$ incorporation into phosphoproteins or phospholipids was to be measured, the tissues were cultured with $[^{32}P]O_4$ (5 μCi/mℓ) during the final 2 hr of incubation. When the rate of $[^{14}C]$-acetate incorporation into triglycerides was to be determined, the tissues were cultured with $[^{14}C]$-acetate (1 μCi/mℓ) for the final 2 hr of incubation.

explants (Figure 1). The onset of this response is between 8 and 12 hr after adding prolactin to the cultured tissues. The onset of the prolactin stimulation of $[^{32}P]O_4$ incorporation into a casein-rich phosphoprotein fraction[17] and $[^{14}C]$-acetate incorporation into triglycerides[18] was found to be between 6 and 12 hr after exposing mouse mammary tissues to prolactin (Figure 1). In view of the packaging process involved with milk fat secretion, it makes intuitive sense that the onset of the prolactin stimulation of phospholipid synthesis should occur in concert with, or immediately subsequent to, the prolactin stimulation of milk product formation. Prolactin has also been shown to stimulate surfactant (phospholipid) formation in cultured organ explants of fetal rat lung.[19] It remains to be determined, however, if the mechanism by which prolactin stimulates phospholipid formation in various types of cells is the same.

Although large amounts of the phospholipids are secreted during lactation, the rate of secretion does not explain the rapid rate of turnover of phosphatidylinositol (PI) which occurs in lactating bovine mammary cells.[20-22] The half-life of PI in those cells was found to be about 1 hr. It is not unlikely, as will be discussed later, that this rapid turnover is related to the mechanism by which hormones regulate lactational processes in mammary tissues.

B. General Role of Phospholipids in Cellular Regulation

There are many hormones and other cellular regulatory agents which bind to specific cell surface receptors and initiate their actions on cells via effects on the metabolism of phospholipids in cell membranes. There are two firmly established pathways by which perturbations of membrane phospholipids culminate in metabolic changes in cells.[7-13] In addition, these pathways may be mutually functional in certain stimuli-response systems.

One of the pathways involving membrane phospholipids entails the cleavage of un-

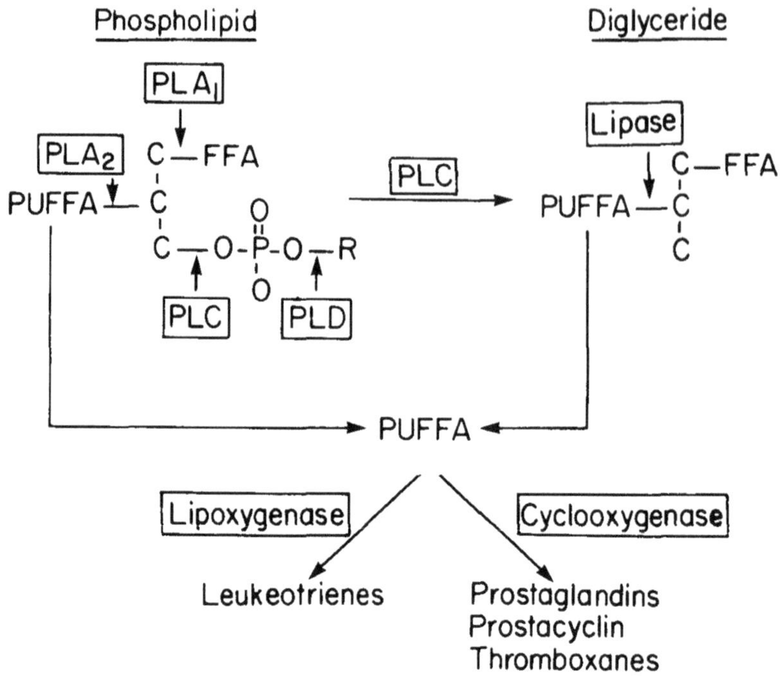

FIGURE 2. Sources of the precursors for the synthesis of prostaglandins and leukotrienes. Abbreviations: PUFFA, polyunsaturated free fatty acids; FFA, free fatty acid; PLA, phospholipase A_1; PLA_2, phospholipase A_2; PLC, phospholipase C; PLD, phospholipase D; A, structural group determining the class of phospholipid.

saturated fatty acids from the 2 acyl position of the glycerol portion of phospholipids or diglycerides (Figure 2). The unsaturated fatty acids then serve as substrates for the cyclooxygenase or lipoxygenase products (prostaglandins or leukotrienes, respectively) which specific cells are programed to produce. The prostaglandins or leukotrienes then carry out the intracellular alterations ascribed to the stimulatory agent. It is well established that the rate-limiting step for the biosynthesis of prostaglandins and leukotrienes in cells is the provision of the polyunsaturated fatty acid substrates. These substrates are provided in cells by at least two different mechanisms. The first involves an activation of phospholipase A_2 (Figure 2) and the direct cleavage of unsaturated fatty acids from the 2-acyl position of membrane phospholipids. A second mechanism (Figure 2) involves an action of two enzymes: first, phospholipase C (PLC) cleaves phospholipids yielding a diacylglycerol plus a phosphate moity; then a specific diacylglycerol lipase cleaves unsaturated fatty acids from the 2-acyl position of diacylglycerols.

Another pathway by which some cell surface regulatory agents effect cellular responses by perturbations of membrane phospholipids is by stimulating the turnover of phosphatidylinositides in cells. These responses are known to be initiated by the activation of PI-specific phospholipase C[23] (Figure 3). The metabolism of PI is complicated by the existence of polyphosphorylated derivatives of PI; these derivatives are present in cell membranes at a fraction of the concentration of PI. Nevertheless, they appear to be very important for the effects of some regulatory agents on cellular responses.[7-13] The formation of the polyphosphatidylinositides involves a phosphorylation of PI on the 4 carbon of the inositol moiety to form monophosphatidylinositol (DPI) (Figure 3). The number 5 carbon on the inositol portion of DPI can be further phosphorylated to yield diphosphatidylinositol (TPI). These inositide derivatives can then be acted on by PLC to yield a 1,2 diacylglycerol plus inositol triphosphate (IP_3),

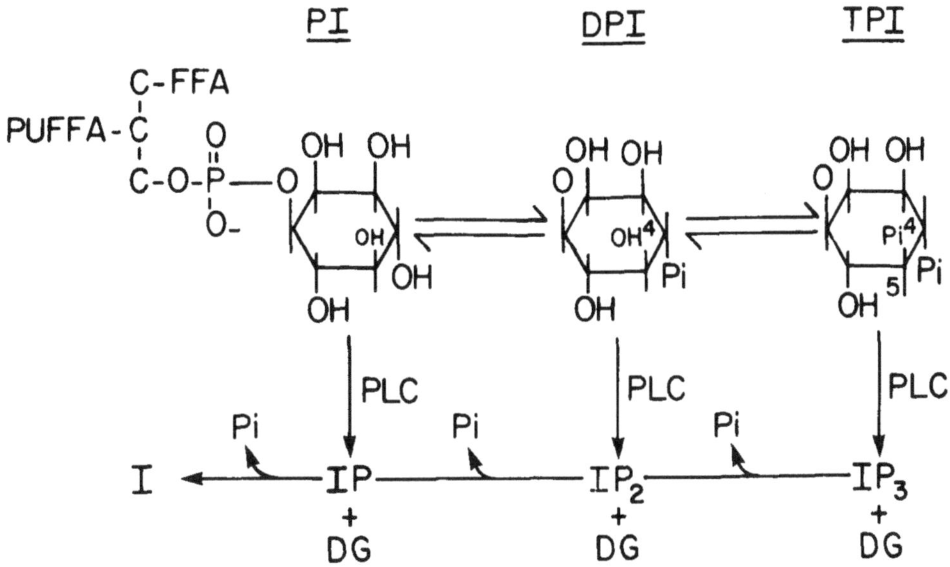

FIGURE 3. Effect of phospholipase C on phosphatidylinositol derivatives. Abbreviations: PI, phosphatidylinositol; DPI, phosphatidylinositol monophosphate; TPI, phosphatidylinositol bisphosphate; PLC, phospholipase C; DG, diglyceride; IP, inositol phosphate; IP_2, inositol bisphosphate; IP_3, inositol triphosphate; FFA, free fatty acid; PUFFA, polyunsaturated free fatty acid; and Pi, phosphate group.

inositol diphosphate (IP_2), or inositol monophosphate (IP_1). IP_3, IP_2, and IP_1 can be progressively dephosphorylated to form free inositol. Several intermediates in the metabolism of the inositol phosphatides may function to carry out the actions of hormones and other regulatory agents on cells. The 1,2 diacylglycerides are known to be activators of a membrane-bound, calcium-dependent kinase enzyme called protein kinase C.[10] IP_3 functions (perhaps as a calcium ionophore) to increase the concentration of calcium ions inside cells.[10] The 2 acyl free fatty acids on the 1,2 diacylglycerides may be released by a lipase as discussed above, and serve as substrates for the prostaglandins and/or the leukotrienes. Finally, the 1,2 diacylglycerides may be phosphorylated to phosphatidic acid which may also function to increase the intracellular concentration of calcium ions. It is further possible that a combination of the above changes may be required for the actions of certain stimuli on cells.

C. Effects of Phospholipase Inhibitors on Prolactin Responses

One way by which one can accrue evidence that the action of a stimuli on a cell is carried out via a perturbation of membrane phospholipids is to show that phospholipase inhibitors will abolish the stimuli responses. Two drugs which are known to inhibit the phospholipases include parabromphenacyl bromide (BPB) and quinacrine (mepacrine). BPB is known to inhibit phospholipase A_2 and quinacrine inhibits both phospholipases A_2 and C.[24-27] These inhibitors have been shown to abolish the prolactin actions on DNA synthesis,[28] RNA synthesis,[29,30] ornithine decarboxylase activity,[31] phosphoprotein (casein) synthesis,[29,30] and triglyceride synthesis[49] in cultured mouse mammary gland explants. The minimum concentrations of these drugs which abolish various prolactin responses in cultured mouse tissues are shown in Table 1.

Employing mammary explants from pseudopregnant rabbits, Dusanter-Fourt et al.[32] reported that BPB at a concentration of 50 μM did not inhibit the prolactin stimulation of casein synthesis; casein was isolated in those studies via immunoprecipitation. We[50] have subsequently cultured mammary tissues derived from 12- to 14-day pregnant rab-

Table 1
MINIMUM CONCENTRATIONS OF PHOSPHOLIPASE INHIBITORS REQUIRED TO INHIBIT PROLACTIN RESPONSES IN THE MAMMARY GLAND

Metabolic response	Drug employed: *Para* bromphenacyl bromide	Drug employed: Quinicrine	Ref.
DNA synthesis	—	1 μM	28
RNA synthesis	50 μM	50 μM	29,30
Ornithine decarboxylase activity	100 μM	50 μM	31
Casein synthesis	50 μM	10 μM	29,30
Triglyceride synthesis	50 μM	250 μM	Unpublished

Note: Methods employed in these studies are detailed in the reference.

bits and found that 5O μM BPB indeed does not abolish the prolactin stimulation of casein synthesis; an isoelectric precipitation method was employed in our studies to assess the rate of casein synthesis. When employing BPB at concentrations of 100 μM and above, however, we found that this drug does abolish the effect of prolactin on casein synthesis in cultured rabbit tissues. Thus, the phospholipase inhibitors have been found to abolish all the prolactin responses in which these drugs have been tested. Although possible nonspecific effects of the phospholipase inhibitors must be considered, the currently available data allows us to conclude that ongoing phospholipase activity in cells is essential for prolactin to express its actions on a number of metabolic processes in mammary tissues.

D. Effects of Phospholipases on Metabolic Processes in Mammary Cells

1. Phospholipase A_2

Several years ago we[15] observed that exogenously added phospholipase A_2 (PLA_2), like prolactin, causes a stimulation of the rate of [^{3}H]-uridine incorporation into RNA in cultured mouse mammary gland explants. In order for PLA_2 to be efficacious in these studies, it is assumed that the PLA_2 embeds itself in the membranes of the mammary cells and has its action on the phospholipids of the membrane structures. The time course and magnitude of the PLA_2 responses in the cultured mammary gland explants were the same as that of prolactin. In addition, prolactin and PLA_2 produced nonadditive responses when these agents were tested together at maximal stimulatory concentrations. These observations support the conclusion that prolactin and phospholipase A_2 are functioning via the same metabolic pathway. Also, these observations are compatible with the thesis that prolactin may be effecting an enhanced rate of RNA synthesis by stimulating PLA_2 activity in its target cells. In subsequent studies, we observed that exogenously added PLA_2 would also stimulate the rate of [^{3}H]-leucine incorporation into a casein-rich phosphoprotein fraction in cultured mouse mammary tissues; PLA_2 was only efficacious, however, if the tissues were also exposed to 0.5 mM spermidine added to the cultured medium.[33] More recently, we have shown that PLA_2 will stimulate ornithine decarboxylase (ODC) activity in cultured mouse mammary tissues, but the magnitude of a maximal stimulatory concentration of PLA_2 is only a fraction of that elicited with prolactin.[31] The prolactin and PLA_2 effects on ODC activity are, however, nonadditive when maximum stimulatory concentrations of these agents are tested together. PLA_2, in the presence and absence of spermidine, has no effect on the rate of [^{14}C]-acetate incorporation into triglycerides in cultured mouse mammary tissues.[49] Thus, exogeneously added PLA_2 will reproduce entirely, or in part, at least certain of the metabolic actions of prolactin in mammary tissues.

In another series of experiments, we found that PLA_2 stimulates guanylate cyclase activity in the 150,000-g pellet fraction of mouse mammary gland homogenates.[34] Since guanylate cyclase catalyzes the conversion of GTP to cyclic GMP, it was of great interest that PLA_2 stimulates guanylate cyclase activity, since in earlier studies we found that cyclic GMP, like PLA_2 will reproduce the prolactin stimulation of RNA synthesis in cultured mouse mammary tissues. These observations are therefore compatible with the postulate that the actions of prolactin in the mammary gland may be mediated, at least in part, by an enhanced activity of PLA_2 and a consequential enhanced production of cyclic GMP.

We next attempted to show a direct effect of prolactin on phospholipase A_2 activity in membrane preparations from mouse mammary tissues.[35] The substrate employed for the enzyme assay was phosphatidylcholine containing radiolabeled arachidonic acid (20:4) in the 2 acyl position of the glycerol moiety. PLA_2 activity was thus assessed by the rate of [^{3}H]-arachidonic acid released from the radiolabeled phosphatidylcholine. In our initial experiments, we observed a stimulation of PLA_2 activity with one prolactin preparation (NIH-PS-13) and reported the results in a rapid communication in 1978;[35] in subsequent studies, however, we tested other more highly purified prolactin preparations (including NIH-PS-14) and found no effect on PLA_2 activity. To date, there is therefore no conclusive evidence that prolactin has a direct action on PLA_2 activity in its target tissues.

2. Phospholipase C

Like PLA_2, PLC also stimulates ODC activity in cultured mouse mammary tissues with response characteristics similar to that of prolactin. Both the maximal responses and time courses of response are similar.[31] When maximal stimulatory concentrations of PLC and prolactin were tested together, however, a synergistic response was observed, i.e., the response was far in excess of simple additivity. Taken together, these observations suggest that the PLC and prolactin responses may be carried out via common biochemical pathways. More recently, we have further observed that PLC also has a prolactin-like action on stimulating the rate of [^{3}H]-uridine incorporation into RNA in cultured mouse mammary tissues.[36] Both the time course of response and maximal responses were the same with prolactin and PLC. In addition, when tested in concert, the responses with maximal stimulatory concentrations of PLC and prolactin are nonadditive. PLC also enhances the rate of [^{3}H]-leucine incorporation into a casein-rich phosphoprotein fraction; however, this response is only observed if the tissues are cultured at the same time with 0.5 mM spermidine. In unpublished studies, we[49] found that PLC had no effect on the rate of [^{14}C]-acetate incorporation into triglycerides in cultured mouse mammary tissues; this was true both in the presence and absence of 0.5 mM spermidine. The responses to PLC are therefore very similar to those elicited with PLA_2, except that PLC was far more potent than PLA_2 in stimulating ODC activity.

E. Possible Mechanism by Which Phospholipase C Affects Mammary Cells

As was pointed out above, the PLC that is activated by a variety of stimuli in membranes of target cells is specific for the inositol phosphatides.[10] The products of this enzymatic reaction, i.e., diglycerides and inositol phosphates, are known to have profound effects in cells. The diglycerides are known to activate a protein kinase in the plasma membrane (protein kinase C) which catalyzes the phosphorylation of specific tyrosyl residues on certain proteins. Certain of the phorbol esters, including 4β-phorbol 12-myristate 13 acetate (TPA), are structural analogs of the diglycerides and also are

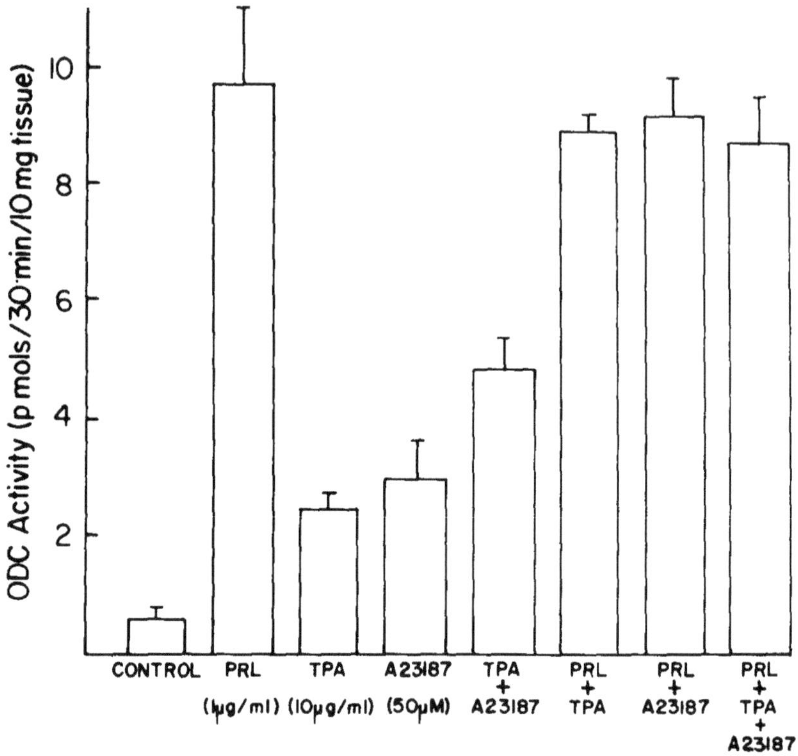

FIGURE 4. Effect of TPA, A23187, and prolactin, on ornithine decarboxylase activity in mouse mammary gland explants. Mammary tissues from 12 to 14-day pregnant mice were initially cultured for 24 hr with 1 μg/mℓ insulin plus 10^{-7} *M* cortisol. The substances listed were then added to the cultures and incubations continued for 6 hr; ODC activity determined as described earlier.

known to stimulate protein kinase C.[36] Since PLC reproduces the prolactin stimulation of ODC activity in cultured mouse mammary tissues, TPA was tested in order to determine whether it would also effect an enhanced activity of ODC in these tissues. In addition, since protein kinase C is a calcium-dependent enzyme, the activity of ODC was tested in tissues that were also exposed to the calcium ionophore A23187. Some of the results of these studies are shown in Figure 4. TPA indeed does stimulate ODC activity in cultured mammary tissues; its effect is maximal with TPA concentrations above 1 μg/mℓ.[37] In addition, the TPA response is nonadditive to that elicited with a maximal stimulatory concentration of prolactin, thus suggesting that these agents are stimulating ODC activity via a common pathway. It was also observed that the calcium ionophore A23187 at concentrations between 10 and 50 μ*M* effected, by itself, an increased ODC activity. When tested together at maximum stimulatory concentrations, TPA and A23187 elicited an additive response (Figure 4). This observation makes intuitive sense, since protein kinase C is a calcium-dependent enzyme and the A23187 may function to increase the calcium ion concentration in the cells, and thus potentiate the TPA activation of protein kinase C. Alternatively, the provision of calcium ions via A23187 could alter other metabolic processes, aside from an activation of protein kinase C, in potentiating the effect of TPA on stimulating ODC activity. It is of great importance that when TPA, A23187, and prolactin were tested together at maximal stimulatory concentrations, the magnitude of response was not greater than that elicited by prolactin alone (Figure 4); this observation is compatible with the conclusion that the prolactin, TPA, and A23187 responses occur via a common pathway. It is not known why optimal stimulatory concentrations of TPA plus A23187 produce a response which is less that that produced by a maximal stimulatory concentration of prolactin; however, the reduced potency could be explained by a reduced efficacy of TPA, as vs. the endogenous diglycerides, in activating protein kinase C. It remains to be demonstrated, however, that prolactin does in fact activate PLC, which then makes diglycerides available for the activation of protein kinase C in mammary tissues. It also remains to be shown that TPA and TPA plus A23187 will effect changes in metabolic processes other than the stimulation of ODC activity in mammary tissues. Similarly, the mechanism by which A23187 enhances ODC activity remains to be established.

F. Possible Relationship of Phospholipase A_2 to Phospholipase C

When PLA_2 and PLC were tested together at maximal stimulatory concentrations, they produced nonadditive effects on ODC activity and RNA synthesis in cultured mouse mammary cells. These observations suggest that PLA_2 and PLC may be acting on mammary cells via a common biochemical pathway. One tenable hypothesis that can explain how these enzymes may interact is that the inositol phosphates released in response to PLC may increase the intracellular concentration of calcium ions; the calcium ions, perhaps in conjunction with calmodulin, may then activate phospholipase A_2, which is a calcium-dependent enzyme. Further experimentation will obviously be required, however, to define if and how prolactin activates the phospholipase enzymes in mammary cells.

III. CALCIUM AND PROLACTIN

Calcium is known to be involved in a multiplicity of cellular functions.[1-3] Some of these include its role as a cofactor for enzyme activities, the regulation of secretory processes, the control of muscle cell activity, and the regulation of membrane transport phenomena. In the mammary gland, massive quantities of calcium are taken up from the plasma into the alveolar cells during lactation. Most of this calcium is then secreted into the milk, primarily associated with casein micelles. Calcium is also essential for prolactin (and perhaps other hormones) to express its actions on a number of metabolic processes in mammary tissues. Fragmentary evidence suggests that calcium may also be involved in the primary mechanism by which prolactin has its actions on target cells, perhaps by activating enzymes such as protein kinase C as was pointed out above.

A. Calcium Concentrations and Prolactin Responses

Removing calcium from the culture medium has been shown to abolish the effects of prolactin in mammary tissues on several metabolic processes, including RNA synthesis, casein synthesis, and triglyceride synthesis.[38,39] In more recent studies, we have also observed that removal of external calcium abolishes the prolactin stimulation of ODC activity (Table 2) in cultured mouse mammary tissues. It was essential in these later studies, however, to preculture the tissues in the absence of calcium for 2 hr prior to adding prolactin in order to abolish the effect of prolactin on ODC activity. When a preculture period was not employed, i.e., prolactin and EGTA were added at the same time, the magnitude of the prolactin effect on ODC activity was attenuated, but not abolished. This makes intuitive sense, since the ODC response to prolactin is the most rapid of those we have studied; the onset of the ODC response is between 30 and 60 min after exposing mammary tissues to prolactin.[40] Thus, when prolactin and EGTA are added at the same time to cultured mammary cells, it is likely that prolactin initiates this metabolic response prior to the chelation of the required amount of calcium ions. Although these studies clearly show that calcium ions are required for pro-

Table 2
EFFECT OF EXTRACELLULAR CALCIUM ION CONCENTRATIONS ON THE PROLACTIN STIMULATION OF ODC ACTIVITY

Calcium ion concentration (μM)	ODC activity (pmols/30 min/10 mg tissue)		P
	Without PRL	With PRL (1 μg/mℓ)	
0	0.04 + 0.04	0.07 + 0.07	N.S.
1	0.04 + 0.02	0.70 + 0.70	N.S.
10	1.00 + 0.40	20.7 + 2.00	< 0.05
100	0.89 + 0.33	18.5 + 1.2	< 0.05
1600	0.69 + 0.30	19.1 + 1.5	< 0.05

Note: Mammary tissues from 10 to 14-day pregnant mice were initially cultured for 36 hr with 1 μg/mℓ insulin plus 10^{-7} M cortisol. The tissues were next cultured with calcium ions at the concentrations indicated in the table. Finally, 1 μg/mℓ prolactin was added to some of the tissues for the final 4 hr of culture. ODC activity was then determined by methods described earlier.[31]

lactin to express its actions on mammary cells, this information does little to shed light on whether alterations of intracellular calcium ion concentrations may be involved in the mechanism by which prolactin acts on its target cells. In fact, it has been reported that EDTA chelation of calcium in medium bathing cultured rabbit mammary gland explants causes both a reduced rate of protein synthesis and a reduced number of prolactin receptors per cell;[41] calcium ions have also been shown to enhance prolactin binding to mammary gland membranes.[42] Any or all of these changes could be the cause of the failure of prolactin to function in the absence of calcium ions.

In the same studies (Table 2), we have titrated extracellular calcium ion concentrations to determine what calcium ion concentrations are required for prolactin to express its actions in mammary cells. As is apparent from the data in Table 2 and in earlier studies,[39] 5 to 10 μM calcium is the minimal extracellular calcium ion concentration required for prolactin to express its actions on ODC activity, triglyceride synthesis, casein synthesis, and RNA synthesis. Studies showing a minimal extracellular calcium ion requirement are compatible with, but in no way prove, a role for calcium as an intracellular mediator for actions of prolactin.

B. Studies with A23187, a Calcium Ionophore

To implicate calcium influx as a trigger for stimuli responses, many investigators have employed calcium ionophores in attempts to mimic or modulate the actions of various stimuli. In one report,[43] A23187 and increasing calcium ion concentrations were found to enhance the rate of protein synthesis and secretion from isolated acini derived from mammary glands of lactating rats. Protein synthesis and secretion rates were found to increase as calcium ion concentrations were increased from 1 to 18 mM. Although these are rather high extracellular calcium ion concentrations, the fact that A23187 was efficacious in eliciting these cellular responses suggests that calcium ions could participate in regulating lactogenic processes.

In a series of studies in both our laboratory[39] and that of Houdabine,[41] A23187 at several concentrations would not substitute for prolactin in stimulating RNA synthesis, triglyceride synthesis, or casein synthesis in cultured rabbit or mouse mammary tissues. The efficacy of this ionophore in increasing calcium ion permeability in cultured mammary tissues is indicated by the experiments in which A23187 was found to have a relatively small but significant stimulatory effect on ODC activity, both in the presence and absence of the phorbol ester TPA (see above). Also, the fact that the TPA plus A23187 effect on ODC activity is nonadditive to that of the maximal stimulatory concentration of prolactin is compatible with the thesis that the action of prolactin may be carried out, at least in part, by increasing the intracellular concentration of calcium ions. The possible mechanism by which prolactin could function in this regard was discussed in detail above.

Further experiments are obviously needed to prove that calcium ions are linked to the prolactin action on ODC activity and other metabolic processes. In addition, it remains to be established whether TPA plus A23187 will stimulate metabolic processes other than a stimulation of ODC activity in the mammary gland.

C. Calmodulin

Calmodulin is a cellular protein which combines with calcium and reduces the calcium ion concentration required to activate certain calcium-dependent enzymes.[44-46] The fact that calmodulin is contained in mammary tissues was suggested in a report by Pizarro et al.[47] These investigators reported that a protein derived from lactating rat mammary tissues functions in concert with 10 μM calcium to activate cyclic AMP- and cyclic GMP-dependent phosphodiesterase; this protein was identified as calmodulin by showing that trifluoperazine abolished its activity. Calmodulin content was shown to increase about twofold in rat mammary tissues at the time of parturition and its concentration remained the same through 20 days of lactation. In a more recent study, Mullaney and Clegg[48] reported the existence of a specific calcium-calmodulin-dependent cyclic GMP phosphodiesterase obtained from mammary tissues of rats in mid-lactation. It thus seems clear that calmodulin does exist in mammary tissues. It is also likely that calmodulin activates more enzymes in mammary cells other than just the cyclic nucleotide phosphodiesterases. In addition, if prolactin does act on cells to increase intracellular calcium ion concentrations, at least certain of the actions of prolactin may be carried out by the activation of calcium-calmodulin-dependent enzymes.

IV. SUMMARY

The phospholipids and calcium ions have been clearly implicated regarding their involvement in the prolactin regulation of metabolic processes in the mammary gland. Many aspects, however, about the possible prolactin regulation of phospholipid metabolism and calcium ion concentrations remain to be elucidated. Further investigations are needed to determine intracellular calcium ion concentrations in prolactin-stimulated cells, the effects of prolactin on the phospholipase enzymes, and the turnover rates of phospholipids in response to prolactin. Our further understanding of the fundamental actions of prolactin on cells will likely evolve from these types of studies.

REFERENCES

1. Rasmussen, H. and Goodman, D. B. P., Relationships between calcium and cyclic nucleotides in cell activation, *Physiol. Rev.*, 57, 421, 1977.
2. Whitfield, J. F., Boynton, A. L., Macmanus, J. P., Sikorska, M., and Tsand, B. K., The regulation of cell proliferation by calcium and cyclic AMP, *Mol. Cell. Biochem.*, 27, 155, 1979.
3. Williamson, J. R., Cooper, R. H., and Hoek, J. B., Role of calcium in the hormonal regulation of liver metabolism, *Biochim. Biophys. Acta*, 639, 243, 1981.

4. Verheij, H. M., Slotboom, A. J., and de Haas, G. H., Structure and function of phospholipase A_2, *Rev. Physiol. Biochem. Pharmacol.*, 91, 91, 1981.
5. Van Den Bosch, H., Intercellular pholipases A, *Biochim. Biophys. Acta*, 604, 191, 1980.
6. Roelefson, B., Phospholipases as tools to study the localization of phospholipids in biological membranes: a critical review, *J. Toxicol. Toxin Rev.*, 1, 87, 1982.
7. Putney, J. W., Jr., Recent hypothesis concerning the phosphatidylinositol effect, *Life Sci.*, 29, 1183, 1981.
8. Farese, R. V., Phosphoinositide metabolism and hormone action, *Endocr. Rev.*, 4, 78, 1983.
9. Best, L. and Malaisse, W. J., Phospholipids and islet cell function, *Diabetologica*, 25, 299, 1983.
10. Nishizuka, Y., Turnover of inositol phospholipids and signal transduction, *Science*, 225, 1365, 1984.
11. Marx, J. L., A new view of receptor action, *Science*, 224, 271, 1984.
12. Lapetina, E. G., Metabolism of inositides and the activation of platelets, *Life Sci.*, 32, 2069, 1983.
13. Fain, J. N., Lin, S. -H., Litosch, I., and Wallace, M., Hormonal regulation of phosphatidylinositol breakdown, *Life Sci.*, 32, 2055, 1983.
14. Nicoll, C. S., Physiological actions of prolactin, in *Handbook pf Physiology, Section 7, Endocrinology*, Vol. 4 (Part 2), Greep, R. O. and Astwood, E. B., Eds., *American Physiological Society*, Washington, D.C., 1974.
15. Rillema, J. A., and Anderson, L. D., Phospholipases and the effect of prolactin on uridine incorpation into RNA in mammary gland explants of mice, *Biochim. Biophys. Acta*, 428, 819, 1976.
16. Kinsella, J. E. and Infante, J. P., Phospholipid synthesis in the mammary gland, in *Lactation*, Vol. 4, Larson, B. D., Ed., Academic Press, New York, 1978, 475.
17. Rillema, J. A., Foley, K. A., and Etindi, R. N., Temporal sequences of prolactin actions on phospholipid biosynthesis in mouse mammary gland explants, *Endocrinology*, 116, 511, 1985.
18. Cameron, C. M., Linebaugh, B. E., and Rillema, J. A., Hormone control of lipid metabolism in mouse mammary gland explants, *Endocrinology*, 112, 1007, 1983.
19. Mullon, D. K., Smith, Y. F., Richardson, L. L. Hamosh, M., and Hamosh, P., Effect of prolactin on phospholipid synthesis in organ cultures of fetal rat lung, *Biochim. Biophys. Acta*, 751, 166, 1983.
20. Kinsella, J. E., The incorporation of [^{14}C] glycerol into lipids by dispersed bovine mammary cells, *Biochim. Biophys. Acta*, 164, 540, 1968.
21. Kinsella, J. E., Stearyl CoA as a precursor of oleic acid and glycerolipids in mammary microsomes from lactating bovine: possible regulatory step in milk triglyceride synthesis, *Lipids*, 7, 349, 1972.
22. Patton, S., McCarthy, R. D., Plantz, P. E., and Lee, R. F., Phospholipid incorporation into plasma membrane of lactating mammary cell with special reference to sphingomyelin, *Nature (London) New Biol.*, 241, 241, 1973.
23. Shulka, S. D., Phosphatidylinositol specific phospholipases C, *Life Sci.*, 30, 1323, 1982.
24. Roberts, M. F., Deems, R. A., and Dennis, E. A. J., Spectral perturbation of the histidine and tryptophan in cobra venom phopholipase A_2 upon metal ion and mixed micelle binding, *J. Biol. Chem.*, 252, 6011, 1977.
25. Markus, H. B. and Ball, E. G., Inhibition of lipolytic processes in rat adipose tissue by antimalarial drugs, *Biochim. Biophys. Acta*, 187, 486, 1969.
26. Vigo, C., Lewis, G. P., and Piper, P. J., Mechanisms of inhibition of phospholipase A_2, *Biochem. Pharmacol.*, 29, 623, 1980.
27. Hofmann, T., Prescott, S. M., and Majerus, P. W., The effects of mepacrine and p-bromophemacyl bromide on arachidonic acid release in human platelets, *Arch. Biochem. Biophys.*, 215, 237, 1982.
28. Rillema, J. A. and Foley, K. A., Characteristics of the action of prolactin on [^{3}H]-thymidine incorporation into DNA in mammary gland explants from virgin mice, *Horm. Metab. Res.*, 15, 385, 1983.
29. Rillema, J. A., Inhibition of prolactin actions in mouse mammary gland explants by *p* bromphenacyl bromide, a phospholipase A_2 inhibitor, *Proc. Soc. Exp. Biol. Med.*, 161, 355, 1979.
30. Rillema, J. A., Actions of quinacrine on RNA and casein synthesis in mouse mammary gland explants, *Prostaglandins Med.*, 2, 155, 1979.
31. Rillema, J. A., Wing, L. -Y. C., and Foley, K. A., Effects of phopholipases on ornithine decarboxylase activity in mammary gland explants from mid-pregnant mice, *Endocrinology*, 113, 2024, 1983.
32. Dusanter-Fourt, I., Djiane, J., and Houdebine, L. M., Effects of transglutaminase or phospholipase A_2 inhibitors on down-regulation of prolactin receptors and stimulation of casein and DNA synthesis in mammary gland explants, *Mol. Cell. Endocrinol.*, 31, 287, 1983.
33. Rillema, J. A., Linebaugh, B. E., and Mulder, J. A., Regulation of casein synthesis by polyamines in mammary gland explants from mice, *Endrocinology*, 100, 529, 1977.
34. Rillema, J. A. and Linebaugh, B. E., Effects of phospholipase A_2 and triton X-100 on guanylate cyclase activity in mammary gland homogenates from mice, *Horm. Metab. Res.*, 10, 331, 1978.
35. Rillema, J. A. and Wild, E. A., Prolactin activation of phospholipase A activity in membrane preparations from mammary glands, *Endocrinology*, 100, 219, 1977.

36. Wenner, C. E., Tomei, L. D., and Leister, K. J., Tumor promoters: an overview of membrane-associated alterations and intracellular events, *Transplant. Proc.*, 16, 381, 1984.
37. Rillema, J. A., Prolactin stimulation of ornithine decarboxylase activity in the mammary gland may involve an activation of protein kinase C, *Proc. Soc. Exp. Biol. Med.*, 178, 490, 1985.
38. Rillema, J. A., Requirement of calcium ions for prolactin actions on RNA and casein synthesis in mouse mammary gland explants, *Endocrinology*, 106, 1360, 1980.
39. Cameron, C. M. and Rillema, J. A., Extracellular calcium ion concentration required for prolactin to express its actions on casein, RNA and lipid biosynthesis in mouse mammary gland explants, *Endocrinology*, 113, 1596, 1983.
40. Rillema, J. A., Linebaugh, B. E., and Mulder, J. A., Regulation of casein synthesis by polyamines in mammary gland explants from mice, *Endocrinology*, 100, 529, 1977.
41. Houdebine, L. -M., Role du calcium dous l'induction de la synthese des caseines par la prolactine dous la glande mammaire de lapine, *Biol. Cell.* 40, 129, 1981.
42. Shiu, R. P. C. and Friesen, H. G., Properties of a prolactin receptor from the rabbit mammary gland, *Biochem. J.*, 140, 301, 1974.
43. Smith, J. J., Park, C. S., and Kennan, T. W., Calcium and calcium ionophore A23187 alter protein synthesis and secretion by acini from rat mammary gland, *Int. J. Biochem.*, 14, 573, 1982.
44. Means, A. R., Calmodulin: properties, intercellular localization, and multiple roles in cell regulation, *Rec. Prog. Horm. Res.*, 37, 333, 1981.
45. Moore, P. B. and Dedman, J. R., Calcium binding proteins and cellular regulation, *Life Sci.*, 31, 2937, 1982.
46. Manalan, A. S. and Klee, C. B., Calmodulin, in *Advances in Cyclic Nucleotide and Protein Phosphorylation Research*, Vol. 18, Greengard, P. and Robison, G. A., Eds., Raven Press, New York, 1984, 227.
47. Pizarro, M., Puente, J., and Sapag-Hagar, M., Calmodulin and cyclic nucleotide-phosphodiesterase activities in rat mammary gland during the lactogenic cycle, *FEBS Lett.*, 136, 127, 1981.
48. Mullaney, I. and Clegg, R. A., Cyclic AMP phosphodiesterase and cyclic GMP phosphodiesterase activities of rat mammary tissues, *Biochem. J.*, 219, 801, 1984.
49. Cameron, C. M. and Rillema, J. A., unpublished studies.
50. Rillema et al., unpublished observations.

Chapter 7

SYNERGISTIC ACTIONS OF GLUCOCORTICOID AND PROLACTIN IN MURINE MILK-PROTEIN GENE EXPRESSION

Mihir R. Banerjee and Ravi S. Menon

TABLE OF CONTENTS

I. Introduction ... 122

II. Culture Model of the Whole Mammary Organ ... 122

III. Dissociation of Morphogenesis and Lactogenesis Stages ... 122

IV. Utilization of the cDNA Hybridization Probe ... 124

V. Influence of the Steroid or the Polypeptide Hormones ... 125

VI. Residual Steroid Hormone Effect ... 126

VII. Nuclear Interaction of Glucocorticoid Receptors in Mammary Cells ... 126

VIII. Transcriptional Control ... 128

IX. Progesterone, an Antagonist to Lactogenesis ... 129

X. Milk-Protein Gene Expression During Mammary Gland Development In Vivo ... 130

XI. Summary and Comments ... 131

Acknowledgments ... 133

References ... 133

I. INTRODUCTION

Expression of functional differentiation of the mammary epithelial cells is accompanied by the appearance of several specific gene products, such as the caseins, α-lactalbumin (α-la), and the whey-acidic proteins (WAP). In murine mammary glands, prolactin and a glucocorticoid are the principal hormones required for lactogenesis.[1,2] Progesterone acts as an antagonist in this complex multiple hormone regulatory process.[3] Thus, the mammary system presents a suitable model for studies on the mechanism(s) of regulation of several genes during the developmental processes of a major endocrine target organ. During the past several years, we have studied the regulatory role(s) of the different steroid and polypeptide hormones on the expression of the murine milk-protein genes in an isolated whole mammary organ in a unique serum-free culture model. This chapter presents a brief account of the recent findings. The diverse problems associated with the multiple hormone-dependent differentiation of the mammary cells have been discussed at regular intervals,[4-7] and the reader is referred to an extensive recent review[8] which principally concerns the findings about the regulatory processes of milk-protein gene expression.

II. CULTURE MODEL OF THE WHOLE MAMMARY ORGAN

The procedures for culture of the whole mammary organ have been extensively described.[9,10] Briefly, 3 to 4-week-old BALB/c female mice are primed by daily injections of a mixture of estradiol-17β and progesterone for 9 days.[11] On the 10th day, the whole second thoracic mammary glands are excised, and the isolated whole mammary organs resting on a dacron raft are then transferred into a plastic culture dish containing Waymouth's synthetic medium (MB/751). The glands are incubated at 37°C in the serum-free medium supplemented with the appropriate combinations of the different steroid and polypeptide hormones in a humidified atmosphere of 95% oxygen and 5% carbon dioxide. (Figure 1 illustrates that initial culture of the glands for 6 days in medium containing a mammogenic hormone mixture to induce the lobuloalveolar secretory structures, which are reminiscent of the similar structures present in the mammary glands of pregnant mice.) The glands, during subsequent culture in a lactogenic hormone-containing medium (prolactin, cortisol, and insulin), accumulate abundant caseins, reflecting expression of functional differentiation.[12] An additional period of culture in a prolactin-free medium containing insulin and/or aldosterone results in a complete regression of the alveolar structures, leaving a ductal parenchyma.[13] Continued culture of the ductal glands in the mammogenic medium supplemented with epidermal growth factor stimulates a second round of lobuloalveolar development.[14] Furthermore, exposure of the glands to chemical carcinogens during lobuloalveolar morphogenesis induces preneoplastic and neoplastic transformation of the epithelial cells.[15-17] Thus, the isolated whole organ in vitro mimics the entire hormone-dependent developmental stages of the mammary gland in vivo, including the carcinogenesis process. This in vitro model was used for the studies described in this chapter.

III. DISSOCIATION OF MORPHOGENESIS AND LACTOGENESIS STAGES

In the animal, the stages of lobuloalveolar morphogenesis and functional differentiation of the mammary gland overlaps during the second half of the gestational period.[4] Therefore, the two developmental stages needed to be dissociated to obtain a reliable assessment of the role of the individual hormones regulating the expression of the milk-protein genes. This was accomplished by allowing morphogenesis and lacto-

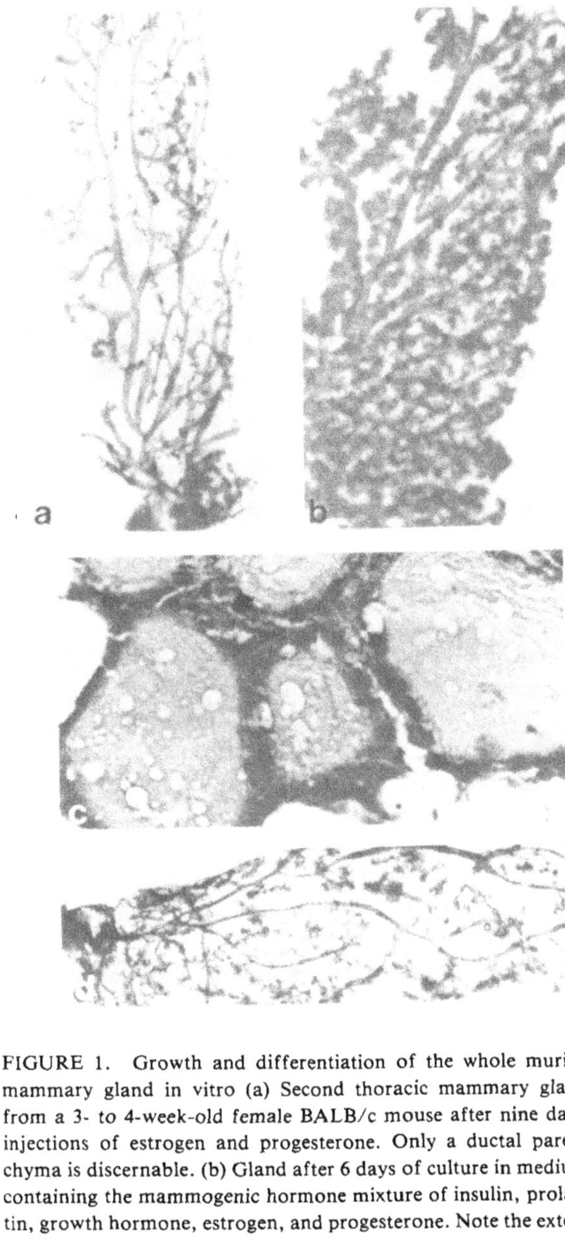

FIGURE 1. Growth and differentiation of the whole murine mammary gland in vitro (a) Second thoracic mammary gland from a 3- to 4-week-old female BALB/c mouse after nine daily injections of estrogen and progesterone. Only a ductal parenchyma is discernable. (b) Gland after 6 days of culture in medium containing the mammogenic hormone mixture of insulin, prolactin, growth hormone, estrogen, and progesterone. Note the extensive development of alveolar epithelial structures. (c) Histological section of a gland that was initially cultured as in panel b, and then incubated for a further 6 days in the presence of insulin, prolactin, and cortisol. The lumens of the alveoli are full of a milk-like secretion, containing proteins and fat droplets. (d) Gland cultured as in b, followed by 12 days in medium containing insulin and aldosterone. The alveolar structures can be seen to have regressed, leaving only a ductal parenchyma.[9]

genesis to occur in a two-step culture model.[18,19] In the step I culture, the glands were incubated in a corticosteroid-free medium containing prolactin, growth hormone, insulin, estrogen, and progesterone (mammogenic hormone mixture) to obtain full lobuloalveolar development. Step II culture then was done in medium containing prolactin, cortisol, and insulin (lactogenic hormone mixture). Insulin is needed for viability of the parenchyma in the serum-free medium. Translation assays of the RNA from the glands after 6 days of step I culture failed to show any casein mRNA (mRNAcsn) activity in a cell-free protein synthesis system derived from heterologous ribosomes; also, no casein was detectable in these glands.[18,20] In contrast, both caseins and the respective mRNAs were abundant in the glands after step II incubation in the medium containing prolactin, cortisol, and insulin. Thus, unlike the mammary glands in the animal, the isolated mammary gland in a corticosteroid-free mammogenic medium completely dissociated the developmental stages of morphogenesis and lactogenesis in the glands in vitro. The two-step culture model thus provides the suitable in vitro model for reliable determination of the role of prolactin and glucocorticoid in expression of the casein genes in the mammary glands not pre-exposed to the steroid hormone.

IV. UTILIZATION OF THE cDNA HYBRIDIZATION PROBE

While the results of the translational assays mentioned above provided a qualitiative determination of the accumulation of the mRNAcsn, the possibility remains that the translational assay may not detect the presence of low levels of mRNAcsn in the glands. Therefore, a quantitative measurement of the concentrations of the mRNA using a sensitive method was needed. Accordingly, a 15S RNA fraction enriched in casein mRNA activity (95%) was isolated from the lactating mammary glands. The 15S RNA was used as a template to synthesize a cDNA (cDNAcsn), which was then used as a sensitive molecular hybridization probe for measuring the concentrations of the mRNAcsn in total RNA from the mammary glands in vitro.[21] It needs to be pointed out that the 15S RNA fraction from lactating mammary tissue includes the mRNAs for the different caseins; consequently, the cDNAcsn represents a hybridization probe for the total casein mRNAs. Caseins are a group of acidic phosphoproteins coded by different mRNAs. Hybribizations analysis of the RNA by the sensitive cDNA probe showed that the concentration of mRNAcsn in the mammary glands after 6 days of step I culture in the corticosteroid-free medium remains as low as 0.00067% of the total RNA mass, representing only 147 molecules of mRNAcsn per epithelial cell. These results also revealed that in the mouse mammary gland in vitro, 147 molecules per cell constitute the basal level of murine mRNAcsn concentration. This indicates that the casein genes in the lobuloalveolar glands after step I culture remain in an uninduced state.[22] This finding is also consistent with the virtual absence of any casein mRNA activity in the glands after step I culture as determined by the translational assays. On the other hand, the same studies also showed (Table 1) that accumulation of the mRNAcsn increased 9-fold over the basal level after only 24 hr culture in the step II medium containing prolactin, cortisol, and insulin, and the level then increased progressively, rising 255-fold over the basal level after 9 days of step II culture with an accumulation of 37,524 molecules of mRNAcsn per epithelial cell. Since there was virtually no rise in cell number in the gland during the 9 days of step II culture, the results reflect a net increase of the mRNAcsn concentration in the glands in vitro. It is of interest to mention that addition of cortisol into the medium during the step I culture promotes a twofold increase of mRNAcsn concentration after 3 days of culture.[23] Thus, the results strongly suggest that maintenance of the uninduced state of the casein genes in the step I culture is due to the absence of glucocorticoid in the medium. The expression of the casein genes during morphogenesis in the step I culture after addition of cortisol in the medium further demonstrates that, as in the pregnant animal, milk-protein gene expression can occur simultaneously with morphogenesis in vitro, provided that glucocorticoid is present in the step I medium.[24] In this context, in the animal, circulating glucocorticoid rises with advanced stages of pregnancy. Thus, the find-

Table 1
NUMBER OF EPITHELIAL CELLS AND CASEIN mRNA MOLECULES PER CELL AT DIFFERENT STAGES OF MORPHOGENESIS AND DIFFERENTIATION OF THE MAMMARY GLAND IN CULTURE

Culture stage	No. cells per gram of tissue[a] × 10^6 ± S.D.	Percent casein mRNA	No. casein fold mRNA molecules per cell	Increase in molecules per cell over IPrlEPGH
Primed glands	0.54 ± 0.13[b]			
Step I				
6-day IPrlEPGH[c]	5.32 ± 0.31[d]	0.00067	147	
Step II				
6-Day IPrlEPGH + 1-day IPRLF	4.72 ± 0.39	0.0033	1,366	9.3
6-Day IPrlEPGH + 3-day IPrlF	5.95 ± 0.50	0.029	4,826	32.8
6-Day IPrlEPGH + 6-day IPrlF	6.09 ± 0.71	0.068	23,699	161.2
6-Day IPrlEPGH + 9-day IPrlF	5.43 ± 0.62	0.09	37,524	255.2

[a] The values reported are a mean of five determinations.
[b] About 27 primed glands (wet weight = 0.5 g) were pooled for each determination.
[c] I, insulin; Prl, prolactin; E, estradiol-17β; P, progesterone; GH, growth hormone; and F, cortisol.
[d] About 12 to 16 glands (wet weight = 0.2—0.3 g) from either IPrlEPGH or IPrlF were pooled for each determination.

From Mehta, N. M., Ganguly, N., Ganguly, R., and Banerjee, M. R., *J. Biol. Chem.*, 255, 4430, 1980. With permission.

ings discussed above indicate that glucocorticoid is needed for stimulation of expression of the casein genes in the presence of prolactin.

V. INFLUENCE OF THE STEROID OR THE POLYPEPTIDE HORMONES

The complete dissociation of the stages of morphogenesis and functional differentiation of the mammary gland in the two-step culture model for the first time made it feasible to design experiments for testing the ability of prolactin or glucocorticoid alone for stimulating expression of the casein genes. In these studies, step I incubation was done for 6 days in the corticosteroid-free mammogenic medium to obtain the lobuloalveolar secretory structures, and step II incubations were then done as follows.[25]

In one set of experiments, step II incubation of the glands was done in medium containing prolactin or cortisol in the presence of insulin. As measured by the cDNAcsn probe, virtually no mRNAcsn was present in these glands after 3 days of incubation. In another set of experiments, a combination of cortisol and insulin failed to raise mRNAcsn above the base level, indicating that neither prolactin nor glucocorticoid alone is capable of stimulating casein mRNA accumulation in the mammary glands in vitro. In contrast, when the initial 3 days of incubation with prolactin or cortisol was followed by an additional 3 days of incubation in medium containing the complete lactogenic hormone mixture of prolactin, cortisol, and insulin, abundant casein mRNA was measurable in these glands. These results clearly show that both prolactin and cortisol are essential for expression of the milk-protein genes. Thus, the milk-protein genes are regulated by a synergistic action of the polypeptide and the steroid hormone.

In a second set of experiments in these studies,[25] the lobuloalveolar glands after step I culture were incubated first in the medium containing prolactin and insulin, and then in medium containing cortisol and insulin. RNA from these glands analyzed by the cDNAcsn probe showed a near basal level of mRNAcsn. In contrast, when the lobuloalveolar glands were first incubated with cortisol and insulin and then in medium containing prolactin and insulin, mRNAcsn in these glands increased 18-fold over the basal level after 3 days. This marked increase is not likely due to a stimulatory action of prolactin in the medium, because the experiments described above clearly showed that prolactin alone is incapable of stimulating the mRNAcsn accumulation. This raises the possibility whether the stimulation of mRNAcsn is due to a synergistic action of prolactin present in the medium and residual cortisol retained in the glands, pre-exposed to the steroid hormone.

VI. RESIDUAL STEROID HORMONE EFFECT

Mammary cells pre-exposed to steroid hormones can retain the same hormone for a prolonged period of time.[26] Therefore, the possibility whether the significant accumulation of mRNAcsn in the mammary glands incubated first with cortisol and insulin, and then in medium containing prolactin and insulin was due to a synergistic action of residual steroid hormone with prolactin was determined.[25] Figure 2 shows that the lobuloalveolar glands preincubated with cortisol indeed retain the steroid hormone. Furthermore, depletion of the residual steroid hormone in the cortisol-free medium corresponds with a loss of mRNAcsn concentration in the glands in the medium containing prolactin and insulin. Addition of cortisol in the medium at the time when the residual cortisol and the mRNAcsn reached a basal level stimulated a 25-fold increase in mRNAcsn accumulation after 3 days of incubation. These results demonstrate that a continuous presence of the glucocorticoid in the medium is obligatory for sustained expression of the casein genes. Under similar conditions, a synergistic action of prolactin and residual cortisol is also needed for casein production in rat mammary tissue explants in culture.[26] Thus, contrary to the conclusion[27] that the expression of the casein genes are dependent on prolactin stimulation, the findings described above provide unequivocal evidence that expression of the casein genes in mice and rats requires a synergistic action of both the steroid and the polypeptide hormone. A synergistic action of glucocorticoid and thyroid hormone has also been reported to be required for growth hormone gene expression.[28] The findings described above thus permit the conclusion that the steroid hormone is essential for induction of the casein genes in the presence of prolactin.

VII. NUCLEAR INTERACTION OF GLUCOCORTICOID RECEPTORS IN MAMMARY CELLS

Steroid hormone-inducible gene expression is believed to require specific interaction of the hormone receptor with a nuclear acceptor site.[29] Evidence presented in the preceding section strongly supports the concept that glucocorticoid as a component of the lactogenic hormones plays a major role in regulation of milk-protein gene expression. Thus, it needed to be ascertained whether an interaction of the glucocorticoid-receptor complex with the nuclear acceptor site in the mammary cells is associated with the expression of the casein genes.

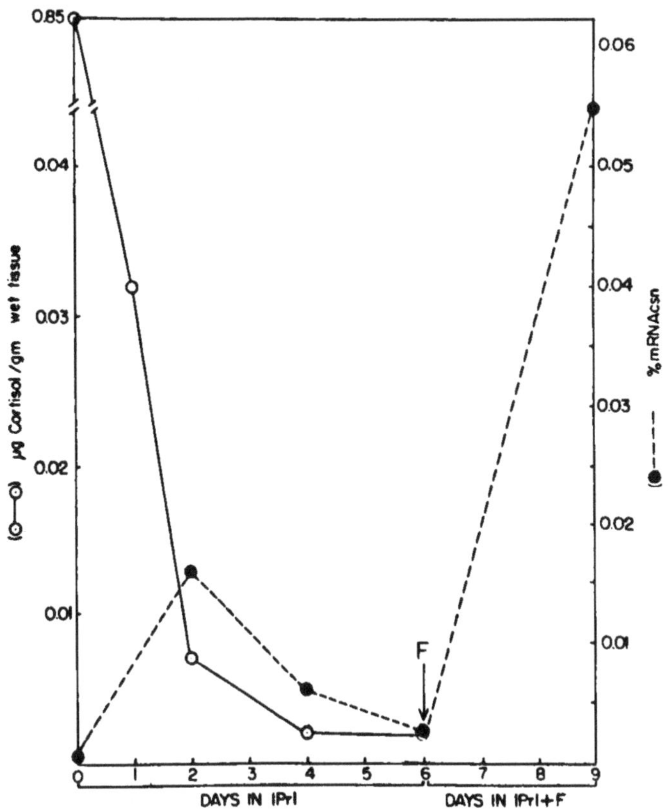

FIGURE 2. Relationship between residual cortisol retained and the mRNAcsn accumulation in the glands.[25]

Pyridoxal-5′-phosphate (PALP), the vitamin B_6 analog, blocks nuclear binding of the steroid hormone-receptor complex, apparently by interacting with the lysine residues in the receptor protein.[30] PALP-mediated inhibition of nuclear binding of steroid receptor also corresponds with a loss of glucocorticoid stimulation of tyrosine amino transferase synthesis.[31] This physiological modulator of steroid hormone regulation of specific gene expression was used to analyze whether the action of PALP on nuclear interaction of the glucocorticoid receptor complex can influence the expression of the casein genes. As usual in these studies,[32] the mammary glands were initially incubated for 6 days in the corticosteroid-free step I mammogenic medium to obtain the lobuloalveolar structures. The glands were then incubated in the step II medium containing the lactogenic hormone combination of prolactin, cortisol, and insulin. Addition of 2 or 5 m*M* PALP in the medium raised its mammary tissue concentration 4- and 12-fold, respectively, over the basal level. The increased tissue concentration of PALP was accompanied by a 50% reduced nuclear binding of ^{3}H-dexamethasone in the glands exposed to 2m*M* PALP. At a 5-m*M* concentration, PALP virtually abolished the nuclear binding of the glucocorticoid receptor complex.

In concurrent studies, RNA from mammary glands in parallel cultures not treated with PALP showed a casein mRNA concentration of 0.03%. In contrast, mRNAcsn concentrations in the glands exposed to 2 and 5 m*M* PALP were reduced to levels of only 0.0041 and 0.0026%. A comparative analysis of the data on ^{3}H-dexamethasone nuclear binding and mRNAcsn concentration in the glands revealed that the loss of nuclear binding of the glucocorticoid receptor complex is accompanied by a corresponding reduction of mRNAcsn accumulation in the glands incubated for 3 days in the step II culture medium containing prolactin, cortisol, and insulin. The same studies also showed that withdrawal of PALP from the medium after 3 days of culture restores the levels of nuclear binding of the glucocorticoid receptor along with a corresponding rise of mRNAcsn concentration in the glands after an additional 3 days of culture in PALP-free medium. This finding indicates that the PALP-mediated inhibition of nuclear binding glucocorticoid receptor and decrease of mRNAcsn concentration does not reflect an altered hormone responsiveness of the mammary cells in organ culture. Therefore, these findings permit the conclusion that the action of the glucocorticoids as a component of the lactogenic hormone mixture is receptor mediated, and the stimulatory action of the steroid hormone on casein gene expression is dependent on the specific nuclear interaction of the glucocorticoid receptor complex with the nuclear acceptor site in the mammary cells. The fact that the level of nuclear binding of the steroid hormone receptor modulates the cellular concentration of mRNAcsn further suggests that glucocorticoid, as a component of the lactogenic hormones, acts at the genomic level of control of milk-protein genes expression.

VIII. TRANSCRIPTIONAL CONTROL

Consistent with the findings in the mammary glands in organ culture, studies in the animal showed that surgical removal of the endogenous source of glucocorticoid from lactating mice by bilateral adrenalectomy causes loss of mRNAcsn activity[33] and also reduces the concentration of the mRNAcsn from 1.5 to 0.25% after 5 days.[21,34] A single injection of cortisol to the adrenalectomized animals raised mRNAcsn concentration four fold over the level in the adrenalectomized animal after 12 hr. This observation confirms that the mammary cells in vivo respond to the regulatory action of glucocorticoids in a manner similar to the glands in organ culture. Moreover, the marked increase in mRNAcsn accumulation in the adrenalectomized animals after a single injection of the steroid hormone also suggests an enhancement of transcription of the casein genes. This possibility was then tested in the following experiments. In these studies,[35] mammary cell nuclei isolated from lactating mice were allowed to synthesize RNA in a cell-free system containing mercurated CTP (Hg-CTP). Mercurated RNA (Hg-RNA) synthesized in vitro was then isolated from the endogenous RNA in the reaction mixture by SH-agarose affinity column chromatography.[36] The purified Hg-RNAs which represent the newly synthesized transcripts were then hybridized to the ^{3}H-cDNAcsn probe.[21] Results showed that adrenal ablation of the lactating animals causes a 75% reduction of mRNAcsn transcripts after 5 days. A single injection of cortisol to the 5-day adrenalectomized mice enhanced the mRNAcsn transcription two-fold over that in the adrenalectomized animal.

Attempts[21] to measure the influence of prolactin in lactating mice treated with 2-α-bromoergocryptine (CB-154), a specific prolactin inhibitor,[37] failed to show any alteration of the level of mRNAcsn transcription in mice with 80% suppression of endogenous prolactin level caused by CB-154. This finding, however, may not reflect a lack of prolactin action at the transcriptional level of casein gene expression, because the 20% level of circulating prolactin may be sufficient to maintain the high rate of transcription in the presence of the high level of circulating glucocorticoid in the lactating animal. Nevertheless, the results of the transcriptional measurements described above clearly demonstrate that maintenance of the high level of transcription of the casein genes in the lactating animal is dependent on the enriched glucocorticoid environment in the animal in the presence of endogenous prolactin. Thus, we conclude that glucocorticoid as a component of the lactogenic hormones does modulate milk-protein gene expression, acting at the genomic level.

IX. PROGESTERONE, AN ANTAGONIST TO LACTOGENESIS

Progesterone acts as an inhibitor of lactogenesis.[3] Thus, it was of interest to know whether the ovarian steroid hormone exerts its influence at the regulatory level of expression of the casein genes. Progesterone is known to compete for the cytoplasmic glucocorticoid receptors in mouse mammary cells.[38] Thus, attempts were made to ascertain the biochemical pathway of action of the ovarian steroid hormone and the possible mechanism of the inhibitory action. As usual, in these studies,[23] the mammary glands were incubated first in the corticosteroid-free step I mammogenic medium for 6 days to obtain the lobuloalveolar structures, and only a basal level of mRNAcsn was present in these glands. Subsequent incubation of the lobuloalveolar glands in a progesterone-free medium containing 1 or 5 μg/mℓ, cortisol plus prolactin and insulin promoted mRNAcsn accumulation 17- and 19-fold, respectively, over the basal level after 6 days. However, addition of 1 μg/mℓ progesterone in the same medium after the initial 3 days of incubation blocked the increase in mRNA accumulation. Interestingly, however, 1 μg/mℓ each of progesterone and cortisol enhanced the mRNAcsn level only three fold, whereas 5 μg/mℓ cortisol in the presence of 1 μg/mℓ progesterone stimulated mRNA accumulation tenfold in medium containing prolactin and insulin (5 μg/mℓ).[23] The results thus suggest that the inhibitory action of progesterone is related to a progesterone (P) to glucocorticoid (F) ratio in the medium. This possibility was confirmed by the observation that at increased P:F molar ratios in the medium, mRNAcsn concentrations decreased progressively. At an equimolar ratio of progesterone to cortisol (P:F = 1), mRNA concentration in the glands was reduced 84% after 3 days of culture in medium containing prolactin and insulin.

At this point, it is of interest to note that in the pregnant animal, the P:F ratio is likely to be reduced because during gestation serum progesterone remains relatively constant, whereas circulating glucocorticoid rises progessively.[39,40] This relationship between the levels of progesterone and glucocorticoid may also account for the increased casein production in the mammary glands during advanced stages of pregnancy in an enriched endogenous prolactin environment in vivo.[8] The findings described above thus strongly suggest the presence of some antagonistic interaction between progesterone and glucocorticoid, and this may prevent the maximal expression of the milk-protein genes in the pregnant animal. Thus, the findings in the organ culture system seem to mimic the events in vivo.

The ability of progesterone to competitively displace glucocorticoid binding to its cytosol receptors in the mammary cells is well established.[41] Thus, it is conceivable that the inhibitory action of progesterone may be mediated at the level of glucocorticoid interaction with its receptor in the mammary cells. Our studies on glucocorticoid binding to mammary cytosol receptor have shown that increased molar ratios of P:F in the culture medium can cause a 60% inhibition of glucocorticoid binding to mammary cytosol receptors. Molecular hybridization of the RNAs from the glands in parallel cultures showed a corresponding loss of mRNA accumulation in the glands in the presence of prolactin, cortisol, and insulin in the medium.[23] Thus, progesterone appears to act as an antagonist to milk-protein gene expression by displacement of glucocorticoid from its receptor in the mammary cells. In contrast to our findings, progesterone inhibition of casein gene expression in rabbit and rat mammary gland has been postulated as due to an antagonistic influence of progesterone on the stimulatory action of prolactin.[6,42] However, no mechanism for a similar interaction between the polypeptide and the steroid hormone has been presented. Moreover, the biochemical pathway(s) of action of prolactin in regulation of expression of the milk-protein genes as yet remains unknown.

X. MILK-PROTEIN GENE EXPRESSION DURING MAMMARY GLAND DEVELOPMENT IN VIVO

Burditt et al.[43] first reported the use of milk-protein cDNA clones to study the expression of the milk-protein genes during different stages of development of guinea pig mammary gland. Radiolabeled cDNA probes for the caseins and α-LA were used to measure the rate of accumulation of the respective mRNAs in the guinea pig mammary gland during pregnancy and lactation. Casein mRNAs were not detectable at late pregnancy. However, concentrations of the mRNAcsn rose sharply at parturition to levels that were present throughout lactation. The α-LA mRNA sequences were present during late pregnancy, and at parturition the expression reached a maximum level. In addition to mRNA accumulation, a differential rate of casein and α-LA synthesis and secretions was also observed in these studies. The authors postulate that milk production may be regulated at several subcellular sites. The caseins and the whey-protein genes may not express in a coordinate manner.

Hobbs et al.[44] noted a steady increase in the levels of α-, β-, and γ- caseins and the WAP mRNAs in rats during pregnancy. A sharp rise in the rate of accumulation of the milk-protein mRNAS occurred at parturition, which continued through to 18 days of lactation. In mammary explant cultures from mid-pregnant rats, all four milk-protein mRNAs required both prolactin and cortisol for maximum accumulation. Cortisol alone had little effect on the accumulation of three casein mRNAs, but caused a 68-fold increase in WAP mRNA during the 24-hr culture period.

Kulski et al.[45] used mammary explants from adrenalectomized virgin rats in order to determine the importance of glucocorticoid for casein gene expression measured by cloned cDNA probes to the respective mRNAs. In these studies, only a basal level of α-casein mRNA was observed in the presence of insulin and prolactin. However, addition of cortisol in the medium raised the mRNA concentration 20-fold over the basal level. Likewise, β-casein gene expression was virtually absent in the explants in the presence of prolactin and insulin, but addition of cortisol in the same medium promoted abundant accumulation of the specific mRNAs. These authors have concluded that expression of the casein genes are greatly, if not entirely, dependent on glucocorticoid stimulation.

Expression of the milk-protein genes during development of the mouse mammary gland both in vivo and in vitro was studied in our laboratory using the different cloned cDNA probes. The two-step culture model of the whole murine mammary gland described earlier was to study the hormonal regulation of WAP and α-LA gene expression. Mammary glands were initially cultured for 6 days in a mammogenic medium containing insulin, prolactin, estrogen, progesterone, and growth hormone. The glands were then cultured in the presence of various combinations of insulin, prolactin, and hydrocortisone for up to 6 days. The WAP and α-LA mRNA levels were analyzed by the "dot-blot" hybridization technique.[46] The cloned cDNAs were radiolabeled with ^{32}P by nick-translation.[47] The autoradiograms were scanned and integrated for quantitation. The expression of both the WAP and α-LA gene were hormone dependent (Table 2). Only a trace amount of both WAP and α-LA mRNA was detectable at the end of the initial 6 days of alveolar morphogenesis. A maximum increase in both the WAP and α-LA mRNAs accumulation occurred during the subsequent functional differentiation of the glands in the lactogenic medium containing insulin, prolactin, and hydrocortisone. However, a significant amount of the α-LA mRNA accumulation occurred also in a prolactin-free medium containing cortisol, indicating that the whey-protein gene is inducible in the absence of prolactin.

Accumulation of the milk-protein mRNA was also measured during exposure of the mammary glands to the endogenous hormones at different stages of pregnancy and

Table 2
CHANGES IN THE CONCENTRATION OF MILK-PROTEIN mRNAs DURING GROWTH AND DIFFERENTIATION OF THE MURINE MAMMARY GLAND

Stage of development	Ratio of mRNA concentration (normalized with respect to virgin)			
	α_1-Csn	β_2-Csn	WAP	α-LA
Virgin	1	1	1	1
5 Days pregnancy	2	1	1	1
11 Days pregnancy	4	2	1	1
18 Days pregnancy	38	22	5	1
Parturition	200	95	35	20
8 Days lactation	800	350	160	1

Note: The mRNA concentration per gram wet weight of murine mammary tissue for α_1-casein (α_1-Csn), β_2-casein (β_2-Csn), whey-acidic protein (WAP), and α-lactalbumin (α-LA) were calculated by multiplying the increase in RNA levels demonstrated by the dot-blot data by the rise in the yield of total RNA. The increase in RNA yield over that obtained from sexually mature virgin females was 1.6-, 2-, 2.4-, 8-, and 23-fold per gram of tissue, respectively, for mammary gland at 5, 11, 18 days pregnancy, parturition, and 8 days of lactation.

lactation (Table 2). Despite the extensive lobuloalveolar morphogenesis by mid-pregnancy, no significant accumulation of caseins, α-LA, and WAP mRNAs occurred until 11 days of gestation. Subsequently, the levels of the mRNAs for caseins and WAP rise until mid-lactation. The pattern of α-LA gene expression, on the other hand, was markedly different in that only a low level of the mRNA was present through the gestation period. However, around parturition, the concentration of the α-LA mRNA increased rapidly, reaching a maximum level and then decreasing to the prepartum level by 8 days of lactation.

The results described above indicate a noncoordinate expression of the milk-protein genes during mammary gland development. In different species, this characteristic appears to be particularly more evident in expression of the mouse α-LA gene.

Studies in explant cultures of mammary tissues have shown that in the presence of both prolactin and glucocorticoids in insulin-containing medium, casein mRNAs express in a coordinate manner. However, the WAP[4] and β-casein[45] in the rat, and the α-LA in the mouse, appear to be particularly more sensitive to the steroid hormone.

The availability of the milk-protein cDNA clones now have made it possible to isolate and characterize genomic milk-protein genes and identify whether distinct methylation sites are involved in the expression of γ-casein during lactation.[48] With the detailed structure of the milk-protein genes uncovered, it should become possible to determine the intricate molecular mechanisms of the hormonal regulation of mammary differentiation.

XI. SUMMARY AND COMMENTS

It is evident from the discussion in this chapter that significant advances have been accomplished in the understanding of the multiple hormonal regulation of the milk-protein gene associated with the expression of functional differentiation of the mammary cells. Of note is the finding that neither prolactin nor glucocorticoid, the two principal components of the lactogenic hormones, individually are capable of stimulating the genes encoding the major milk proteins, the caseins. The whey proteins, α-LA, and WAP genes also seem to exhibit a similar response to the individual components of the lactogenic hormones. Thus, it is now reasonable to conclude that initiation of lactogenesis is dependent on expression of the milk-protein genes which are regulated by a synergistic action of prolactin and glucocorticoid. The role of the glucocorticoid, which remained neglected during the past decades, has been elucidated. It is now established that glucocorticoid is an obligatory requirement for stimulation of expression of the milk-protein genes, acting at the transcriptional level of control. The elucidation of the important role of the glucocorticoid has been possible because of the development of the unique two-step culture model of the whole mammary organ of the mouse. The isolated mammary organ in vitro mimics the developmental stages of the parenchyma in the serum-free medium, appropriately supplemented with the combinations of the steroid and the polypeptide hormones. Studies in the mouse mammary gland in vitro also revealed that the stimulatory action of the glucocorticoid is correlated with interaction of the hormone receptor with the nuclear acceptor site in the mammary cells. The knowledge about the important role of the glucocorticoid obtained from studies with the organ culture model also prompted the development of another experimental model which combines the in vivo system with the explant culture system. This new approach permits measurement of responses of mammary explants from glucocorticoid-deprived virgin mice to exogenous hormones in the culture medium.[45] Based on the results of the studies using this modification of the explant culture model, the authors made the important observation that the expression of the rat casein genes are largely, if not entirely, dependent on the glucocorticoid hormone.[45] This finding in the rat is consistent with similar observations in the mouse mammary gland in culture. It is also important to mention that most of the findings discussed in this chapter were possible due to the prudent application of specific cDNA probes in molecular hybridization analysis, which provides a quantitative measurement of cellular accumulation of the respective mRNA or transcription of the mRNA in a cell-free nuclear RNA synthesis system. However, despite these advances, our present knowledge about the expression of the milk-protein genes is based mostly on the measurement of the cellular accumulation of the mRNAs. The question remains as to whether the synergistic action of prolactin and glucocorticoid is required for stimulation of the transcription of the different genes, or if the steroid hormone acts alone at the transcriptional level and the polypeptide hormone exerts its influence on some post-transcriptional event. At this time, there is no clue about a possible pathway of interaction of prolactin directly with the genes with resultant onset of its expression. Moreover, it is important to know whether glucocorticoids as a component of the lactogenic hormones directly interact with the mammary cell genome, possibly by binding of the receptor to the putative promoter regions of the genes.[49]

Although it appears that progesterone acts as an antagonist by interrupting the binding of the glucocorticoid to its receptor in mammary cells, the endocrinological significance of this phenomenon remains intriguing. The antagonistic influence of progesterone is likely to remain limited to the gestational period because, at parturition, circulating progesterone levels drop to a basal level.[50] Results of the studies on the expression of milk-protein genes at different stages of development of the mammary gland strongly suggest that the expression of the casein and whey-protein genes is noncoordinate, and these genes exhibit a differential sensitivity to the glucocorticoid. This provokes the question as to whether the site-specific DNA interaction of the hormone receptor is variable among the different genes.

Finally, evidence at this time indicates that prolactin, the classical lactogenic hormone, is essential for glucocorticoid stimulation of the milk-protein genes. However, to date, little is known about a biochemical pathway of action of this important polypeptide hormone which is required for both morphogenesis and expression of functional differentiation of the parenchyma. Numerous attempts to find a mechanism of action of prolactin, mediated through the cyclic nucleotide system, polyamine, and prostaglandins have failed to provide any clues[51,52] with respect to the action of the polypeptide hormone on gene expression in mammary cells, although cyclic nucleotide-mediated actions of prolactin have been observed in other tissues.

Cell differentiation presents a number of challenges. With advances of knowledge and technology, it is anticipated that elucidation of the problems enumerated above will be forthcoming in the near future.

ACKNOWLEDGMENTS

This work was supported by USPHS grants CA11058 and CA25304 from the National Cancer Institute and NIH and Nebraska Department of Health grants 84-054 and 85-03. The cloned cDNA used for mouse α-lactalbumin is a gift from Dr. Pradman Qasba, National Institutes of Health, Bethesda, Maryland. We thank Arvilla Kirchhoff for secretarial assistance.

REFERENCES

1. Nandi, S., Endocrine control of mouse mammary gland development and function in the C3H/H3Crgl mouse, *J. Natl. Cancer Inst.*, 21, 1039, 1958.
2. Lyons, W. R., Li, C. H., Cole, R. D., and Johnson, R. E., The hormonal control of mammary growth and lactation, *Recent Prog. Horm. Res.*, 14, 219, 1958.
3. Assairi, L., Delouis, C., Gaye, P., Houdebine, L. M., Olliver-Bousquet, M., and Denamur, R., Inhibition by progesterone of the lactogenic effect of prolactin in the pseudopregnant rabbit, *Biochem. J.*, 144, 245, 1974.
4. Craig, R. K. and Campbell, P., Molecular aspects of milk protein biosynthesis, in *Lactation, a Comprehensive Treatise*, Vol. 4, Larson, B. L., Ed., Academic Press, New York, 1978, 387.
5. Topper, Y. J. and Freeman, C. S., Multiple hormone interactions in the developmental biology of the mammary gland, *Physiol. Rev.*, 60, 1049, 1980.
6. Houdebine, L. M., Role of prolactin, glucocorticoids and progesterone in the control of casein gene expression, in *Hormones and Cell Regulation*, Vol. 4, Dumont, J. and Nunez, J., Eds., Elsevier/North-Holland, Amsterdam, 1980, 175.
7. Banerjee, M. R., Responses of mammary cells to hormones, *Int. Rev. Cytol.*, 4, 1, 1976.
8. Banerjee, M. R. and Antoniou, M., Steroid and polypeptide hormone interaction in milk-protein gene expression, in *Biochemical Actions of Hormones*, Vol. 12, Litwack, G., Ed., Academic Press, New York, 1985, 237.
9. Banerjee, M. R., Wood, B. G., Lin, F. K., and Crump, L. R., Organ culture of the whole mammary gland of the mouse, in *Tissue Cult. Assoc. Man.*, Vol. 2, Sanford, K. K., Ed., Tissue Culture Association, Rockville, Md., 1976, 457.
10. Banerjee, M. R. and Antoniou, M., Serum-free culture of the isolated whole mammary organ of the mouse: a model for the study of differentiation and carcinogenesis, in *Methods for Serum-Free Culture of Cells of the Endocrine System*, Vol. 2, Barnes, D. W., Sirbasku, D. A., and Sato, G. H., Eds., Alan R. Liss, New York, 1984, 143.
11. Ichinose, R. R. and Nandi, S., Lobuloalveolar differentiation in mouse mammary tissue *in vitro*, *Science*, 145, 496, 1964.
12. Terry, P. M., Ball, E. M., Ganguly, R., and Banerjee, M.R., An indirect radioimmunoassay for mouse casein using ^{125}I-labeled antigen, *J. Immunol. Methods*, 9, 123, 1975.
13. Wood, B. G., Washburn, L. L., Mukherjee, A.S., and Banerjee, M.R., Hormonal regulation of lobuloalveolar growth, functional differentiation and regression of whole mammary gland in organ culture, *J. Endocrinol.*, 65, 1, 1975.

14. Tonelli, Q. J. and Sorof, S., Epidermal growth factor requirement for development of cultured mammary glands, *Nature (London)*, 285, 250, 1980.
15. Lin, F. K., Banerjee, M. R., and Crump, L. R., Cell cycle-related hormone carcinogen interaction during chemical carcinogen induction of nodule-like mammary lesions in organ culture, *Cancer Res.*, 36, 1607, 1976.
16. Telang, N. T., Banerjee, M. R., Iyer, A. P., and Kundu, A. B., Neoplastic transformation of epithelial cells in whole mammary gland *in vitro*, *Proc. Natl. Acad. Sci. U.S.A.*, 76, 5886, 1979.
17. Iyer, A. P. and Banerjee, M. R., Sequential expression of preneoplastic and neoplastic characteristics of mouse mammary epithelial cells transformed in organ culture, *J. Natl. Cancer Inst.*, 66, 893, 1981.
18. Terry, P. M., Banerjee, M. R., and Lui, R. M., Hormone-inducible casein messenger RNA in a serum-free organ culture of whole mammary gland, *Proc. Natl. Acad. Sci. U.S.A.*, 74, 2441, 1977.
19. Banerjee, M. R., Terry, P. M., Sakai, S., and Lin, F. K., Regulation of mRNA and specific milk protein in mammary gland, in *Hormone Research Part III*, Norvell, N. and Shellenberger, T., Eds., Hemisphere, Washington, D.C., 1977, 281.
20. Banerjee, M. R., Terry, P. M., Sakai, S., Lin, F. K., and Ganguly, R., Hormone regulation of casein messenger RNA (mRNA), *In Vitro*, 14, 128, 1978.
21. Ganguly, R., Mehta, N. M., Ganguly, N., and Banerjee, M. R., Glucocorticoid modulation of casein gene transcription in mouse mammary gland, *Proc. Natl. Acad. Sci. U.S.A.*, 76, 6466, 1979.
22. Mehta, N. M., Ganguly, N., Ganguly, R., and Banerjee, M. R., Hormonal modulation of the casein gene expression in a mammogenesis-lactogenesis two-step culture model of whole mammary gland of the mouse, *J. Biol. Chem.*, 255, 4430, 1980.
23. Ganguly, R., Majumder, P. K., Ganguly, N., and Banerjee, M. R., The mechanisms of progesterone-glucocorticoid interaction in regulation of casein gene expression, *J. Biol. Chem.*, 257, 2182, 1982.
24. Ganguly, N., Ganguly, R., Mehta, N. M., Crump, L. R., and Banerjee, M. R., Simultaneous occurrence of pregnancy-like lobuloalveolar morphogenesis and casein gene expression in a culture of the whole mammary gland, *In Vitro*, 17, 55, 1981.
25. Ganguly, R., Ganguly, N., Mehta, N. M., and Banerjee, M. R., Absolute requirement of glucocorticoid for expression of the casein gene in presence of prolactin, *Proc. Natl. Acad. Sci. U.S.A.*, 77, 6003, 1980.
26. Bolander, F. F., Jr., Nicholas, K. R., and Topper, Y. J., Retention of glucocorticoid by isolated mammary tissue may complicate interpretation of results from *in vitro* experiments, *Biochem. Biophys. Res. Commun.*, 91, 247, 1979.
27. Rosen, J. M., Matusik, R., Richards, D. A., Gupta, P., and Rodgers, J. R., Multihormonal regulation of casein gene expression at the transcriptional and posttranscriptional levels in the mammary gland, *Recent Prog. Horm. Res.*, 36, 157, 1980.
28. Spindler, S. R., Mellon, S. H., and Baxter, J. D., Growth hormone gene transcription is regulated by thyroid and glucocorticoid hormones in culture rat pituitary tumor cells, *J. Biol. Chem.*, 257, 11627, 1982.
29. Yamamoto, K. R. and Alberts, B., Steroid receptors: elements for modulation of eukaryotic transcription, *Annu. Rev. Biochem.*, 38, 722, 1976.
30. Litwack, G., Modulator and the glucocorticoid receptor, *Trends Biochem. Sci.*, 4, 217, 1979.
31. Disorbo, D. M. and Litwack, G., Changes in the intercellular levels of pyridoxal 5′-phosphate affect the induction of tyrosineaminotransferase by glucocorticoid, *Biochem. Biophys. Res. Commun.*, 99, 1203, 1981.
32. Majumder, P. K., Joshi, J. B., and Banerjee, M. R., Correlation between nuclear glucocorticoid receptor levels and casein gene expression in murine mammary gland in vitro, *J. Biol. Chem.*, 258, 6793, 1983.
33. Terry, P. M., Lin, F. K., and Banerjee, M. R. Responses of mouse mammary gland casein mRNA to corticosteroid action and suckling, *Mol. Cell. Endocrinol.*, 9, 169, 1977.
34. Banerjee, M. R., Mehta, N. M., Ganguly, R., Majumder, P. K., Ganguly, N., and Joshi, J., Selective gene expression in an isolated whole mammary organ *in vitro*, in *Growth of Cells in a Hormonally Defined Medium*, Vol. 9, Sirbasku, D. A., Sato, G. H., and Pardee, A. B., Eds., Cold Spring Harbor Laboratory, New York, 1982, 789.
35. Ganguly, R. and Banerjee, M. R., RNA synthesis in isolated nuclei of lactating mammary cells in presence of unmodified and mercury-labeled CTP, *Nucl. Acids Res.*, 5, 4463, 1978.
36. Dale, R. M. K., Livingston, D. C., and Ward, D. C., The synthesis and enzymatic polymerization of nucleotides containing mercury: potential tools for nucleic acid sequencing and structural analysis, *Proc. Natl. Acad. Sci. U.S.A.*, 70, 2238, 1973.
37. Welsch, C. W., Squires, M. D., Casseby, E., Chen, C. L., and Meites, J., Median eminence lesions and serum prolactin: influence of ovariectomy and ergocornine, *Am. J. Physiol.*, 221, 1714, 1971.
38. Shyamala, G. and Dickson, C., Relationship between receptor and mammary tumor virus production after stimulation by glucocorticoid, *Nature (London)*, 262, 107, 1976.

39. Tucker, H. A., General endocrinological control of lactation, in *Lactation, a Comprehensive Treatise*, Vol. 1, Larson, B. L. and Smith, V. R., Eds., Academic Press, New York, 1974, 277.
40. Gala, R. R. and Westphal, U., Corticosteroid binding activity in serum of mouse, rabbit and guinea pig during pregnancy and lactation: possible involvement in the initiation of lactation, *Acta Endocrinol.*, 55, 47, 1965.
41. Shyamala, G., Glucocorticoid receptors: mode of action in normal and neoplastic mammary tissue, in *Hormonal Regulation of Mammary Tumors*, Vol. 1, Leung, B. S., Ed., Eden Press, Montreal, 1982, 245.
42. Rosen, J. M., O'Neal, D. L., McHugh, J. E., and Comstock, J. P., Progesterone mediated inhibition of casein mRNA and polysomal casein synthesis in the rat mammary gland during pregnancy, *Biochemistry*, 17, 290, 1978.
43. Burditt, L. J., Parker, D., Craig, R. K., Getova, T., and Campbell, P. N., Differential expression of α-lactalbumin and casein genes during the onset of lactation in the guinea-pig mammary gland, *Biochem. J.*, 194, 999, 1981.
44. Hobbs, A., Richards, D. A., Kessler, D. J., and Rosen, J. M., Complex hormonal regulation of rat casein gene expression, *J. Biol. Chem.*, 257, 3598, 1982.
45. Kulski, J. K., Topper, Y. J., Chomczynski, P., and Qasba, P., An essential role for glucocorticoid in casein gene expression in rat mammary explants, *Biochem. Biophys. Res. Commun.*, 114, 380, 1983.
46. Thomas, P., Hybridization of denatured RNA and small DNA fragments transferred to nitrocellulose, *Proc. Natl. Acad. Sci. U.S.A.*, 77, 5201, 1980.
47. Rigby, P. W. J., Dieckmann, M., Rhodes, C., and Berg, P., Labeling deoxyribonucleic acid to high specific activity *in vitro* by nick translation with DNA polymerase I, *J. Mol. Biol.*, 113, 237, 1977.
48. Johnson, M. L., Levy, J., Supowit, S. C., Yu-Lee, L. Y., and Rosen, J. M., Tissue and cell specific casein gene expression, *J. Biol. Chem.*, 258, 10805, 1983.
49. Pfahl, M., Specific binding of the glucocorticoid-receptor complex to the mouse mammary tumor proviral promoter region, *Cell*, 31, 475, 1982.
50. Tucker, H. A., Endocrinology of lactation, in *Seminars in Perinatology*, Vol. 3, Oliver, T. K. and Kirschbaum, T. H., Eds., Grune and Stratton, New York, 1979, 199.
51. Rilema, J. A., Mode of action of prolactin on normal and neoplastic mammary tissue, in *Hormonal Regulation of Mammary Tumors*, Vol. 2, Leung, B. S., Ed., Eden Press, Montreal, 1982, 77.
52. Oka, T., Perry, J. W., Takemoto, T., Sakai, T., Terada, N., and Inou, H., Polyamines in normal and neoplastic growth of mammary gland, in *Hormonal Regulation of Mammary Tumors*, Vol. 2, Leung, B. S., Ed., Eden Press, Montreal, 1982, 205.

Chapter 8

ROLE OF POLYAMINES IN PROLACTIN ACTIONS

Takami Oka and Michiyoshi Taga

TABLE OF CONTENTS

I. Introduction 138

II. The Mammary Gland System 138

III. Polyamines 139
- A. Polyamine Metabolism 140
- B. Inhibitors of Polyamine Biosynthesis 141

IV. Polyamine Accumulation in the Mammary Gland 141
- A. Studies In Vivo 141
- B. Circulating Polyamines and Mammary Transport System for Polyamines 143
- C. Studies in Organ Culture Systems 143

V. Hormonal Regulation of Polyamine Biosynthetic Enzymes 144
- A. Arginase 144
- B. Ornithine Decarboxylase 144
- C. *S*-Adenosylmethionine Decarboxylase 145
- D. Spermidine Synthase 145
- E. Overview 145

VI. Role of Polyamines in the Action of Lactogenic Hormones 146

VII. Concluding Remarks 148

References 148

I. INTRODUCTION

Prolactin has been shown to exert many different actions in various biological systems (see appropriate chapters in this volume). Although certain polypeptide hormones elicit their biological effects through stimulation of the cyclic AMP system,[1] prolactin belongs to the class of hormone whose action is not mimicked by the cyclic nucleotide. The current model of polypeptide hormone action suggests that the interaction between a polypeptide hormone and its target tissue is mediated by hormone-specific binding sites in the plasma membrane. The membrane-bound receptor molecules for prolactin have been found in a number of its target tissues.[2] This raises the question as to how the prolactin-receptor interaction transmits its signal to elicit intracellular metabolic processes. Since the involvement of intracellular second messengers is known to be required for the action of other polypeptide hormones, the possibility remains that the effect of prolactin involves some intracellular mediators. Over the past several years, the studies in this area have provided several possible candidates, which include the polyamines, cyclic GMP, calcium ions, and prostaglandins (see appropriate chapters in this volume). In this chapter, experimental findings relating to the role of the polyamines in the action of prolactin are reviewed with focus on studies carried out in the mammary gland system. However, comments will also be made on the experimental work with other biological systems.

II. THE MAMMARY GLAND SYSTEM

In order to study the mechanism of hormone action, it is necessary to have a model system in which the effect of hormone can be monitored by specific biochemical markers. In the case of prolactin, the mammary gland provides an excellent model, since mammary epithelial cells are dependent on prolactin to express their differentiated function to synthesize specific milk components, such as milk proteins, milk fats, and lactose.[3-5]

The development of the mammary gland leading to functional differentiation of mammary epithelium occurs following the onset of gestation. During pregnancy, estrogens stimulate growth and development of the mammary duct system, but only when progesterone is also present does full alveolar development take place. In hypophysectomized animals, however, these ovarian steroids fail to stimulate development. Studies involving endocrine ablation and replacement therapy indicated the role of prolactin, growth hormone, and adrenal corticoids in normal lobuloalveolar growth,[6,7] The normal development of the mammary gland during pregnancy involves largely a phase of epithelial cell proliferation, which is followed by functional differentiation at parturition. During the period of lactation, mammary epithelial cells synthesize and secrete milk proteins such as casein and α-lactalbumin, lactose, lipid, and other components of milk.[5] These specific cell products serve as biochemical indicators of the functional differentiation of mammary cells.

Experimentation in vivo allows only limited control of hormonal milieu. Organ cultures of mammary tissue in a chemically defined, serum-free medium have been used successfully to analyze the role of several hormones in the growth and differentiation of mammary epithelial cells.[3-5] These studies have shown that proliferation of mouse mammary epithelium is stimulated by insulin and/or epidermal growth factor.[5,8,9] In tissue explants prepared from virgin mice, the proliferative response is delayed unless the animals are pretreated with prolactin.[10] Since the circulating levels of prolactin (or placental lactogens) increase during pregnancy, it was postulated that this hormone sensitizes mammary epithelial cells to the mitogenic action of insulin.

The addition of glucocorticoid with insulin stimulates the formation of cytoplasmic structures such as rough endoplasmic reticulum that are characteristic of secretory cells.[11,12] The combination of insulin and glucocorticoid has also been shown to increase the prolactin receptors in mammary cells.[13] Thus, these hormones serve to prepare mammary epithelial cells to respond to the stimulation by prolactin. When mammary explants pretreated with insulin and glucocorticoid are exposed to prolactin, the epithelial cells express their differentiated function by producing milk components such as casein, α-lactalbumin, and lactose. These results indicate that insulin, glucocorticoid, and prolactin act in a prescribed sequence to induce the functional differentiation of the mouse mammary gland. This combination of hormones has been shown to be effective in stimulating lactogenic response of cultured mammary tissues from rats, hamsters, rabbits, and dogs, although the rabbit gland can be induced to produce milk components in the absence of glucocorticoid.[14]

These studies of the mammary gland culture system have provided important information concerning the hormonal regulation of lactogenesis. For example, progesterone,[15-17] cyclic AMP,[18-22] and epidermal growth factor[23] have been shown to act as negative regulators of lactogenesis. The concentrations of these agents are elevated during pregnancy when precocious production of milk protein is inhibited and decline after parturition when the synthesis of milk proteins increases.[15-23] Another interesting complexity relating to hormonal regulation of milk protein synthesis is a marked difference in glucocorticoid concentration required for the induction of α-lactalbumin and casein.[24-26] These studies have indicated that casein induction is progressively stimulated by cortisol up to a steroid concentration of 3 μM, whereas α-lactalbumin synthesis is enhanced by 30 nM cortisol and reduced at higher concentrations. Additional studies have shown that cortisol-induced inhibition of α-lactalbumin accumulation can be reversed by prostaglandins, although these agents have no effect on casein synthesis.[27,28] These results suggest that the action of cortisol is mediated by different mechanisms in terms of the induction of the two milk proteins.

In this section, only some features of the mammary gland system pertinent to the present topic have been presented. More extensive accounts of other important findings have been described in other chapters of this monograph as well as in recent reviews.[3-5]

III. POLYAMINES

The polyamines are aliphatic, nitrogenous, nonprotein substances which are found in all living cells and biological fluids. The major polyamines include putrescine, spermidine, and spermine (Figure 1). Because the primary and secondary amino groups of these compounds are positively charged at physiological pH, the polyamines can interact with negatively charged cellular constituents such as nucleic acids, proteins, and phospholipids. This electrostatic binding is presumed to be important for some of the effects that polyamines have in a variety of systems. The history of polyamine research dates back many decades; however, the biological and physiological importance of polyamines have only recently been recognized. Studies in various cell-free systems have shown that the polyamines stabilize membranes, interact with macromolecules such as DNA, RNA, protein, and phospholipid, and stimulate the synthesis of nucleic acids and protein, as well as the activities of many enzymes.[28] In mammalian systems, high concentrations of polyamines are found in tissues that are active in protein synthesis.[29-31] Cellular polyamine levels have also been shown to increase dramatically during the early phase of cell growth.[29-31] In addition, a number of hormones are found to stimulate polyamine biosynthesis in their target tissues.[29-31] These observations have

POLYAMINE	STRUCTURE
Putrescine	$NH_2CH_2CH_2CH_2CH_2NH_2$
Spermidine	$NH_2CH_2CH_2CH_2NHCH_2CH_2CH_2CH_2NH_2$
Spermine	$NH_2CH_2CH_2CH_2NH(CH_2)_4NHCH_2CH_2CH_2NH_2$

FIGURE 1. The chemical structures of the major, naturally occurring polyamines.

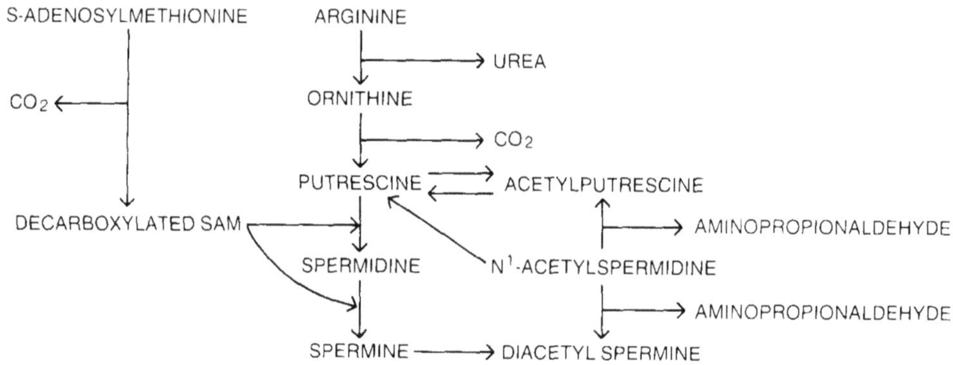

FIGURE 2. Biosynthetic pathway of polyamines in eukaryotic cells.

been explored further by the use of various inhibitors of polyamine biosynthesis to assess the role of the polyamines in cell growth and function. These studies, along with other finding of the effect of the polyamines in vivo and in vitro, suggest an involvement of polyamines in cell proliferation and protein synthesis.

A. Polyamine Metabolism

Polyamine biosynthesis in eukaryotic cells involves several precursors such as arginine, ornithine, methionine, and *S*-adenosylmethionine (Figure 2). Ornithine is mainly formed in the urea cycle through the cleavage of urea from arginine by arginase. Ornithine is then converted to putrescine in a reaction catalyzed by ornithine decarboxylase (ODC).[29-31] *S*-Adenosylmethionine is a universal methyl donor in transmethylation reactions. In polyamine biosynthesis, this compound is first decarboxylated by *S*-adenosylmethionine decarboxylase (*S*-ado-met DC).[29-31] *S*-Ado-met serves as the propylamine donor in the formation of spermidine from putrescine by spermidine synthase as well as in the formation of spermine from spermidine by spermine synthase.[29-31]

Putrescine, spermidine, and spermine can be converted to various metabolites by such enzymatic reactions as acetylation, oxidation, and conjugation in mammalian systems.[29-31] Some of these metabolites are shown in Figure 3. The biological importance of polyamine metabolites and their role in the regulation of polyamine biosynthesis remain to be elucidated.

In the biosynthesis of polyamines, ODC and *S*-ado-met DC are considered to be the key enzymes. These two enzymes exhibit extremely rapid rates of turnover: the half-life is estimated to be 10 to 20 min for ODC and 30 to 60 min for *S*-ado-met DC.[29-31] In contrast, spermidine and spermine synthases are much more stable, and their cellu-

POLYAMINE	METABOLITE
Putrescine	N-acetylputrescine
	γ-aminobutyrate
	2(3)hydroxy putrescine
	hypusine
Spermidine	N^1-acetyl spermidine
	N^8-acetyl spermidine
	Putreanine
	Isoputreanine
Spermine	Spermic acid

FIGURE 3. Some metabolites of polyamines in eukaryotic systems.

lar content is much higher than that of ODC and *S*-ado-met DC.[29-31] Thus, polyamine biosynthesis is mainly regulated through modulation of the activities of the two decarboxylases. Since the activity of *S*-ado-met DC in eukaryotic systems is dependent on the presence of putrescine,[32] the importance of ODC as the regulatory enzyme is further emphasized. The regulation of ODC is apparently complex, and thus the enzyme activity can be modulated by alterations in the rate of synthesis and turnover, conversion of active to inactive forms, and interactions with antizyme, a specific protein inhibitor of ODC.[28-31,33]

B. Inhibitors of Polyamine Biosynthesis

Since ODC and *S*-ado-met DC are the rate controlling enzymes of polyamine biosynthesis, these two enzymes have been the main targets for designing inhibitors of polyamine biosynthesis (Figure 4). Competitive inhibition of ODC can be achieved with various analogs of ornithine such as α-hydrazino-ornithine,[34] α-methylornithine,[35] and 1,4-diamino-2-butene.[36] In addition, α-difluoromethylornithine[37] is found to be a potent and irreversible inhibitor of ODC. *S*-Ado-met DC is competitively inhibited by methylglyoxal *bis*(guanyl-hydrazone) (MGBG),[38] whereas MBAG, 1,1′-(methylethanediylidenedinitrilo)-bis (3-aminoguanidine), inhibits the enzyme irreversibly.[39] These inhibitors can be used as a valuable means of assessing the role of polyamine in cells. Although some caution should be exercised in interpreting the results of experiments with such inhibitors, these studies have provided clear evidence for a causal link between polyamine function and cell growth and differentiation.[34-37]

IV. POLYAMINE ACCUMULATION IN THE MAMMARY GLAND

A. Studies In Vivo

The biosynthesis of polyamines has been shown to vary as a function of mammary

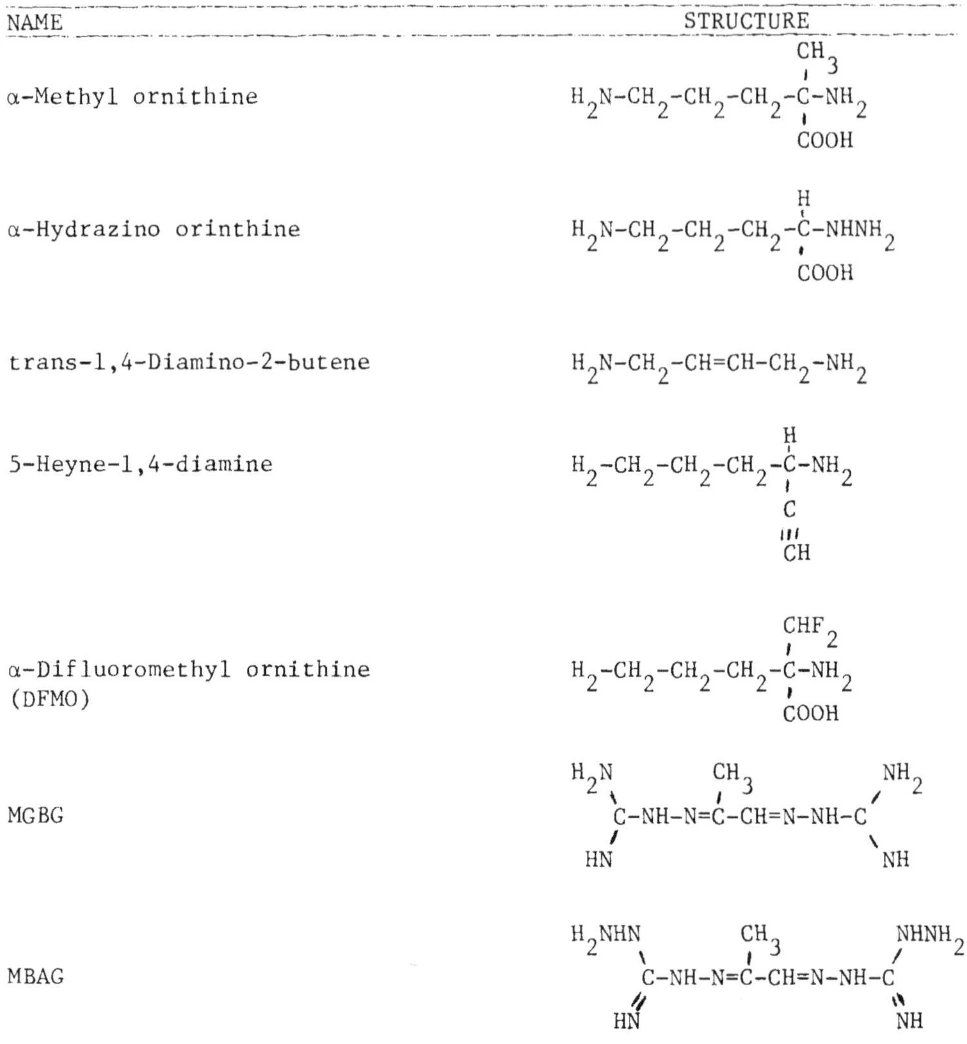

FIGURE 4. Some inhibitors of polyamine biosynthesis in eukaryotic systems.

gland development.[40,41] The levels of spermidine and putrescine are low in mammary tissue from virgin mice and rats, but during pregnancy, when rapid cell proliferation occurs, the concentrations of these polyamines begin to increase. During the period of lactation when the mammary cells have terminally differentiated to produce milk proteins, the polyamine concentrations approach a maximum. The relative increase in spermidine concentration is two- to threefold during pregnancy and six- to eightfold during lactation over those levels in the virgin stage. The concentration of putrescine increases five- and twofold during pregnancy and lactation, respectively. In contrast, there is little change in spermine levels throughout the development of the mammary gland.

The increases in the concentrations of spermidine and putrescine are accompanied by a rise in the activities of their associated enzymes, such as arginase, ODC, *S*-ado-met DC, and spermidine synthase.[40,41] In the lactating rat mammary gland,[42] insulin deficiency has been shown to cause a decrease in the activities of ODC and *S*-ado-met DC, which could be reversed by insulin treatment. Prolactin deprivation resulted in decreases in ODC activity and milk production. In addition, starvation caused a marked decrease in ODC and milk production and a smaller decrease in S-ado-met DC. These results suggest that polyamine biosynthesis may be important for lactogenesis, and its regulation involves the action of insulin and prolactin.

B. Circulating Polyamines and Mammary Transport System for Polyamines

The results presented above indicate that the concentrations of spermidine and putrescine in the mammary gland are elevated during pregnancy and lactation. During this period, however, the levels of circulating spermidine, the predominant polyamine in rat blood, also increase substantially.[43] There is a large increase in spermidine that occurs just prior to parturition, and a rapid fall just after birth, followed by another increase to the highest level at 3 days of lactation. The majority of the circulating polyamine is present in the cellular component of the blood, i.e., erythrocytes, whereas less than 10% is detected in the plasma as free spermidine. The blood concentration of spermine, as opposed to that of spermidine, does not increase during pregnancy or lactation. Although the origin, fate, and role of circulating spermidine during pregnancy and lactation are unknown, it is possible that the blood polyamine may be an important determinant of the cellular polyamine level in the mammary gland. In this regard, it is interesting to note that mammary epithelial cells have been shown to possess a transport system for polyamines, which can be stimulated by insulin and prolactin.[44] The stimulatory effect of prolactin involves enhancement of V_{max} for spermidine, whereas insulin increases the V_{max} for spermidine influx and decreases its efflux. In contrast to insulin, the stimulatory effect of prolactin is manifested after a considerable lag period, and is sensitive to actinomycin D. In view of the fact that prolactin stimulates the transport of the small cations such as Ca^{2+} and Na^{+} in lower invertebrates,[45] and spermidine is also a naturally occurring cation, it is possible that the action of prolactin on the polyamine transport in mammary cells may have some evolutionary counterpart in lower invertebrates.

C. Studies in Organ Culture Systems

The changes which occur during pregnancy and lactation in vivo can be induced in vitro by culturing mammary explants in a chemically defined medium containing appropriate combinations of insulin, prolactin, and cortisol. It has been shown that the concentration of spermidine in mammary epithelium increases by about threefold when mammary tissue explants from mid-pregnant mice are cultured in the presence of insulin, cortisol, and prolactin, the same combination of hormones which induces milk-protein synthesis in this system.[46] Significant increases in the spermidine concentration occur prior to the accelerated synthesis of milk proteins. The combination of insulin and prolactin, which results in a smaller stimulation of milk-protein synthesis, correspondingly elicits a smaller increase in the intracellular concentration of spermidine. The combination of insulin and cortisol caused a slight increase in the polyamine level, whereas insulin alone maintained the initial concentration. By contrast, the concentration of spermine increased only slightly, and no appreciable difference was found among various combinations of the three hormones. These data obtained by studies in vitro are consistent with the findings in vivo, and thus it may be concluded that enhancement of spermidine accumulation by insulin, cortisol, and prolactin in vitro reflects physiological changes in mammary epithelium during lactogenesis.
nesis.

The pattern of polyamine accumulation in cultured mouse mammary gland appears to be influenced by the stage of animals used. Thus, when mammary explants from virgin mice are cultured, insulin alone stimulates markedly the accumulation of spermidine and putrescine prior to the augmentation of DNA synthesis.[47] Studies using several inhibitors of polyamine biosynthesis and various culture conditions suggest that both putrescine and spermidine are important for insulin-stimulated DNA synthesis in mammary explants.[47,48] However, the requirement of the polyamines for DNA synthesis is only apparent in the case of mammary cells of virgin mice, which contain low levels of the polyamines. The mammary cells of mid-pregnant mice, however, are already in a proliferative phase and contain sufficient levels of the polyamines for DNA synthesis. Accordingly, the requirement for putrescine may not be apparent in these cases when the cellular levels of the polyamines are initially high, even though the biosynthesis of the polyamines is blocked by inhibitors. Such considerations may be important in evaluating the role of polyamines in cell growth and differentiation by use of inhibitors of polyamine biosynthesis.

V. HORMONAL REGULATION OF POLYAMINE BIOSYNTHETIC ENZYMES

A. Arginase

Arginase catalyzes the formation of ornithine, and thereby participates in the biosynthesis of polyamine. The activity of arginase in the mammary gland increases markedly during lactation.[49] The hormonal regulation of arginase activity was examined in mammary organ culture systems. It has been shown that prolactin in combination with insulin increases arginase activity by severalfold.[50,51] Stimulation of arginase activity by prolactin is not a consequence of the increase in the number of mammary epithelial cells, since the effect of the hormone can be observed in postmitotic cells.[50,51] The stimulatory effect of prolactin was only apparent after a 2 to 3-day exposure to this hormone, when total arginase activity was assayed.[51] However, mammary cells contain two forms of arginase, a mitochondrial enzyme and a cytosolic enzyme, which differ in terms of size, heat stability, and Mn^{++} dependence.[52] When the effect of prolactin was examined on the activities of the two arginases, it became apparent that prolactin rapidly stimulates the activity of soluble arginase, whereas the activity of mitochondrial enzyme was unaffected.[41] Time course studies revealed a close temporal relationship between increases in the activity of soluble arginase and the biosynthesis of spermidine from arginine.[41] These results suggest that it is the cytosolic form of arginase which participates in polyamine biosynthesis.

B. Ornithine Decarboxylase

Studies using mammary explant cultures have shown that mammary ODC activity increases in response to a variety of stimuli. In many instances, the pattern of increase in ODC is characterized by a rapid, large rise and a rapid fall approaching the initial levels. In some cases, however, a biphasic increase has been observed.[53]

In cultured mammary tissue, the activity of ODC increases in a biphasic manner with the first hormone-independent peak occurring at 3 to 4 hr of culture, and the second hormone-dependent one at later times.[47,54] The appearance of the first peak has been shown to be influenced by the osmolarity in the culture medium,[55] whereas the second peak of activity is dependent on the action of both insulin and prolactin.[47,54,56] Several derivatives of cyclic AMP as well as prostaglandin E_1 also enhance the first peak of ODC activity.[54] Studies with actinomycin D and cycloheximide suggest that the first peak of ODC activity may be regulated at a post-transcriptional level, whereas a second increase may be at both transcriptional and translational levels.

Addition of prolactin stimulates ODC activity in the daughter cells which have been formed in the presence of insulin and cortisol.[51,54] The hormone-stimulated increase occurred rapidly, reaching the maximum at 2 to 4 hr, and declined to the basal level by

24 hr after the addition of prolactin.[51,54] This double incubation system has been useful to study the action of prolactin.[51,54,57-59] In this system, both 3-isobutyl-1-methylxanthine, an inhibitor of cyclic nucleotide phosphodiesterase,[60] and dibutyryl cAMP have been shown to stimulate ODC activity.[61] It is interesting to note that the optimal prolactin and dibutyryl cAMP responses require different concentrations of cortisol, which are greater than $10^{-7}M$ and less than $10^{-7}M$, respectively.[61] These studies have led to the proposal that prolactin and the cyclic nucleotide stimulate ODC via the same mechanism.[61] The mechanism by which prolactin stimulates the activity of ODC in mammary cells is unknown. Recent experiments with various prostaglandins indicate that PGF_1, E_2, and I_2 elicit a concentration-dependent stimulation of ODC activity.[62] Moreover, the prolactin action on ODC activity has been shown to be attenuated by indomethacin, an inhibitor of prostaglandin biosynthesis,[62] or by quinacrine, an inhibitor of phospholipases A_2 and C,[63] whereas these two enzymes enhance ODC activity in cultured mammary explants. These results are compatible with the view that the stimulatory action of prolactin may involve a stimulation of phospholipases A_2 activity, followed by an enhanced synthesis of prostaglandins.[63]

The stimulatory effect of prolactin on ODC activity has been reported in other biological systems such as liver[64] and human peripheral blood lymphocytes.[65] Richards has shown that prolactin stimulates ODC activity in different target tissues, and the magnitude of response in these cells appears to correlate well with the number of receptor sites.[66] The ability of prolactin to stimulate ODC in a variety of tissues in hypophysectomized rats has been recently shown to be blocked by cyclosporine.[67] This effect of cyclosporine appears to be the results of its ability to block the binding of prolactin to its receptors.[66] Studies of prolactin action in cultured rat mammary tumors indicate that prolactin increases ODC activity, and cyclic AMP and/or cyclic GMP may augment the prolactin-induced response by increasing the sensitivity of cells to the hormone.[68]

C. *S*-Adenosylmethionine Decarboxylase

The combination of insulin and glucocorticoid stimulates the activity of *S*-ado-met DC in cultured mouse mammary gland.[46] The increase in enzyme activity occurs in parallel with the rise in the concentration of spermidine in cultured tissue.[46] Studies using a monospecific antibody to the purified enzyme indicate that the increase in *S*-ado-met DC is primarily due to the enhanced level of enzyme protein rather than modulation of the activity of the enzyme.[69,70] *S*-Ado-met DC appears to be the rate-limiting enzyme in the biosynthesis of spermidine in cultured mammary tissue, as judged by the results of experiments with MGBG.[46,51] The activity of *S*-ado-met DC in the gland is not affected by prolactin.[46]

D. Spermidine Synthase

The activity of spermidine synthase in mouse mammary epithelium is stimulated by the action of both insulin and cortisol.[71] Since the hormonal stimulation of *S*-ado-met DC activity can be blocked by actinomycin D or cycloheximide, the increase in enzyme activity may require both new RNA and protein synthesis.[70] Additional studies with cycloheximide suggest that this enzyme has a half-life of at least 12 hr.[71]

E. Overview

The data presented above indicate that the synergistic actions of insulin, glucocorticoid, and prolactin stimulate the activities of enzymes involved in the biosynthesis of spermidine (Figure 5). It is clear that the activities of the first two enzymes in spermidine formation are regulated by prolactin and insulin, whereas the last two enzymes

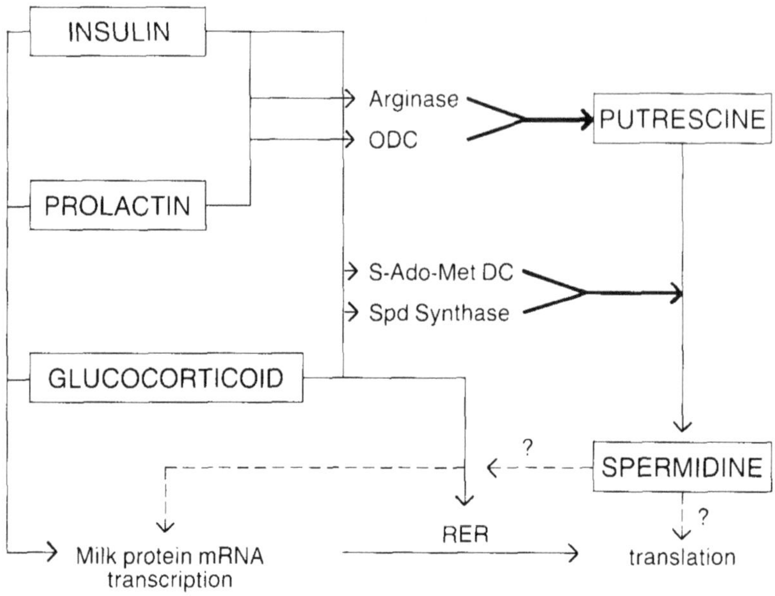

FIGURE 5. The interactions of lactogenic hormones on spermidine biosynthesis in the mouse mammary gland.

are under the control of glucocorticoid and insulin. Moreover, the increase of the four enzymes is asynchronous and differs in magnitude. These results obtained by experiments in vitro are largely in accord with those from studies in vivo.

VI. ROLE OF POLYAMINES IN THE ACTION OF LACTOGENIC HORMONES

Several inhibitors of polyamine biosynthesis have been employed to assess the importance of polyamines in the hormonal induction of lactogenesis. Studies with MGBG, a potent inhibitor of S-ado-met DC, indicate that the addition of the drug in combination with insulin, cortisol, and prolactin, inhibits increases in spermidine accumulation and synthesis of milk proteins, casein, and α-lactalbumin in cultured mouse mammary explants.[46,57] More recently, MGBG has been also shown to abolish the prolactin-stimulated lipid biosynthesis.[72] However, MGBG did not block the synthesis of nonmilk protein,[46] as well as the prolactin stimulation of total RNA synthesis.[58] In addition, the inhibitory effect of MGBG on milk-protein synthesis was found to be specifically reversed by exogenous spermidine. Similar results have been obtained with another inhibitor of polyamine biosynthesis, α-hydrazino-δ-aminovaleric acid, which is a potent inhibitor of ODC.[73] These results suggest that spermidine may be necessary for the hormonal stimulation of lactogenesis. On the other hand, recent studies have shown that α-difluoromethyl ornithine, an irreversible inhibitor of ODC, almost completely inhibits the prolactin-stimulated ODC activity in mammary explants primed with insulin and cortisol, but failed to suppress the effect of prolactin on RNA, casein, or lipid synthesis.[73] These results argue that the stimulation of ODC by prolactin is not necessary for the subsequent lactogenic responses of mammary cells to the hormone, when mammary tissues are pre-exposed to insulin and cortisol.

Several lines of evidence indicate that under certain conditions spermidine can mimic some of the effects of lactogenic hormones, glucocorticoid, and prolactin in organ culture of the mouse mammary gland. Thus, the addition of spermidine in place of cortisol to a medium containing insulin and prolactin stimulates milk-protein synthesis.[46] The increase in α-lactalbumin is similar to that elicited by the combination of insulin, cortisol, and prolactin, whereas the increase in casein production is about half that seen in the triple hormone system. The effect of spermidine was observed at a concentration as low as 0.1 m*M*, which is in the physiological range. The differential stimulation of synthesis of α-lactalbumin and casein by spermidine may be related to the different effects of cortisol concentrations on the production of the two milk proteins.[24-26] The ability of spermidine to mimic the action of glucocorticoid has also been shown by the studies on the synthesis and secretion of plasminogen activator in explants of lactating mammary gland,[74] and the accumulation of glucose-6-phosphate dehydrogenase activity,[75] and the induction of choline kinase activity in cultured mammary explants.[76] In all cases, spermidine appears to act by mimicking the effect of glucocorticoid which augments the action of prolactin and insulin. These results are consistent with the view that glucocorticoid, in combination with insulin and prolactin, facilitates the accumulation of spermidine via stimulation of *S*-ado-met DC, and the polyamine, in turn, mediates the action of glucocorticoid (see Figure 5).[46]

On the other hand, studies using mouse mammary explants primed with insulin and glucocorticoid have shown that spermidine in combination with prostaglandin(s) can mimic the action of prolactin on casein synthesis, whereas spermidine alone or prostaglandin(s) by itself is ineffective.[77] Moreover, the combination of spermidine with other agents such as cyclic GMP, arachidonic acid, or phospholipase A_2 that stimulates RNA synthesis has been shown to enhance the rate of casein synthesis, though the extent of these effects is not quite comparable to that of prolactin.[51] These results have led to the hypothesis that any or all of these agents, including spermidine, may participate in the prolactin-dependent lactogenic processes.[78]

The role of spermidine in milk-protein synthesis by mammary explants appears to be different among various species. In contrast to the mouse system, rabbit mammary explants require only insulin and prolactin for milk-protein synthesis, and both spermidine and diaminohexane, an inhibitor of spermidine synthesis, did not affect milk-protein synthesis in this system.[79] Casein synthesis in cultured rat mammary tissue has been shown to depend on both glucocorticoid and spermidine, but spermidine cannot substitute for glucocorticoid.[80] Studies of casein mRNA accumulation in rat mammary organ culture have shown that MGBG blocks the prolactin stimulation of casein mRNA, but the effect of the drug is not reversed by the addition of spermidine.[81] In this system, however, the addition of spermidine with insulin is able to maintain the level of casein mRNA at a level comparable to that found in the presence of insulin and cortisol.[79] At present, the reason for some of the observed species differences is not known.

Studies with the mouse organ culture system suggest that spermidine is necessary for the hormonal stimulation of lactogenesis. However, the molecular mechanism of spermidine action remains to be elucidated. Recently it has been found that spermidine enhances the efficiency of the translation of the mRNAs for casein in a cell-free system derived from wheat germ.[82] The stimulatory effect of spermidine involves the rates of both initiation and elongation of polypeptide synthesis. In the presence of an appropriate concentration of K^+, spermidine also enhances the relative amount of casein polypeptide synthesized in this system. It would be of interest to assess the physiological relevance of these observations.

More recent studies have shown the presence of a cyclic AMP-independent protein kinase in the mouse mammary gland whose activity can be stimulated by the polyamines.[83] The polyamine-responsive protein kinase activity in the mammary tissue increases during pregnancy and lactation. In an organ culture system, optimal enzyme activity largely depends on the presence of prolactin. Moreover, dibutyryl cAMP inhibits this enzyme in cultured mammary gland. The enzyme utilizes casein as substrate, which is synthesized only in the mammary tissue. It is possible that this enzyme plays a role in the process of lactogenesis and may be physiologically regulated by the polyamines.

VII. CONCLUDING REMARKS

The involvement of the polyamine spermidine in the hormonal regulation of lactogenic processes in the mammary gland has been summarized in this chapter. Several lines of evidence suggest that spermidine plays an important role in the synthesis of milk components such as milk protein and milk fat. Intracellular spermidine concentrations are highly elevated in mammary glands of lactating animals as well as in tissues that are hormonally induced to differentiate in vitro. All of the polyamine biosynthetic enzymes including arginase, ornithine decarboxylase, *S*-adenosylmethionine decarboxylase, and spermidine synthase are stimulated by one or more of the lactogenic hormones, insulin, cortisol, and prolactin in cultured tissues. Inhibitors of polyamine biosynthesis inhibit the hormonal stimulation of milk-protein synthesis, but their effects can be overcome by spermidine. At the present time, however, both the precise mode of spermidine action and its molecular mechanisms remain unclear. Further progress in this area will be greatly enhanced by experimentations using genetic, molecular biological, and cell biological techniques. In addition, it remains to be determined whether the polyamine is involved in the actions of prolactin on other biological systems. It is hoped that future work provides some answers to these questions.

REFERENCES

1. Robinson, G. A., Dutcher, R. W., and Sutherland, E. W., *Cyclic AMP*, Academic Press, New York, 1971.
2. Kelly, P. A., Ferland, L., and Labrie, F., Endocrine control of prolactin receptors, in *Progress in Prolactin Physiology and Pathology*, Robyn, C. and Harter, M., Eds., Elsevier/North-Holland, Amsterdam, 1978, 59.
3. Banerjee, M. R., Responses of mammary cells to hormones, *Int. Rev. Cytol.*, 47, 1, 1976.
4. Rosen, J. M., Matasik, R. J., Richards, D. A., Gupta, P., and Rogers, J. R., Multihormonal regulation of casein gene expression at the transcriptional and post-transcriptional levels in the mammary gland, *Rec. Prog. Horm. Res.*, 36, 157, 1980.
5. Topper, Y. J. and Freeman, C. S., Multiple hormone interactions in the developmental biology of the mammary gland, *Physiol. Rev.*, 60, 1049, 1980.
6. Lyons, W. R., Li, C. H., and Johnson, R. E., The hormonal control of mammary growth and lactation, *Res. Prog. Horm. Res.*, 14, 219, 1958.
7. Nandi, S., Endocrine control of mammary gland development and function in the C3H/He Crgl mouse, *J. Natl. Cancer Inst.*, 21, 1039, 1958.
8. Turkington, R. W., The role of epidermal growth factor in mammary gland development in vitro, *Exp. Cell Res.*, 57, 79, 1969.
9. Taketani, Y. and Oka, T., Epidermal growth factor stimulates cell proliferation and inhibits functional differentiation of mouse mammary epithelial cells in culture, *Endocrinology*, 113, 871, 1983.
10. Oka, T. and Topper, Y. J., Is prolactin mitogenic for mammary epithelium?, *Proc. Natl. Acad. Sci. U.S.A.*, 69, 1693, 1972.
11. Mills, E. S. and Topper, Y. J., Some ultrastructural effect of insulin, hydrocortisone and prolactin on mammary gland explants, *J. Cell Biol.*, 44, 310, 1970.
12. Oka, T. and Topper, Y. J., Hormone-dependent accumulation of rough endoplasmic reticulum in mouse mammary epithelial cells in vitro, *J. Biol. Chem.*, 246, 7701, 1971.
13. Sakai, S., Enami, J., Nandi, S., and Banerjee, M. R., Prolactin receptor on dissociated mammary epithelial cells at different stages of development, *Mol. Cell. Endocrinol.*, 12, 285, 1978.

14. Forsyth, I. A., The endocrinology of lactation, in *Biochemistry of Lactation,* Mephem, T., Ed., Elsevier/North-Holland, Amsterdam, 1983, 309.
15. Rosen, J. M., O'Neal, D. L., McHugh, J. E., and Comstock, J. P., Progesterone mediated inhibition of casein mRNA and polysomal casein synthesis in the rat gland during pregnancy, *Biochemistry,* 17, 290, 1978.
16. Ganguly, R., Majumder, P. K., Ganguly, N., and Banerjee, M. R., The mechanism of progesterone-glucocorticoid interaction in regulation of casein gene expression, *J. Biol. Chem.,* 257, 2182, 1982.
17. Kuhn, N. J., The biosynthesis of lactose, in *Biochemistry of Lactation,* Mephem, T. B., Ed., Elsevier/North-Holland, Amsterdam, 1983, 159.
18. Sapag-Hagar, M. and Greenbaum, A. L., Adenosine 3',5'-monophosphate and hormone interrelationships in the mammary gland of the rat during pregnancy and lactation, *Eur. J. Biochem.,* 47, 303, 1974.
19. Rillema, J. A., Cyclic nucleotides and the effect of prolactin on uridine incorporation in RNA in mammary gland explants of mice, *Horm. Metab. Res.,* 7, 45, 1975.
20. Loizzi, R. F., dePont, J. J. H. H. M., and Bonting, S. L., Inhibition by cyclic AMP of lactose production in lactating guinea pig mammary gland slices, *Biochim. Biophys. Acta,* 392, 20, 1975.
21. Speake, B. K., Dils, R., and Mayer, R. J., Regulation of enzyme turnover during tissue differentiation: studies on the effect of hormones on the turnover of fatty acid synthetase in rabbit mammary gland in organ culture, *Biochem. J.,* 148, 309, 1975.
22. Perry, J. W. and Oka, T., Cyclic AMP as a negative regulator of hormonally induced lactogenesis in mouse mammary gland organ culture, *Proc. Natl. Acad. Sci. U.S.A.,* 77, 2093, 1980.
23. Teketani, Y. and Oka, T., The biological action of epidermal growth factor and its functional receptors in normal mammary epithelial cells, *Proc. Natl. Acad. Sci. U.S.A.,* 80, 2647, 1983.
24. Ono, M. and Oka, T., α-Lactalbumin-casein induction in virgin mouse mammary explants: dose-dependent differential action of cortisol, *Science,* 207, 1367, 1980.
25. Ono, M. and Oka, T., The differential action of cortisol on the accumulation of α-lactalbumin and casein in midpregnant mouse mammary gland in culture, *Cell,* 19, 473, 1980.
26. Nagamatsu, Y. and Oka, T., Differential action of cortisol on the synthesis, turnover of α-lactalbumin and casein and their mRNA accumulation in cultured mammary gland from midpregnant mice, *Biochem. J.,* 212, 509, 1983.
27. Terada, N., Ono, M., Nagamatsu, Y., and Oka, T., The reversal of cortisol-induced inhibition of α-lactalbumin by prostaglandins in the mouse mammary gland in culture, *J. Biol. Chem.,* 257, 11199, 1982.
28. Terada, N., Leiderman, L. J., and Oka, T., Interaction of cortisol and prostaglandin E_2 on the expression of α-lactalbumin gene, *Biochim. Biophys. Res. Commun.,* 80, 2647, 1983.
29. Tabor, C. W. and Tabor, H., 1,4-Diaminobutane (putrescine), spermidine, and spermine, *Annu. Rev. Biochem.,* 45, 285, 1976.
30. Williams-Ashman, H. G. and Canellakis, Z., Polyamines in mammalian biology and medicine, *Perspect. Biol. Med.,* 22, 44, 1979.
31. Pegg, A. E. and McCann, P. C., Polyamine metabolism and function, *Am. J. Physiol.,* 243, C212, 1982.
32. Pegg, A. E., Purification of rat liver S-adenosyl-L-methionine decarboxylase, *Biochem. J.,* 141, 581, 1974.
33. Canellakis, E. S., Kyriakidis, D. A., and Pawlak, J. W., The complexity of regulation of ornithine decarboxylase, *Med. Biol.,* 59, 279, 1981.
34. Inoue, H., Kato, Y., Takigawa, M., Adachi, K., and Takeda, Y., Effect of DL-α-hydrozino-S-aminovaleric acid, an inhibitor of ornithine decarboxylase, on polyamine metabolism in isoproterenol-stimulated mouse parotid glands, *J. Biochem.,* 77, 879, 1975.
35. Mamont, P. S., Bohlen, D., McCann, P. P., Bey, P., Schuber, F., and Tardif, C., α-Methyl ornithine, a potent competitive inhibitor of ornithine decarboxylase, blocks proliferation of rat hepatoma cells in culture, *Proc. Natl. Acad. Sci. U.S.A.,* 73, 1626, 1976.
36. Heby, O. and Janne, J., Polyamine antimetabolites: biochemistry, specificity and biological effects of inhibitors of polyamine synthesis, in *Polyamines in Biology and Medicine,* Morris, D. R. and Marton, L. J., Eds., Marcel Dekker, New York, 1981, 243.
37. Metcalf, B. W., Bey, P., Danzin, C., Jung, M. J., and Vevert, J. P., Catalytic irreversible inhibition of mammalian ornithine decarboxylase by substrate and product analogs, *J. Am. Chem. Soc.,* 100, 2551, 1978.
38. Williams-Ashman, H. G. and Schenone, A., Methyl-glyoxal bis (guanylhydrazone) as a potent inhibitor of mammalian and yeast S-adenosyl-methionine decarboxylases, *Biochem. Biophys. Res. Commun.,* 46, 288, 1972.
39. Pegg, A. E., Inhibition of mammalian S-adenosyl-methionine decarboxylase activity by 1,1'-[(methylethane-diylidene)-dinitrilo]bis(3-aminoguanidine), *J. Biol.Chem.,* 253, 539, 1978.

40. Russell, D. H. and McVicker, T. A., Polyamine biogenesis in the rat mammary gland during pregnancy and lactation, *Biochem. J.*, 130, 71, 1972.
41. Oka, T., Sakai, T., Lundgren, D. W., and Perry, J. W., Polyamines in growth and development of mammary gland, in *Hormones, Receptors, and Breast Cancer*, McGuire, W. L., Ed., Raven Press, New York, 1978, 301.
42. Brosnan, M. E., Ilic, V., and Willianson, D. H., Regulation of the activity of ornithine decarboxylase and S-adenosyl-methione decarboxylase in mammary gland and liver of lactating rats, *Biochem. J.*, 202, 693, 1982.
43. Lundgren, D. W. and Oka, T., Alterations in polyamine levels in rat blood during pregnancy and lactation, *Am. J. Physiol.*, 234, E451, 1978.
44. Kano, K. and Oka, T., Polyamine transport and metabolism in mouse mammary gland, *J. Biol. Chem.*, 251, 2795, 1976.
45. Bern, H. A. and Nicoll, C. S., The comparative endocrinology of prolactin, *Recent Prog. Horm. Res.*, 24, 681, 1968.
46. Oka, T. and Perry, J. W., Spermidine as a possible mediator of glucocorticoid effect on milk protein synthesis in mouse mammary epithelium in vitro, *J. Biol. Chem.*, 249, 7647, 1974.
47. Sakai, T., Lundgren, D. W., and Oka, T., Polyamine biosynthesis and DNA synthesis in cultured mammary gland explants from virgin mice, *J. Cell. Physiol.*, 95, 259, 1978.
48. Inoue, H. and Oka, T., The effect of inhibitors of ornithine decarboxylase on DNA synthesis in mouse mammary gland, *J. Biol. Chem.*, 255, 3308, 1980.
49. Yip, M. C. M. and Knox, W. E., Function of arginase in lactating mammary gland, *Biochem. J.*, 127, 893, 1972.
50. Oka, T. and Perry, J. W., Arginase effects lactogenesis through its influence on the biosynthesis of spermidine, *Nature (London)*, 250, 660, 1974.
51. Rillema, J. A., Linebaugh, B. E., and Mulder, J. A., Regulation of casein synthesis by polyamines in mammary gland explants of mice, *Endocrinology*, 100, 529, 1977.
52. Glass, R. D. and Knox, W. E., Arginase isozymes of rat mammary gland, liver and other tissues, *J. Biol. Chem.*, 248, 5785, 1973.
53. McCann, P. P., Tardif, C., Mamont, P. S., and Schuber, F., Biphasic induction of ornithine decarboxylase and putrescine levels in growing HTC cells, *Biochem. Biophys. Res. Commun.*, 64, 336, 1975.
54. Oka, T. and Perry, J. W., Studies on the regulatory factors of ornithine decarboxylase activity during development of mouse mammary epithelium in vitro, *J. Biol. Chem.*, 251, 1738, 1976.
55. Perry, J. W. and Oka, T., Regulation of ornithine decarboxylase in cultured mammary gland by the osmolarity in the cellular environment, *Biochim. Biophys. Acta*, 629, 24, 1980.
56. Aisbitt, R. P. G. and Barry, J. M., Stimulation by insulin of ornithine decarboxylase activity in cultured mammary tissue, *Biochim. Biophys. Acta*, 320, 610, 1973.
57. Turkington, R. W., Lockwood, D. H., and Topper, Y. J., The induction of milk protein synthesis in post-mitotic mammary epithelial cells exposed to prolactin, *Biochim. Biophys. Acta*, 148, 475, 1967.
58. Rillema, J. A., Action of prolactin on ornithine decarboxylase activity in mammary gland explants of mice, *Endocrine Res. Commun.*, 3, 297, 1976.
59. Rillema, J. A., Wing, L. Y. C., and Camerone, C. M., Effect of various concentrations of prolactin and growth hormone on the magnitude of stimulation of RNA synthesis, casein synthesis, and ornithine decarboxylase activity in mouse mammary gland explants, *Horm. Res.*, 15, 133, 1981.
60. Rillema, J. A. and Wing, L. Y. C., Effect of 3-isobutyl-1-methylxanthine on prolactin actions on RNA synthesis, casein synthesis, lipid synthesis and ornithine decarboxylase activity in mouse mammary gland explants, *Horm. Metab. Res.*, 14, 195, 1982.
61. Wing, L. Y. C. and Rillema, J. A., Effects of cyclic nucleotides on ornithine decarboxylase activity in mammary gland explants from mid-pregnant mice, *Biochim. Biophys. Acta*, 756, 266, 1983.
62. Wing, L. Y. C. and Rillema, J. A., Prostaglandin stimulation of ornithine decarboxylase activity in mammary gland explants from mid-pregnant mice, *Prostaglandins*, 25, 321, 1983.
63. Rillema, J. A., Wing, L. Y. C., and Foley, K. A., Effects of phospholipases on ornithine decarboxylase activity in mammary gland explants from midpregnant mice, *Endocrinology*, 113, 2024, 1983.
64. Hurley, T. W., Kuhn, C. M., Schanberg, S. M., and Handwerger, S., Differential effects of placental lactogen, growth hormone and prolactin on rat liver ornithine decarboxylase activity in the perinatal period, *Life Sci.*, 27, 2269, 1980.
65. Russell, D. H., Matrisian, L., Kibler, R., Larson, D. F., Poulos, B., and Magun, B. E., Prolactin receptors on human lymphocytes and their modulation by cyclosporine, *Biochim. Biophys. Res. Commun.*, 121, 899, 1984.
66. Richards, J. F., Ornithine decarboxylase activity in tissues of prolactin-treated rats, *Biochem. Biophys. Res. Commun.*, 63, 292, 1975.
67. Russell, D. H., Larson, D. F., Cardon, S. B., and Copeland, J. G., Cyclosporin inhibits prolactin induction of ornithine decarboxylase in rat tissues, *Mol. Cell Endocrinol.*, 35, 159, 1984.

68. Frazier, R. P. and Costlow, M. E., Prolactin stimulation of ornithine decarboxylase activity in cultured rat mammary tumors, *Exp. Cell. Res.*, 138, 39, 1982.
69. Sakai, T., Hori, C., Kano, K., and Oka, T., Purification and characterization of S-adenosyl-L-methionine decarboxylase from mouse mammary gland and liver, *Biochemistry*, 18, 5541, 1970.
70. Sakai, T., Perry, J. W., Hori, C., and Oka, T., Putrescine and the regulation of S-adenosyl-L-methionine decarboxylase in cultured mouse mammary gland, *Biochim. Biophys. Acta*, 614, 577, 1980.
71. Oka, T., Perry, J. W., and Kano, K., Hormonal regulation of spermidine synthase during the development of mouse mammary epithelium in vitro, *Biochem. Biophys. Res. Commun.*, 79, 979, 1977.
72. Rillema, J. A. and Cameron, C. M., Early action of prolactin on ornithine decarboxylase activity is not essential for the subsequent actions of prolactin on casein and lipid biosynthesis, *Proc. Soc. Exp. Biol. Med.*, 174, 28, 1983.
73. Oka, T., Perry, J. W., and Terada, N., The regulatory function of spermidine in hormonal control of the development of mouse mammary gland in culture, *Fed. Proc.*, 41, 3073, 1982.
74. Ossowski, L., Biegel, D., and Reich, E., Mammary plasminogen activator: correlation with involution, hormonal modulation and comparison between normal and neoplastic tissue, *Cell*, 16, 929, 1979.
75. Oka, T. and Perry, J. W., Studies on the function of glucocorticoid in mouse mammary epithelial cell differentiation in vitro, *J. Biol. Chem.*, 249, 3586, 1974.
76. Oka, T. and Perry, J. W., Glucocorticoid stimulation of choline kinase activity during the development of mouse mammary gland, *Dev. Biol.*, 68, 311, 1979.
77. Rillema, J. A., Activation of casein synthesis by prostaglandins plus spermidine in mammary gland explants of mice, *Biochim. Biophys. Res. Commun.*, 70, 45, 1976.
78. Rillema, J. A., Mechanism of prolactin action, *Fed. Proc.*, 39, 127, 1980.
79. Houdebine, L. M., Devinoy, E., and Delouis, C., Role of spermidine in casein gene expression in the rabbit mammary gland, *Biochemie*, 60, 753, 1978.
80. Bolander, F. F. and Topper, Y. J., Relationship between spermidine, glucocorticoid and milk proteins in different mammalian species, *Biochem. Biophys. Res. Commun.*, 90, 1131, 1979.
81. Matusik, R. J. and Rosen, J. M., Prolactin regulation of casein gene expression: possible mediators, *Endocrinology*, 106, 252, 1980.
82. Takemoto, T., Nagamatsu, Y., and Oka, T., The study of spermidine-stimulated polypeptide synthesis in cell-free translation of mRNA from lactating mouse mammary gland, *Biochim. Biophys. Acta*, 740, 73, 1983.
83. Leiderman, L. J. Criss, W. E., and Oka, T., The characterization of a polyamine-responsive, cyclic nucleotide-independent protein kinase activity in the mouse mammary gland, *Biochim. Biophys. Acta*, 844, 95, 1985

Chapter 9

PROLACTIN AND THE REGULATION OF SECRETION INCLUDING MEMBRANE FLOW: POTENTIAL ROLES FOR TUBULIN AND MICROTUBULES*

Robert F. Loizzi

TABLE OF CONTENTS

I. Introduction ... 154

II. Secretion ... 155
- A. Secretory Events ... 155
- B. Prolactin and Secretion ... 156

III. Tubulin and Microtubules ... 157
- A. Membrane Tubulin ... 157
  - 1. Plasmalemma ... 157
  - 2. Secretory Vesicles ... 158
- B. Microtubules ... 158
  - 1. Assembly and Disassembly ... 158
  - 2. Regulation ... 159
    - a. Calcium and Calmodulin ... 160
    - b. Cyclic Nucleotides ... 161
  - 3. Microtubules, Secretion, and Transport ... 162
  - 4. Microtubules in Mammary Gland ... 164
    - a. Occurrence ... 164
    - b. Effects of Microtubule-Altering Drugs ... 164
    - c. Tubulin Polymerization During the Lactation Cycle ... 164
      - i. Intact Animals ... 164
      - ii. The Alveolar Cell ... 165
    - d. Tubulin Polymerization In Vitro ... 167
    - e. Tubulin Polymerization During Induced Lactogenesis In Vivo ... 167

IV. Summary: Potential Sites for Prolactin Tubulin/Microtubule Interactions ... 168

References ... 171

* The author's research is supported by NIH grant HD 11601.

I. INTRODUCTION

Prolactin has been associated with the onset of secretory activity in mammary gland and in a few other tissues such as the parathyroid gland[1] and pinealocytes.[2,3] While previous chapters in this volume attest to the current intense interest in mechanisms by which prolactin stimulates cells, the emphasis has been primarily on events at the plasmalemma following hormone-receptor binding, including biochemical cascade reactions and the search for prolactin second messengers. There remains an information gap between these steps and the classical picture of cellular secretion, including widespread activation of membranous organelles, development and translocation of secretory vesicles or granules, and the complex process of exocytosis. Moreover, while much of the current research on prolactin action is concentrated on its role in stimulating the synthesis of secretory products, in mammary cells synthesis and secretion actually may be regulated by separate mechanisms. Ollivier-Bousquet and Denamur[4] observed that in mammary gland slices from lactating ewes and rabbits, the intracellular transit times and secretory rates for milk proteins were not diminished either by the absence of prolactin or treatment with a protein synthesis inhibitor, but both were accelerated by dibutyryl 3′,5′-cyclic adenosine monophosphate. This raises the possibility that while prolactin does participate in the differentiation of mammary cells to the secretory state and the manufacture of secretory products, this hormone may not directly regulate secretion in the narrow sense of exo- or emiocytosis. However, one is then left with the questions, "What does regulate secretion?" and "Does prolactin influence this process even indirectly?"

One approach to answering these questions is provided by the observation that synthesis and secretion in many cell types are differentially affected by compounds that alter microtubule structure. Colchicine, vincristine, vinblastine, and taxol[5] all bind to tubulin, the primary microtubule protein, and reduce the quantity of functional microtubules by either promoting their disassembly or causing them to form aggregates such as paracrystals or bundles. These compounds also tend to inhibit secretion to a greater extent than synthesis of secretory product. In addition to their effects on exocytosis, microtubule-altering drugs also have an effect on the endocytic uptake of prolactin itself. Ollivier-Bousquet showed that colchicine inhibited not only the release of casein from incubated rabbit mammary gland fragments, but also diminished the endocytic vesicles which were also a result of prolactin stimulation.[6] Similarly, Houdebine demonstrated that a series of five different microtubule-disrupting drugs were able to inhibit the initiation of casein and DNA synthesis by prolactin, presumably by inhibiting internalization of the hormone via endocytosis.[7] Finally, Posner et al. observed that colchicine injected into rats inhibited the internalization of ^{125}I-labeled prolactin into liver cells and also the uptake of this hormone into vesicular elements of the Golgi complex.[8] Accordingly, microtubules may be involved in the bulk transport or motility aspect of secretion, both intracellular and across the plasmalemma, in either direction. This is significant for two reasons. First, most theories of prolactin regulation of secretion involve only the synthesis of the secretory product, not its transport. Thus, rather than stimulus-secretion coupling, such models stop at stimulus-synthesis coupling. Yet, as stated above, drugs such as colchicine will allow synthesis stimulation to continue, while blocking exocytosis. Regulation of synthesis, therefore, does not automatically explain regulation of secretion. Secondly, bulk movement itself is a form of information transfer. Plasmalemmal and other membranes, with a complement of membrane-bound receptors, enzymes, and other complexes, shuttle between the exterior of the cell and internal sites.[9] Moreover, internalized prolactin is also transported to those sites.[10-12] This raises the interesting, albeit speculative possibility, that bulk transport is an intermediate step in the regulation not only of exocytosis, but also of synthesis itself. These concepts justify a review of potential sites for involvement of microtubules and tubulin in prolactin-induced secretory processes. It is an incomplete picture since many other aspects of bulk transport, such as the role of microfilaments, are being omitted. Furthermore, this review will focus primarily on mammary gland for its prolactin target cell. Despite these limitations, it is hoped that juxtaposing these concepts will provide clues to potential sites of prolactin influence in the secretory process.

II. SECRETION

A. Secretory Events

Few cellular processes integrate structure and function in the number and variety of ways as does secretion. Advances in almost any area of cell biology such as receptor physiology, membrane excitability, peptide synthesis, fluid and electrolyte transport, membrane-bound enzyme complexes, cell ultrastructure, energy metabolism, gene expression, motility and contractility, protein phosphorylation, calcium and cyclic nucleotide second messengers, etc. have found some application in the secretory process, thus accounting for rapid advances in our understanding. The Latin root for secretion means "to separate" and reflects the original concept of a process in which glands simply filter substances from the blood. We now know that most types of secretions are a mix of epithelial cell products synthesized *de novo,* and plasma substances transported across the epithelial sheet either transcellularly or between the cells, paracellularly. In exocrine cells, contributions may occur in gland regions where secretory cells are clustered into designated secretory sites (acini or alveoli) or along ducts.

The events of secretion have been reviewed extensively.[13-15] In mammary epithelial cells and other cells producing a proteinaceous secretory product, one of the earliest signs of preparation for secretion includes development and proliferation of the rough endoplasmic reticulum (RER), the ribosome-studded membrane system of channels or cisternae in which large quantities of exportable, nascent peptides are first formed. These molecules are transported to the Golgi apparatus, probably via smooth vesicles, where further processing such as glycosylation and packaging into secretory vesicles takes place.[9,16] In mammary cells, the Golgi apparatus is also the site of lactose synthesis by a complex consisting of the ubiquitous Golgi enzyme galactosyltransferase and the milk protein α-lactalbumin.[17] In late pregnancy, glycogen synthesis decreases, and the increased availability of glucose facilitates increased lactose synthesis.[18] Production of the osmotically active lactose within the Golgi saccules[19] is thought not only to carry water transcellularly from blood to lumen, but to cause a dilatation[20] of the usually flattened Golgi saccules arranged in a cup-shaped stack. Swelling of the Golgi vesicles has been quantitated and correlated with milk yield.[21] Material arriving from the RER enters the Golgi complex at its outer, convex, or *cis* face, while finished products exit from the inner, concave, or *trans* face. Thus, evidence of secretory products in mammary cells begins with the appearance of dense, amorphous material within vesicles exiting the Golgi apparatus. This material, which contains milk proteins, including caseins, plus calcium and phosphate, gradually condenses into a dense, micellar granule with a characteristically periodic substructure. Aqueous secretory vesicles containing one or more granules as well as lactose[19,22] and water are transported from the Golgi complex to the apical cell border. The vesicular contents are released to the lumen[23] by a process of reverse pinocytosis or emiocytosis which includes fusion of the vesicle membrane with the plasmalemma and then fission of the fused membranes and expulsion of the vesicular contents. Alternatively, chains of vesicles may fuse and their contents may be released via compound exocytosis.[24] Milk lipids, which arise in the cytosol as droplets, also move to the apical surface, but are then released to the lumen in an apocrine fashion[25] taking along some apical cell membrane and forming the membrane-covered milk fat globules found in milk. It can be seen that all of the above secretory functions involve membranes, and that membranes differ functionally among organelles, including the plasmalemma. The concept that intracellular membranes undergo differentiation, as to their enzymatic and biochemical composition, as they move among various sites within the cell is referred to as "endomembrane flow." See Keenan's review on specialization and function of the plasmalemma and endomembranes for more details.[26] An important application to this discussion is that in functions involving membrane cycling, membrane availability may be a limiting factor, particularly in secretion and endocytosis. Thus, observation of inhibition at one site may be related to blockage of donor membrane from a distant site. The results of several studies indicate that maintenance of epithelial cell-cell contacts[13,27] and cell polarity, which is aided by growing cells on or in collagen gels,[28-37] are necessary prerequisites for differentiation of secretory cells and the vectorial, basoapical translocations inherent in complex, secretory activity *in situ*.

B. Prolactin and Secretion

The role of prolactin in some of the above processes has been examined. For example, the possible mechanisms of prolactin action on the molecular biology of casein and α-lactalbumin synthesis, including its binding to the prolactin receptor and endocytosis of the hormone-receptor complex, have been reviewed elsewhere.[38-40] Prolactin stimulates a wide variety of activities associated with milk secretion. In explants and cultured cells of mammary gland from several species, prolactin induces the synthesis and secretion of the milk-sugar lactose[41] and α-lactalbumin, the specifier protein for lactose synthetase,[38,40,42] casein,[43-46] and milk fatty acids.[47] Lipid accumulation is induced by prolactin in human breast cancer cells.[48] Prolactin increases the uptake of amino acids by mammary gland in vivo[49] and it also increases glycogen metabolism in mammary gland, preparing it for anaerobic metabolism.[50] Two different nanogram concentration ranges of prolactin optimally stimulate synthesis of RNA and casein, and ornithine decarboxylase activity in mouse explants.[51] The activity of ouabain-sensitive Na^+, K^+-ATPase in mammary gland slices is increased by prolactin[52] as is the short circuit current across mammary epithelial cells growing on floating collagen gels[53,54] which, in concert with "tighter" tight junctions, account for an increased active sodium reabsorption during milk secretion. Within 15 min of adding prolactin to rabbit mammary gland fragments, the cells undergo significant changes in morphology which are indicative of milk secretion; in particular, the Golgi complex undergoes gross differentiation, including a large increase in volume.[39] Within 1 hr, active casein synthesis is underway. Many of these responses are gradually lost by cells in culture with aging.[55] Similarly, rat mammary cells obtained from ovariectomized donors lose their potential to differentiate in response to prolactin.[56] Morphological changes at the ultrastructural level are also produced in pinealocytes,[2] resulting in increased peptide secretion by these cells.[3] For a more complete listing of secretion-related effects of prolactin in mammary gland and other tissues and species, see reviews by Topper and Freeman,[57] Cowie et al.,[58] and Ensor.[59]

The studies of Topper and others[57] indicate that while development of the synthetic machinery of the mammary epithelial cell is dependent on the presence of hydrocortisone, and inhibition of secretion a function of progesterone, it is the presence of prolactin which stimulates the secretory processes. Virtually no attention, however, has been paid to examining whether prolactin in some manner controls the various types of transport associated with and probably essential for secretion. These include: uptake of the hormone-receptor complex, transport of nascent peptides from RER to Golgi apparatus, movement of substances through the Golgi during processing, transport of secretory vesicles and lipid droplets to the apical surface, and the two types of exocytosis for fat droplets and vesicles described above. These can be viewed from the standpoint of organelle or secretory product transport through the cell or as transport of a functional enzymatic complex via endomembrane flow. Although there is no evidence for a specific prolactin role in these examples of intracellular transport, cellular differentiation is essential for secretion, and this includes the *in situ* secretory cells. Moreover, microtubules or their principal protein tubulin have been implicated in a large number of secretion-related processes. Therefore, in searching for evidence of cytoskeletal participation in a prolactin-dependent secretory cell, the mammary gland alveolar cell will be examined. This will be limited primarily to tubulin, either in relation to microtubule function or as unpolymerized tubulin, and some discussion of actin microfilaments.

III. TUBULIN AND MICROTUBULES

A. Membrane Tubulin

Tubulin is a constituent of many cell membranes. When tubulin is measured by the ^3H-colchicine binding assay (described below), a significant amount of binding activity is associated with the final pellet, even after extraction of free tubulin and solubilization of microtubules.[60,61] A major portion of this activity probably includes molecular tubulin which is incorporated into various cell membranes, as well as some adsorbed tubulin and small, microtubule fragments. Dustin[62] has reviewed the localization of tubulin in the plasmalemma and other cell membranes, as well as the association of microtubules with these structures. The techniques have included extraction and electrophoretic analysis of tubulin-like proteins from isolated membranes, binding of ^3H-colchicine to isolated membranes, electron microscopic (EM) observations of whole microtubules or autoradiography of labeled colchicine in various cell membranes, immunofluorescent concanavalin A observations of colchicine-induced capping in lymphocytes, and similar studies using fibroblasts and polymorphonuclear leukocytes. Tubulin localization in membranes has been observed in brain cells and synaptosomal membrane preparations from brain, avian erythrocytes, as well as the leukocytes mentioned above. Bhattacharyya and Wolff[63] isolated tubulin from guinea pig brain plasmalemma fractions, which electron microscopic examination had shown to be microtubule-free, and were able to polymerize this tubulin into microtubules. Zisapel et al.[64] identified tubulin in synaptic vesicle membrane by gel electrophoresis and peptide mapping. They concluded that this membrane contains more α subunit than β, and that the former is an integral vesicle membrane protein, while the latter is peripherally attached and easily dissociated from the membrane. Intracellularly, tubulin has been associated with pores of the nuclear membrane in mastocytes of rat intestine, and a variety of cell organelles, particularly the Golgi vesicles. Membrane tubulin may function as a nucleating site for the growth of microtubules or it may have a more physiological, regulatory function.

1. Plasmalemma

One aspect of the role of tubulin in membranes is its postulated function of regulating the association of subunits within plasmalemmal receptor-enzyme complexes. Zor[65] has reviewed and attempted to integrate two dichotomous theories for the role of cyclic nucleotides in hormone action: one originating in Ralls' observations[66] and proposing that hormones stimulating adenylate cyclase cause an increase in cyclic AMP concentrations in the cell resulting in a variety of actions including cytoskeletal changes and the second stemming from theories of Peters[67] in which hormones are thought to act through a reorientation of the cytoskeletal system, which in turn stimulates the adenylate cyclase system. A key point in either theory is that the binding of a hormone to its receptor in the cell membrane promotes, first, a binding of the guanyl nucleotide with the GTP regulatory protein or G-unit, followed by a coupling of the G-unit[68] with a catalytic moiety of adenylate cyclase. It has been shown that increasing the mobility of proteins within the cell membrane by a variety of means increases the interaction of the G-unit with the catalytic moiety. One method is to use microtubule-altering agents such as colchicine.[69] The explanation for this effect is that the ability of the G-unit to diffuse laterally within the membrane is one limiting factor in cyclase activation, and that colchicine and membrane-fluidizing agents such as free fatty acids and certain local anesthetics all increase the freedom of protein mobility within the membrane and, thus, cyclase activation. Since similar effects are seen with other microtubule-altering agents such as vincristine, it is assumed that they act on microtubules attached to the membrane or on nonpolymerized tubulin molecules within it, and that either action would increase planar mobility of subunits within the plasmalemma. Similar effects have been seen in broken cell preparations of cerebral cortex[69] and in cultured S49 lymphoma cells in which adrenergic- and PGE_1-stimulated accumulation of cyclic AMP is enhanced by colchicine and vinblastine.[70]

2. Secretory Vesicles

Tubulin in vesicle membranes may have a role in the aggregation of vesicles or fusion with the plasmalemma which, in turn, may be essential to exocytosis. Later in this article, the role of tubulin and microtubules in the intracellular transport of secretory vesicles will be discussed. However, the possibility arises that incorporation of tubulin within secretory vesicle membranes may offer an alternate explanation for the inhibitory effects of microtubule-altering drugs on secretion. Namely, drugs such as colchicine may bind to molecular tubulin which is incorporated into secretory vesicle membrane, thereby interfering with some necessary step in their transport or exocytosis. Just this suggestion has been raised by Patton and his groups for mammary gland[71,72] and by Busson-Mabillot et al. regarding secretion by lacrimal glands.[73] Experiments have been carried out with artificial membrane vesicles or liposomes in which tubulin has been incorporated into the membrane.[74] When tubulin is inserted into a membrane consisting of bilayers of dipalmitoyl phosphatidylcholine and vesicles are held at a temperature below phase transition for 10 to 20 min, in this case 28 to 30°C, the vesicles have a tendency to aggregate and to fuse. Addition of calcium and other divalent cations, with the exception of magnesium, also caused vesicles to fuse into larger structures. Microtubule-altering drugs, however, had no effect on liposomal aggregation or fusion. Moreover, vesicle-bound tubulin did not associate with microtubules when tubulin was assembled in vitro.

B. Microtubules

Microtubules have been characterized in lactating mammary epithelial cells by several investigators.[75-77] They are elongated, cylindrical, apparently hollow structures, approximately 25 nm in diameter and up to several μm long, composed of polymerized α and β tubulin dimers, each with a molecular weight of 55,000, and several associated proteins, including tau protein and the high molecular weight MAPs. See reviews by Soifer[78] and Dustin[62] for more details.

1. Assembly and Disassembly

The cytoskeletal and transport functions of microtubules depend to some extent on the ability of 6S dimeric tubulin, 110,000 mol wt to polymerize into microtubules under physiological conditions.[62,78,79] Study of tubulin assembly and disassembly received an impetus from Weisenberg's observations that tubulin solutions isolated from brain would form microtubules in vitro providing several factors, including tubulin and calcium concentrations, ionic strength, and temperature,[80,81] were controlled. Purified tubulin has certain minimal requirements for polymerizing, including maintenance of low calcium concentration with a chelator such as EGTA, temperature close to 37°C, guanine nucleotides, particularly 1 m*M* GTP, and a critical tubulin concentration estimated at 0.2 mg/mℓ. The latter is lower if the assembly solution contains high concentrations of sucrose or glycerol. The opposite conditions, namely cold temperature, elevated calcium, and lack of GTP result in disassembly. Tubulin solutions of relatively high purity can be produced from brain homogenates by alternating the two sets of conditions several times, i.e., successive polymerization-depolymerization cycles and separation by ultracentrifugation. Under these conditions, polymerization is aided by yet another factor, a family of microtubule-associated proteins (MAPs) which copurify with tubulin and which constitute about 10 to 15% of the weight of thrice-cycled tubulin. These include both low molecular weight proteins, such as tau proteins,[82] which have molecular weights of about 70,000, and MAPs with molecular weights of approximately 300,000 to 350,000.[79] The MAPs can be separated from tubulin by their adherence to a phosphocellulose column, and they can be released by the addition of KCl to the elution buffer. Electron micrographs of microtubules polymerized in the presence of MAPs show short, filamentous projections extending laterally from the microtubules at regular intervals. Sloboda et al.[79] reported that the polymerization process involves phosphorylation of one of the high molecular weight MAPs, MAP-2.

2. *Regulation*

In vivo regulation of microtubule assembly and disassembly is likely to involve a great many factors, including some of those mentioned above. Schliwa et al.[83] using detergent-extracted monkey cells, observed that microtubule assembly and disassembly is very sensitive to calcium concentrations. Microtubules were completely stable when calcium concentrations were below 1 μM, but concentrations greater than 1 to 4 μM induced microtubule disassembly. This began in the cell periphery and proceeded toward the cell center and, in concentrations up to 500 μM, this pattern and time course of disassembly was not markedly altered, suggesting that, within this concentration range, calcium effects are catalytic rather than stoichiometric. In the millimolar range, calcium results in rapid destruction of microtubules. The conclusion is that microtubules are very sensitive to physiological concentrations of calcium within the cell.

Initiation of microtubule assembly in vivo may require special sites called microtubule organizing centers (MTOCs)[84] which are associated with a variety of cell organelles such as membranes, the kinetochores of chromosomes, centrioles, or existing microtubules. Since physiological regulation of polymerization is determined to some extent by the level of unpolymerized tubulin available for assembly into microtubules, the factors governing tubulin synthesis are relevant to microtubule formation. Ben-Ze'ev et al.[85] observed that when microtubules in cultured fibroblasts were depolymerized with colchicine or nocadazole, there was a rapid inhibition of tubulin synthesis and a reduction in the level of translatable tubulin mRNA. Vinblastine, which also disrupts microtubules, but leads to the aggregation of tubulin into large paracrystals without increasing free tubulin, resulted in an enhancement of tubulin synthesis rather than an inhibition.

The picture emerges that within the cytoplasm, free tubulin dimers and microtubules are in a state of dynamic equilibrium and that under certain physiological conditions there is a shift in the direction of polymerized tubulin, and the resulting structures then appear in electron micrographs as normal-looking microtubules. For example, during early mitosis, the appearance of spindle fibers is due to the polymerization of tubulin into microtubules of which the spindle is composed. Drugs such as colchicine are thought to inhibit spindle formation and other types of tubulin polymerization by binding to the tubulin dimer at the growing end of the potential microtubule and thus blocking addition of the next dimer. Disappearance of formed microtubules after treatment with these drugs would then require that such "capped" microtubules disassemble at the opposite end, gradually shortening until only molecular tubulin remained. Inherent in this scheme is an assumption that assembly takes place primarily at one end of microtubules and disassembly at the other end. Margolis and Wilson[86] in 1978 reported evidence for just such specialization. They pulse labeled bovine brain microtubules with ^{3}H-GTP and observed that, under steady state conditions, microtubules maintained a constant length while adding tubulin dimers at one end (indicated by GTP binding) and removing tubulin at the other end (GTP release) at equal rates. The turnover rate for tubulin in this system was rather slow, about 0.69 μm/hr, which would account for the relative insensitivity of bovine brain microtubules to drug-induced depolymerization in vitro. Since the microtubules in this steady-state situation did not change length, but rather the GTP pulse label traveled down the length of tubules from the assembly to the disassembly ends, this concept has given rise to the term "treadmilling" and is the basis for one theory of how microtubules affect transport within cells.

a. Calcium and Calmodulin

Calcium concentrations above micromolar levels promote depolymerization of microtubules, while lowering calcium levels promotes tubulin polymerization. Cytosolic calcium levels are kept at low levels by various organelles. Mitochondrial uptake of calcium, for example, has been used in an in vitro system to stimulate tubulin polymerization.[87] In mammary gland cells, the Golgi complex contains a Ca^{2+}-ATPase system which transports calcium from the cytosol into Golgi vesicles.[88-90] Besides reducing cytosolic calcium levels, particularly in the region of the Golgi complex, calcium is packaged into secretory vesicles containing milk proteins. Uptake of calcium probably stimulates lactose synthesis within the Golgi saccules via stimulation of UDPase.[91,92] The ability of calcium ion to inhibit microtubule formation is increased significantly in the presence of calmodulin or calcium-dependent regulator protein,[93,94] a major calcium-binding protein in nonmuscle cells analogous to troponin in muscle cells. Evidence that calmodulin may have a physiological role in regulating microtubule assembly and disassembly includes its localization on the mitotic spindle.[95-97] It should be pointed out that calcium-calmodulin need not serve simply to achieve overall disruption or overall shortening of microtubules but may also carry out more subtle and precise activities, such as increasing the rate of disassembly at one end which is balanced by assembly at an equal rate the opposite end, resulting in increased tubulin turnover rate and treadmilling without a change in microtubule length. This may explain the common observation of calcium as a stimulator of secretion in mammary gland[98] and most other secretory cells.[99] Calmodulin is an intracellular receptor for calcium and a modulator for its actions as a second messenger analogous to cyclic AMP. The role of calmodulin in stimulus-secretion coupling by calcium has been described for a wide variety of exocrine- and endocrine-secreting cells[100-102] and is the basis for a theory of neurotransmission and synaptic modulation.[103] Means et al.[100,101] have reviewed the various fundamental cellular activities which are known to be regulated by calcium-calmodulin. These include cyclic nucleotide and glycogen metabolism, intracellular motility via microtubules and microfilaments, calcium transport, and a group of less well-defined Ca^{2+}-dependent protein kinases. The stimulatory action of calcium-calmodulin on cyclic AMP phosphodiesterase not only has provided a convenient assay for calmodulin, but is an important point of interaction between the regulatory systems utilizing these two second messengers, calcium and cyclic AMP.[104,105]

These universal processes are found in all animal species and are central for normal secretory cell function. Calmodulin has been measured in mammary gland and its concentration is related to the lactation cycle. Riss and Baumrucker purified calmodulin from bovine tissue and reported that its concentration approximately doubled 1 week prepartum.[106] Similarly, Pizarro et al.[107] observed in rats a sharp rise in calmodulin activity from days 15 through 20 of pregnancy with the levels during lactation approximately double those of early pregnancy. This was accompanied by a similar rise in high K_m cyclic AMP phosphodiesterase. These studies do not reveal whether the increased calmodulin activity functions in microtubule assembly/disassembly and/or Golgi calcium uptake or still other processes; however, the coincident calmodulin increase with the prepartum increase in prolactin secretion, at least in the rat,[58] and rapid alterations in mammary gland cyclic AMP concentrations (see below) do suggest a regulatory role for calcium-calmodulin and cyclic nucleotides in lactogenesis.

b. Cyclic Nucleotides

The effects of cyclic nucleotides on microtubule formation are less clear than that of calcium although current evidence seems to favor the concept that cyclic AMP promotes assembly under certain conditions. Earlier studies using cultured cells indicated first, that as cells reached confluency and growth slowed, cyclic AMP levels rose and second, that exogenous cyclic AMP or its analogs decreased cell proliferation.[108-110] These effects were usually accompanied by increased cell differentiation, including greater numbers and organization of microtubules.[111] In transformed cell lines, this response was referred to as "reverse transformation".[112] Conversely, the concept that cyclic AMP had a negative influence of polymerization arose in part from observations that induced lysosomal enzyme secretion by polymorphonuclear leukocytes was inhibited by colchicine and cyclic AMP and stimulated by cyclic GMP.[113,114] Finally, in several types of secretory cells, cyclic AMP is reported to stabilize microtubules and/or increase the rate of secretion.[115] Garland[116] reported that cyclic AMP inhibited assembly of microtubules in vitro when using "crude" 100,000 g supernatant tubulin, but not twice cycle-purified tubulin. One approach, therefore, has been to examine the various proteins which copurify with tubulin during repeated depolymerization-polymerization cycling. These include, among others, the MAPs described above and protein kinases, both cyclic AMP-dependent (cAMPdPK) and independent. Cyclic AMP-dependent phosphorylation of MAPs has been reported to promote,[79] have no effect on,[117] or inhibit[118] tubulin polymerization.

In order to examine *in situ* factors affecting assembly, Brinkely et al.[119] devised a lysed cell system for measuring microtubule growth. It consisted of cultured cells grown on coverslips which had been stabilized and permeabilized and could serve as templates for initiating microtubule assembly with the addition of soluble, exogenous factors such as 6S tubulin and nucleotides. Using antitubulin immunofluorescence, this system permitted quantitation of the number of organizing centers (MTOCs) and the number and length of microtubules. It was found that SV-3T3 cells have less ability to initiate microtubule assembly than 3T3 cells; specifically, while SV-3T3 cells had the same number of organizing centers, the number of microtubules growing from each center was about half that in 3T3 cells. This difference, however, was eliminated by the addition of cyclic AMP which stimulated assembly by SV-3T3 cells, but not in 3T3 cells. These results suggested that cyclic AMP-stimulated protein phosphorylation of an unknown substrate was involved in tubulin assembly and may be retarded in SV-3T3 cells due to lack of a required factor. The same group showed that both cell lines had the requisite enzymes for carrying out cyclic AMP-dependent protein phosphorylation,[120] and that endogenous proteins and also the exogenous tubulin were phosphorylated.[80] The difference between the two cell lines was in the degree of phosphorylation of the substrates. Extensive debate is being carried out concerning the mechanism and sites of protein phosphorylation in tubulin polymerization and its regulation. Dallas and Solomon[121] identified two phosphorylated polypeptides isolated from extracted cytoskeletal material with molecular weights of 69 and 80 kdalton. Soifer et al.[122] used photoaffinity labeling of cyclic AMP-binding proteins to localize the regulatory subunit for the cAMPdPK which copurifies with tubulin. While some activity was associated with the high molecular weight MAPs and tubulin, the major cyclic AMP-binding protein was distinct from both.

Tissue changes in mammary gland cyclic nucleotides and related enzymes suggest they have important regulatory roles in lactogenesis and lactation. Cyclic AMP levels in rats,[123-125] mice,[126] and guinea pigs[127] increase progressively during pregnancy, reaching a peak shortly before parturition, at which time they drop abruptly, are maintained at low levels during lactation, and then rise again during weaning. Conversely, cyclic GMP levels undergo the reverse, a decrease during pregnancy and a rapid increase following parturition. These changes in cyclic nucleotides are accompanied or preceded by changes in their respective cyclases and phosphodiesterases, which could explain the fluctuations of the nucleotides. In vitro studies with dibutyryl cyclic AMP caused inhibition of nucleic acid and fatty acid synthesis,[128] casein synthesis,[129] and lactose synthesis.[130,131] We[131] observed that inhibition of lactose synthesis in guinea pig mammary gland slices occurred consistently with phosphodiesterase inhibitors, such as methyl xanthines, and approached 100% in dose-response studies, while inhibition with cyclic AMP and various analogs usually reached a plateau at 30 to 40%. Wilde and Kuhn[132] reported similar inconsistent inhibition of lactose synthesis and glucose uptake by isolated mammary gland acini. Nevertheless, the rapid drop of cyclic AMP at parturition combined with the in vitro inhibitory effects of cyclic AMP suggested a physiological, inhibitory role for this nucleotide with respect to synthesis of milk constituents. Inhibition was also observed in explant cultures of midpregnant mouse mammary glands. Perry and Oka[133] observed that raising intracellular cyclic AMP levels by a variety of means resulted in greater inhibition of hormonally induced synthesis of α-lactablumin than casein, also suggesting a negative regulatory role to cyclic AMP in milk protein synthesis. Cyclic AMP is also excreted from mammary tissue via milk.[125] We observed that the content of cyclic AMP in guinea pig milk is highest in early lactation when lactose production is maximal, i.e., lactose synthesis is highest when cyclic AMP removal is greatest.[127] Conversely, cyclic AMP is reported to have a stimulatory effect on intracellular protein transport and exocytosis in rabbit[4] and bovine[134] mammary gland slices. This discrepancy points out another example of apparently separate regulation or uncoupling of milk-protein synthesis from secretion which had been mentioned at the beginning of this article; with respect to the role of microtubules in secretion, it revives conflicting observations on the effects of cyclic nucleotide on tubulin polymerization.

3. Microtubules, Secretion and Transport

In 1968, Lacy et al.[135] reported that insulin secretion by β cells in response to glucose stimulation occurs via emiocytosis and, based on inhibition with antimicrotubular agents, transport of the β granule to the plasmalemma is dependent on a microtubular-microfilamentous system. Since then, the cytoskeleton and microtubules have been implicated in secretion for a growing list of exocrine and endocrine cells.[115,136] Lacy's group later observed with cinemicrography saltatory β-granule movements in cultured β cells, and their inhibition by microtubule-altering drugs, which supported the original concept.[137] Similar studies using time-lapse cinematography confirmed these observations.[138] Assays of the tubulin pools (cf. Section III. B. 4. C.) in islet cells showed that glucose stimulation increases the size of the polymerized pool, suggesting increased microtubules.[139-141] However, E. M. morphometric analyses of β cell microtubules in pancreases from in vivo experiments in which glucose was administered to vincristine-treated rats showed no change in either microtubule length or numbers compared with controls.[142] The authors concluded that glucose-induced secretion did not utilize microtubules. Similar findings were reported in immunofluorescence studies by Boyd et al.[143] which indicated that microtubules facilitate sustained secretion of insulin (e.g., over a 2-hr period), but probably were not involved in immediate secretion. In view of the dynamic nature of microtubules and their ability to assemble and disassemble at opposite ends without an observable change in length, morphometric data must be carefully interpreted. Such findings suggest that the relationship between microtubules and secretion may be more complex than first supposed. Assuming that secretion is dependent on microtubules, by what mechanism are secretory granules transported? One mode[79] utilizes MAPs as attachment points or bridges to microtubules. Again, using pancreatic β cells, Suprenant and Dentler observed β secretory granules binding to microtubules with darkfield microscopy.[144] Binding increased in the presence of cyclic AMP, and granules were released by Mg-ATP. Electron micrographs confirmed that microtubule granule associations were mediated by lateral projections of high molecular weight MAPs from the surface of microtubules assembled in vitro.

With regard to other cell types, microtubules are reported to take part in anterior pituitary secretion of growth hormone and prolactin[145-147] and to bind membranes of granule-containing secretory vesicles.[148] Colchicine also inhibits protein[149] and lipoprotein[150] secretion by hepatocytes and shifts the tubulin pool to free, dimeric tubulin in perfused liver.[60] Finally, in mammary gland, colchicine and vinblastine inhibit the secretion of each of the primary constituents of milk, namely, fat, casein, lactose, and exocytosis of secretory vesicles,[71,151-156] to a much greater extent than synthesis, as well as prolactin-induced endocytosis[8] and internalization of the prolactin-receptor complex.[7,10-12] Evidence has also been reported suggesting that microtubules translocate newly synthesized materials from the RER to the Golgi complex, including proinsulin in β cells,[157] proparathyroid hormone,[158,159] and salivary protein in the lacrimal gland.[73] This type of transport inhibition, in secretory cells where protein-dependent synthesis of the final product occurs in the Golgi complex, such as lactose synthesis in mammary cells, would be considered a synthetic block rather than a secretory block.

Secretion involves not only the exocytic transfer of condensed secretory product to the cell exterior, but usually there is an accompanying flow of water in which the secretory solids are suspended. Therefore, it is interesting to note that microtubules have also been implicated in the transport of water across vasopressin-sensitive cells, including toad urinary bladder[7,160] and mammalian kidney collecting duct.[161] Vasopressin or antidiuretic hormone (ADH) on the blood or basal side stimulates cyclic AMP production in these cells, leading to increased water permeability on the luminal or apical membrane, thus permitting a flow of water along the osmotic gradient from lumen to blood. Colchicine and other antimitotic drugs inhibit this action of vasopressin on transepithelial water transport. There is no evidence that a similar mechanism exists in secretory cells. However, postsecretory transport of secretory material through ducts requires a fluid medium, and alteration of the permeability of the membranes of the secretory cells to fluids and electrolytes might utilize tubulin within the membrane, if not microtubules. The variety of cells and tissues in which secretion is inhibited by microtubule-altering agents is compelling; however, each drug has its own set of side effects which should be thoroughly tested before assuming that the observed effect, such as inhibition of secretion, is due to a specific blockage of microtubule functioning.

4. Microtubules in Mammary Gland

a. Occurrence

Electron microscopic studies show that microtubules are located in almost every plant and animal cell examined.[75] In rat[76] and bovine[77] mammary cells, microtubules have a vectorial distribution in which their concentration increases in a basal to apical direction. They are oriented longitudinally along the same axis, with a higher concentration just beneath the apical membrane.

b. Effects of Microtubule-Altering Drugs

Administration of colchicine and other microtubule-altering drugs to lactating goats and rodents in vivo, or incubating tissue fragments or cells with the same drugs, inhibits secretion of milk components. Electron microscopic autoradiographic studies by Ollivier-Bousquet and Denamur[162] showed that colchicine inhibits exocytosis of milk protein in incubated mammary gland fragments. Patton and co-workers[163] carried out a series of in vivo studies testing the effects of microtubule-altering drugs administered to goats via retrograde infusion through the teat canal. Milk flow was suppressed maximally up to 69% by 36 hr following infusion of colchicine or vincristine, the latter requiring one tenth the dose of colchicine, and normal flow resumed after 72 to 96 hr. In a later study,[164] using this technique with rats and goats, inhibition was related to events in the mammary alveolar cell which included cytoplasmic disorganization, loss of cell polarity, and an accumulation of secretory vesicles and lipid droplets of increased size, suggesting interference with exocytosis. However, the paucity of observable microtubules in alveolar cells and their lack of association with vesicles suggests that the suppressive action of these drugs may be due to interference with vesicular membrane fusion and cytoplasmic organization rather than a direct role of microtubules in exocytosis. We reported that colchicine decreased lactose secretion by guinea pig mammary gland slices[165,130] and that similar inhibition was observed with vincristine, but not with lumicolchicine.[156] The reduced secretion of lactose into the media was shown to be due to a suppression of exocytosis, rather than decreased synthesis. Conversely, cytochalasin B, a drug which alters microfilaments and inhibits glucose transport, inhibits lactose synthesis in guinea pig slices.[166] While both drugs inhibit prolactin-stimulated endocytosis in rabbit mammary gland slices, cytochalasin B actually stimulates casein secretion.[6] Nikerson et al. compared in vivo and in vitro effects of colchicine[151] in rat mammary gland and observed cytoplasmic disorganization and loss of microtubules in both cases. In a similar study with vinblastine,[152] these morphological changes were accompanied by decreased protein synthesis, as well as secretion. The same workers further differentiated among the actions of these drugs using isolated acini from mammary gland.[155] While colchicine caused an accumulation of protein and, presumably, lactose, synthesis of both was inhibited by cytochalasin B and vinblastine, the latter interfering with protein transport from the rough endoplasmic reticulum.

c. Tubulin Polymerization During the Lactation Cycle

i. Intact Animals

The inhibitory effects of microtubule-altering drugs on the secretion of milk components as evidence for microtubule involvement is somewhat compromised to the extent that they exhibit other sometimes toxic effects unrelated to microtubules. While quantitative analysis of microtubules from electron micrographs showed a vectorial cytoplasmic gradient supporting a secretory role[151] and a good correlation with mammary gland function and drug treatment,[77,151] often their appearance in micrographs are too few in number to carry out meaningful morphometric analysis. Our laboratory, therefore, sought to quantitate microtubules biochemically using a colchicine binding assay based on several sources[167-169,140] in which the free and polymerized tubulin pools are first separated by ultracentrifugation in different buffers.

Tissues are first homogenized in a microtubule-stabilizing (MTS) buffer which, together with a calcium chelator and warm temperature, protect existing microtubules against disassembly[170,171,140] during cell disruption. MTS buffer contains 50% glycerol, 10% DMSO, GTP, $MgCl_2$, and EGTA in phosphate buffer, pH 6.95, at room temperature. Assembly of tubulin into microtubules in vitro is not likely because homogenization in MTS buffer dilutes the tubulin precursors below the required critical concentration for polymerization.[140] Ultracentrifugation of this homogenate (100,000 × g for 45 min) separates endogenous, polymerized (i.e., microtubular) tubulin in the pellet (P_1) from endogenous dimeric (free) tubulin. This first, "MTS" supernatant is reserved for the assay and the pellet is resuspended in a tubulin-solubilizing (TS) buffer and incubated for 30 min on ice to depolymerize the microtubules. TS buffer differs from MTS in that it contains 0.25 *M* sucrose instead of glycerol and no EGTA. Following a second ultracentrifugation, the S-2 "TS" supernatant is reserved and assayed for polymerized tubulin. In our earlier studies, MTS and TS fractions were incubated in the same colchicine concentrations containing ^{3}H-colchicine tracer, the MTS for 150 min and the TS for 90 min, and the free colchicine removed by incubating with activated charcoal for 10 min. This was modified later by using a higher concentration for MTS (42 μM vs. 11 μM) based on different binding characteristics in the two buffers, and small Sephadex G-100® columns for better separation of bound from free colchicine. The size of the polymerized tubulin pool is expressed as a percent of the total, i.e., free plus polymerized, tubulin-bound colchicine measurements. The colchicine binding assay, with some variations, has been used to measure free and polymerized tubulin in brain,[140,168] liver,[60,168,172] isolated pancreatic islets,[139,141,172] leukocytes,[169,172] parotid gland,[168] kidney,[168] spleen,[168] muscle,[168] lung,[168] pancreas,[172] platelets,[172] cultured bone cells,[173] mammary gland,[2,55,56,61,76,174] and cell-free assembly of brain tubulin.[116]

In our first study,[61] free and polymerized tubulin pools were measured in multiple mammary gland biopsies, an average of four per animal, from 23 guinea pigs over the period from 3 weeks prepartum through weaning. Values were grouped into 5-day periods in either direction from birth and averaged. The results indicate that the relative size of the polymerized pool remains low, 9 to 15% of the total, through 1 week prepartum, at which time it begins a gradual increase to a peak of 35 to 40% during mid-lactation, and then returns to low levels at weaning. The absolute increase in polymerized tubulin is approximately sevenfold; however, since the total tubulin increases nearly threefold from 0.87 to 2.34 μg/mg tissue, indicating net tubulin synthesis, this results in a smaller change in percent polymerization. We observed a similar increase in rat mammary gland from day 20 of pregnancy, 6.0% polymerization, to day 9 of lactation, 21.8%.[174] Glands from lactating rats have 50% more total tubulin than the pregnant animals.

ii. The Alveolar Cell

While the biopsy study suggests a correlation between increased polymerized tubulin in mammary gland tissue and the onset and maintenance of lactation, such tissue changes might be due to changes in cellular composition (e.g., relative contributions of acinar, duct, stromal, leukocyte, and adipocyte cell populations) rather than epithelial intracellular changes associated with milk secretion. Two approaches were used to study tubulin changes in the alveolar cells: immunofluorescent localization of tubulin and quantitation of tubulin pools in isolated mammary gland cell populations. In the first approach,[175] tubulin antibody was produced in rabbits against guinea pig brain tubulin which had been purified by three cycles of polymerization-depolymerization followed by polyacrylamide gel electrophoresis. Cryostat sections of mammary glands from late pregnant (1 week prepartum), early lactating (2 to 3 days post-partum), and peak lactating guinea pigs were exposed first either to antiserum or affinity column-purified antitubulin and then to fluorescein-labeled goat antirabbit IgG. Sections from pregnant animals contained small alveoli, composed of nonsecreting cuboidal cells surrounding a lumen and myoepithelial cells, capillaries, and sparse connective tissue. Acinar cells contained a diffuse fluorescence throughout the cytoplasm. Myoepithelial cells fluoresced brightly even in control sections stained with nonimmune serum, possible due to the presence of endogenous, fluorescent neurotransmitters. In tissues from early lactation, bright concentrations of fluorescence started appearing in the apical regions of some acinar cells, but not in other cell types. At peak lactation, antitubulin was present in nearly every acinar cell, usually appearing as a bright, sometimes punctate layer directly below the apical membrane and appeared to surround lipid droplets and other inclusions in this region. The immunofluorescence results, therefore, suggest a developmental change in tubulin distribution from pregnancy to lactation in which tubulin becomes more concentrated in the apical portions of alveolar secretory cells but is not present in appreciable quantities in other cell types. These fluorescent images of antitubulin in sections of three-dimensional tissues differ from the usual lacy network seen in relatively flat, cultured cell preparations in which microtubules are easily identified.[176,177] The tubulin distribution, however, does correspond with electron microscopic descriptions of apically concentrated microtubules in mammary gland cells[76,88] and suggests that the tubulin and microtubule increases during lactogenesis and lactation observed in the biopsy studies may reflect changes primarily within alveolar cells.

In the second type of study,[178] cells were isolated from mammary glands of late pregnant (1 wk or less prepartum) and peak lactating guinea pigs by enzymatic digestion with collagenase, hyaluronidase, α-chymotrypsin, and elastase, followed by centrifugation of a continuous density gradient of 3.5 to 5% bovine serum albumin.[179,180] Pregnant and lactating animals usually yielded six to eight bands of mammary gland cells. Light[181] and electron microscopy,[182] including L.M. morphometric analyses of cross-sectional areas and other parameters, revealed a variety of cell types with secretory alveolar cells concentrated in the intermediate bands. Tubulin assays were done on each band. Bands 4 to 5, which were least contaminated with cell types other than alveolar secretory cells, contained twice as much tubulin as the two extreme bands, 1 and 7, (8 to 11 pmol/10^6 cells vs. 2 to 5 pmol in cells from lactating animals) and more than twice as much of it in the polymerized state (16 to 24 vs. 55%). The differences between lactating and pregnant animals were smaller than expected, but still significant (32% more tubulin per cell in the lactating and 17% greater polymerization). Two factors may account for this. Firstly, lactogenesis-related differentiation in guinea pig mammary gland begins much earlier in pregnancy than in many other species such as rat and mouse. Thus, animals in this study, within 1 week of partus, had already undergone significant alveolar cell differentiation. Secondly, it is likely that the largest secretory cells, those laden with secretory vesicles and lipid droplets, were the most fragile and the least likely to survive enzymatic treatment and repeated centrifugation. Like those of the immunofluorescence study, the results of this study indicated that alveolar secretory cells have a considerably richer tubulin supply than other cell types in mammary gland, as well as a much larger polymerized pool. Thus, the increased polymerization with lactation observed in the biopsy study described earlier could be due to both intra-alveolar cell increases as well as an enrichment of these cells within the gland as lactation approaches.

d. Tubulin Polymerization In Vitro

Our laboratory has carried out preliminary studies on the effects of drugs and hormones on tubulin polymerization in mammary tissues and cells in vitro.[183] Short-term effects were studied in mammary gland slices from 18-day pregnant and peak lactating rats incubated for 2 hr as described previously.[48,184] Free and polymerized tubulin pools were assayed using the ^{3}H-colchicine binding assay. In mammary tissues from both the pregnant and lactating rats, the size of the polymerized tubulin pool was increased significantly by DB-cyclic AMP and cyclic GMP, both in micromolar quantities and taxol. The net tubulin content did not change, thus indicating lack of stimulation of tubulin synthesis in this short time interval. The control polymerized pool in lactating tissue was about four times that in the pregnant tissue, but the absolute increases in percent polymerization due to the drugs was approximately the same in the two groups. Electron microscopic examination of these slices showed few microtubules which were located mainly beneath the apical membrane, often parallel to it, and in close proximity to secretory vesicles and lipid droplets undergoing exocytosis. Long-term in vitro effects of drugs and hormones were studied in mammary cells isolated from 15-day pregnant rats and grown for 1 week on floating collagen gel membranes[29,30] with drug exposure during the last 5 days. The polymerized pool in control cultures was about double that in control lactating slices, which reflects the extensive microtubule complex observed in normal, cultured cells. Experiments with cyclic AMP resulted in very low cell numbers and tubulin assays were not possible. However, all three steroid hormones used, hydrocortisone, estradiol-17β, and progesterone, significantly increased percent polymerization beyond the control values. In addition, the total tubulin concentration also markedly increased with the steroids. Dose-response studies with taxol and hydrocortisone yielded increases of two- to threefold in both polymerization and tubulin content. We did not observe additivity when the hormones were used in combination, suggesting that each was maximally stimulating tubulin or generalized protein synthesis. Ultrastructurally, taxol yielded bundles of microtubules in various locations, especially in the peripheral cytoplasm adjacent to the plasmalemma between the RER and Golgi complex. We speculate that the increases in polymerization observed were due to different causes. The cyclic nucleotide effects in mammary gland slices were probably the result of direct stimulation of polymerization via the microtubule-associated protein kinase. Taxol stimulation was probably the result of decreasing the microtubular pool by formation of taxol complexes[5] and, thus, shifting the equilibrium toward polymerization. In cultured cells, the resulting decrease in free tubulin would then induce tubulin synthesis, resulting in an increased total tubulin. Finally, the steroids, all of which are known to induce mammary cell development, may have stimulated tubulin synthesis directly as well as other proteins. A large free tubulin pool would stimulate polymerization, thus shifting the equilibrium toward microtubules again, but for a different reason. Polymerization may also have been stimulated directly by the steroids, perhaps reflecting a mitogenic effect. The results also caution against direct application of tissue culture observations to *in situ* cells. The large size of the polymerized tubulin pool in cultured cells compared to that in slices reflects the large microtubule complex observed in cultured cells with immunofluorescence and electron microscopy and which appears to be related to the flat shape of these cells and the solid substratum. An *in situ* epithelial cell, polarized in a sheet supported by neighboring cells and a tight junctional collar, probably utilizes much less tubulin for support function and only a fraction of the total tubulin for the rapid, dynamic processes involved in transport and secretion.

e. Tubulin Polymerization During Induced Lactogenesis In Vivo

Since the amount of polymerized tubulin in mammary gland appeared to increase with lactation, we examined the possibility that such an increase might be induced artificially by stimulating lactogenesis in the intact animal.[174] Pregnant rats were ovariectomized (OVX) at 18 days utilizing the procedures described by Kuhn[185] and others[186-189] and mammary glands were examined for lactose and tubulin at intervals over the following 48 hr. In chronically cannulated rats, this procedure decreased blood progesterone levels to very low values within 6 hr while increasing prolactin levels at 16 to 18 hr.[186,187] Normally, prolactin levels rise 1 to 2 days prior to parturition.[187,190] Between 18 and 24 hr lactose content rose from 0.5 to 3.4 μm/mg tissue, about 700%, indicating the onset of induced lactogenesis, and up to 5.9 μg/mg by 48 hr. Intact rats had a mammary gland tubulin content of 10.4 pmol/mg protein of which 3.6% was polymerized. During the first 18 hr following surgery, the average size of the polymerized pools for the OVX and sham-operated groups was 5.4 and 4.7% of the total tubulin. Between 18 and 24 hr however, that of the OVX group increased 2.4-fold. Mean values for the period 24 to 48 hr were 13.0% for the OVX and 6.0% for the sham. Lactating rats had a polymerized pool, which was 21.8% of the total. Total tubulin content, however, did not change significantly. Mean values over the entire 48-hr period were 14.3 pmol/mg protein for the OVX and 14.7 for the sham-operated rats. A significant increase was observed in mammary glands of lactating rats which contained 21.0 pmol tubulin per milligram of protein. The results indicate that in rats, the increase in the size of the polymerized tubulin pool in mammary gland occurs simultaneously with initiation of lactose synthesis. In addition, the net increase in tubulin synthesis observed in lactating animals was not present in the OVX-induced rats. Increased polymerization, therefore, had been stimulated in OVX rats without a significant increase in tubulin content, although the latter does increase with lactation.

In rats, induction of lactogenesis by OVX was shown by Kuhn to be the result of progesterone withdrawal.[185,191] Raising prolactin levels in late pregnant rats with perphenazine alone does not induce lactogenesis.[192] In the above study, the major decrease in progesterone levels had occurred 6 hr following surgery, and 18 hr preceding the rise in tubulin polymerization. The question arises, therefore, was the increase in polymerization also the result of progesterone withdrawal? A second ablation study[193] was carried out examining the effects of estrogen and progesterone replacement on OVX-induced changes. Implants containing estradiol-17β, progesterone, or both were inserted subcutaneously at the time of surgery in 18-day pregnant rats and mammary glands were collected 24 hr later. As before, there was an increase in the polymerized pool in OVX rats, 10.6 vs. 4.4% in the sham-operated concomitant with the initiation of lactose synthesis, while total tubulin did not change. In rats fitted with progesterone implants or progesterone plus estradiol, the increases in tubulin polymerization and lactose were both blocked. Estradiol alone had no effect on polymerization and slightly reduced lactose content compared to OVX. Hormone replacements did not affect total tubulin content. Thus, stimulation of polymerization and lactose synthesis in these animals were both related to progesterone withdrawal.

IV. SUMMARY: POTENTIAL SITES FOR PROLACTIN TUBULIN/MICROTUBULE INTERACTIONS

As stated previously, there is no direct evidence that prolactin exerts some of its secretory influence through the tubulin/microtubule components of the cytoskeleton. Moreover, the involvement of these components in secretion is based on two types of evidence: first, a correlation between events of the lactation cycle with either tubulin polymerization or apical tubulin concentrations in alveolar cells (neither of which necessarily reflects microtubule formation) and, second, the inhibitory actions of a variety of microtuble-altering drugs, whose side effects create problems which are only partially offset by the different ways in which many of them disrupt microtubules. Nevertheless, the striking information gap between prolactin-induced secretion, particularly in mammary gland, on the one hand, and the myriad of evidence, albeit largely circumstantial, of cytoskeletal involvement in secretion, is itself sufficient justification for hypothesizing potential sites where prolactin might exert its effects through tubulin or microtubules. We catalog some of these below. The observation by Ollivier-Bousquet[6] that cytochalasin B and colchicine inhibited prolactin-induced endocytosis suggests that the effects of these drugs on alveolar cell secretory activity may be due in part to their interference in the uptake of necessary precursors, membrane, membrane components such as adenylate cyclase, cAMPdPK, and transport ATPases, or informational molecules such as prolactin itself. Along this line, Houdebine[7] observed that several microtubule-altering agents inhibited the induction of casein and lactose synthase in rabbit mammary gland explants, and this did not involve lysosomal degradation of the prolactin receptor complex. Thus, microtubules or tubulin may function in the transmission of prolactin information responsible for initiation of milk synthesis.

Our rat ablation studies on mammary gland tubulin polymerization[174,193] suggest that progesterone withdrawal in late pregnancy induces tubulin polymerization, but not net tubulin synthesis, as part of the lactogenesis complex. Lactating mammary cells, however, do have higher tubulin contents. Therefore, applying these observations to normal pregnancy brought to partus, we speculate that progesterone withdrawal would tend to reduce the concentrations of free tubulin dimers in alveolar cells, which in turn would stimulate tubulin synthesis.[85] The increase in prolactin following progesterone withdrawal in rats allows the possibility that prolactin may also promote tubulin synthesis. To accomplish this, prolactin may initiate a cascade of reactions at the cell membrane or it may be internalized into the cell as described by Nolan[10,11] for mammary gland and Josefsberg for liver,[194] via an endocytic process[6,7] and act at the Golgi complex or other cell organelle. Either process may involve tubulin and/or microtubules. Action at the plasmalemma may involve a process analogous to that proposed for the adenyate cyclase system whereby enzyme activation occurs when subunits within the receptor-enzyme complex are allowed to diffuse together by altering either tubulin embedded within the membrane or microtubules attached to it. Endocytic internalization of prolactin might involve membrane tubulin in the budding of vesicles from the plasmalemma and fusion with organelles as well as microtubules aiding intracellular transport of the vesicles. Furthermore, endocytosis would not only serve to internalize prolactin, it could also serve to carry plasmalemmal enzymes and adsorbed calcium to the same organelles. Thus, transport ATPases such as Ca^{2+}-ATPase and adenylate cyclase may be transferred from plasmalemma to the Golgi complex, while extracellular Ca^{2+} would be delivered to the Golgi vesicles where it would stimulate lactose synthetase, resulting in lactose production and osmotic flow of water across the epithelial cells toward the lumen and, therefore, milk flow.

New tubulin arising from prolactin stimulation and not incorporated into membranes would raise the cytosolic concentration of molecular tubulin and shift the equilibrium toward polymerized tubulin, thus stimulating microtubule formation, and this would continue through peak lactation. In our studies, the size of the polymerized pool in lactating rats was considerably larger than that in OVX, pregnant rats. Newly synthesized tubulin would also be available for incorporation into membranes, particularly that of Golgi vesicles, and eventually added to the plasmalemma as vesicles are transported to the cell apex, perhaps with the aid of newly formed microtubules. Besides synthesis of tubulin itself, prolactin may stimulate production of other proteins such as the MAPs and calmodulin, which in turn, would help regulate microtubule formation. The fact that mammary gland calmodulin undergoes a significant increase in late pregnancy in cow[106] and rat[107] makes calmodulin production another potential site for prolactin regulation, not only of microtubule formation, but also of Ca^{2+}-activation of microfilaments in endocytosis and other forms of membrane flow and calcium uptake by the Golgi complex. Finally, prolactin may increase water flow across the mammary cell by interacting with tubulin or microtubules associated with either the plasmalemma or the Golgi complex membrane, and thus increase the permeability of these membranes similar to the action of vasopressin in toad bladder. The driving force for this water movement is the increased concentration of osmotically active lactose within the Golgi saccules and vesicles. Increased water flow is a necessary complement to solute synthesis and secretion to facilitate transport of secretions through ducts over finite distances.

Finally, such speculations must take into account the probable influence of placental lactogen on any of these processes. In the rat, low molecular weight rPL peaks in late pregnancy at 17 to 21 days[195] and binds equally well to mammary gland and ovarian PRL receptors.[196]

Potential sites for prolactin acting through tubulin and/or microtubules are summarized here and illustrated in Figure 1. They include:

1. Stimulation of plasmalemmal tubulin leading to endocytosis, internalization of hormone as well as enzyme complexes, including adenylate cyclase, and possibly translocating the latter to the Golgi complex where increased cyclic AMP levels induce tubulin polymerization beginning at the Golgi complex and moving toward the plasmalemma.
2. Stimulation of synthesis of tubulin, MAPs, or calmodulin, the first leading to an increased free tubulin pool, the other two to increase ability for regulating microtubule formation; new dimeric tubulin would also be available for insertion into vesicles and plasmalemma, thereby increasing their ability for fusion and allowing greater regulation of intramembrane enzyme subunit movements.
3. Activation of calcium pumps in the plasmalemma, the Golgi complex, or other organelles, reducing cytosolic calcium and inducing microtubule formation and/or treadmilling, resulting in intracellular, endocytic, or exocytic translocations; Ca^{2+} uptake by the Golgi saccules would also stimulate lactose synthesis and initiate osmotic water transport
4. Stimulation of calmodulin, adenylate cyclase, or cAMP-dependent protein kinase, leading to microtubule treadmilling and/or formation, the latter then stimulating tubulin synthesis by reducing free tubulin levels
5. Increased stimulation of plasmalemmal tubulin and/or microtubules resulting in increased membrane permeability to water and water influx, thus aiding lactose-initiated osmotic water flow across the epithelial cell from base to apex, increased transport of glucose to Golgi complex, accelerated lactose synthesis, and increased milk secretory flow rate.

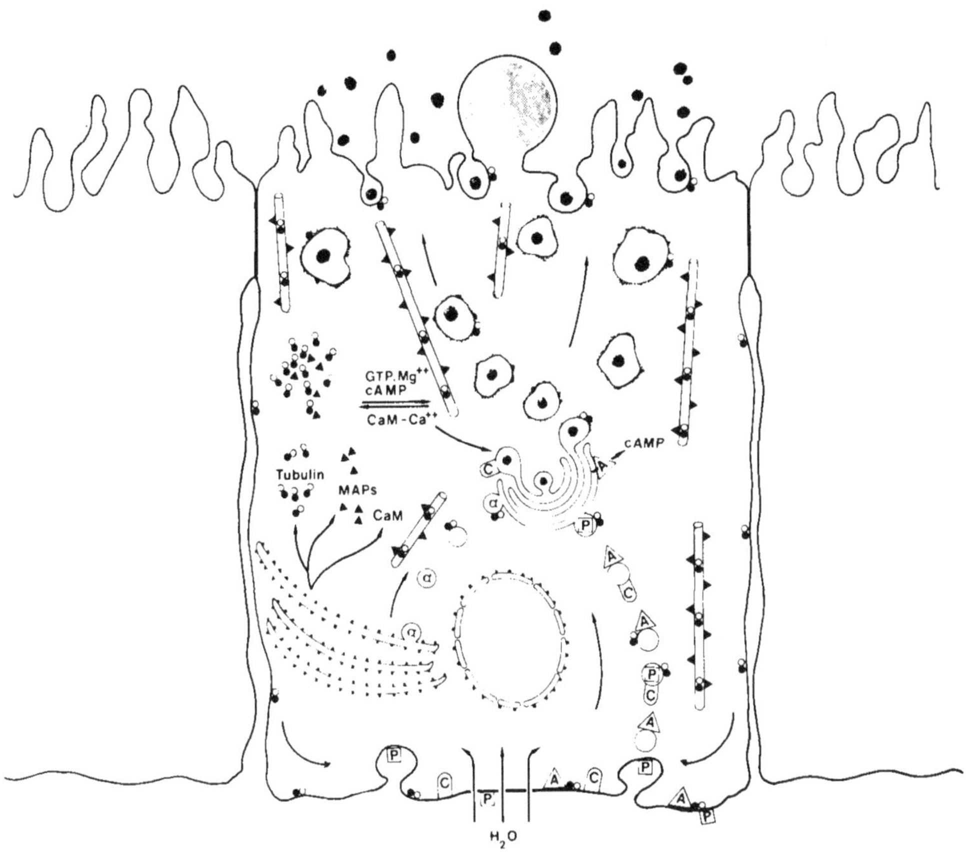

FIGURE 1. Potential sites of interaction between prolactin (P) and the tubulin-microtubule system during milk secretion by the mammary gland alveolar cell. These include transport of endocytic vesicles to Golgi complex, increased water permeability across basolateral plasmalemma, and stimulation of protein synthesis in RER including α,β dimeric tubulin (open-closed twin circles), MAPs (closed triangles), calmodulin (CaM), and α-lactalbumin (α) and other milk proteins. Endocytic vesicles translocate prolactin, adenylate cyclase (A), and transport enzyme complexes such as Ca^{2+}-ATPase (C) to Golgi complex where they participate in regulating microtubule opposite ends assembly-disassembly. Microtubules then function in transport and exocytosis of milk constituents such as lipids (large, shaded globes), proteins (dark granules), and lactose (in vesicles with proteins and calcium). Refer to Section IV for more details.

REFERENCES

1. Magiola, L. and Forte, L. R., Prolactin stimulation of parathyroid hormone secretion in bovine parathyroid cells, *Am. J. Physiol. (Endocrinol. Metabl. 10)*, 247, E675, 1984.
2. Karasek, H., Bartke, A., and Hansen, J. T., Influence of prolactin on pinealocytes of the mouse with hereditary hypopituitarism: a quantitative ultrastructural study, *Mol. Cell. Endocrinol.*, 29, 101, 1983.
3. Haldar-Misra, C. and Pevet, P., Influence of prolactin on the processes of protein and/or peptide secretion in mouse and rat pinealocytes. An *in vitro* study, *J. Neural Transm.*, 58, 245, 1983.
4. Ollivier-Bousquet, M. and Denamur, R., Effect de l'etat physiologique et du 3'5' adenosine monophosphate cyclique sur le transit intracellulaire et l'excretion des proteines du lait. Etude autoradiographique en microscopie electronique, *J. Microsc. Biol. Cell.*, 23, 63, 1975.
5. Manfredi, J. J., Parness, J., and Horwitz, S. B., Taxol binds to cellular microtubules, *J. Cell Biol.*, 94, 688, 1982.

6. Ollivier-Bousquet, M., Effects de la cytochalasine B et de la colchicine sur l'action rapide de la prolactine dans la glande mammarie de lapine en lactation, *Eur. J. Cell Biol.*, 19, 168, 1979.
7. Houdebine, L.-M., Effect of various lysosomotropic agents and microtubule disrupting drugs on the lactogenic and the mammogenic action of prolactin, *Eur. J. Cell Biol.*, 22, 755, 1980.
8. Posner, B. I., Verma, A. K., Patel, B. A., and Bergeron, J. J. M., Effect of colchicine on the uptake of prolactin and insulin into Golgi fractions of rat liver, *J. Cell Biol.*, 93, 560, 1982.
9. Farquhar, M. G., Multiple pathways of exocytosis, endocytosis, and membrane recycling: validation of a Golgi route, *Fed. Proc.*, 42, 2407, 1983.
10. Nolan, J. and Witorsch, R., Detection of endogenous immunoreactive prolactin in rat mammary epithelial cells during lactation, *Endocrinology*, 99, 949, 1976.
11. Nolan, J. M., Intracellular prolactin in rat corpus luteum and adrenal cortex, *Endocrinology*, 102, 402, 1978.
12. Malven, P. V. and Keenan, T. W., Immunoreactive prolactin in subcellular fractions from bovine mammary tissue, *J. Dairy Sci.*, 66, 1237, 1983.
13. Pitelka, D. R., Hamamoto, S. T., Duafala, J. G., and Nemanic, M. K., Cell contacts in the mouse mammary gland. 1. Normal gland in postnatal development and the secretory cycle, *J. Cell Biol.*, 56, 797, 1973.
14. Pitelka, D. R. and Hamamoto, S. T., Form and function in mammary epithelium: the interpretation of ultrastructure, *J. Dairy Sci.*, 60, 643, 1977.
15. Teyssot, B. and Houdebine, L.-M., Induction of casein synthesis by prolactin and inhibition by progesterone in the pseudopregnant rabbit treated by colchicine without any simultaneous variations of casein mRNA concentration, *Eur. J. Biochem.*, 117, 563, 1981.
16. Rothman, J. E., The Golgi apparatus: two organelles in tandem, *Science*, 213, 1212, 1981.
17. Nicholas, K. R., Hartmann, P. E., and McDonald, B. L., α-Lactalbumin and lactose concentrations in rat milk during lactation, *Biochem. J.*, 194, 149, 1981.
18. Emerman, J. T., Bartley, J. C., and Bissell, M. J., Interrelationship of glycogen metabolism and lactose synthesis in mammary epithelial cells of mice, *Biochem. J.*, 192, 695, 1980.
19. Sasaki, M., Eigel, W. N., and Keenan, T. W., Lactose and major milk proteins are present in secretory vesicle-rich fractions from lactating mammary gland, *Proc. Natl. Acad. Sci. U.S.A.*, 75, 5020, 1978.
20. Sasaki, M. and Keenan, T. W., Membranes of mammary gland. XV. 5-Thio-D-glucose decreases lactose content and inhibits secretory vesicle maturation in lactating rat mammary gland, *Exp. Cell Res.*, 111, 413, 1978.
21. Holt, C., Swelling of Golgi vesicles in mammary secretory cells and its relation to the yield and quantitative composition of milk, *J. Theor. Biol.*, 101, 247, 1983.
22. White, M. D., Kuhn, N. J., and Ward, S., Permeability of lactating-rat mammary gland Golgi membranes to monosaccharides, *Biochem. J.*, 190, 621, 1980.
23. Cline, P. R., Zamora, P. O., and Hosick, H. L., Morphology and lactose synthesis in tissue culture of mammary alveoli isolated from lactating mice, *In Vitro*, 18, 694, 1982.
24. Dylewski, D. and Keenan, T. W., Compound exocytosis of casein micelles in mammary epithelial cells, *Eur. J. Cell Biol.*, 31, 114, 1983.
25. Murakami, M., Shimada, T., Nishida, T., and Sakima, M., Scanning electron microscopic study of the mammary gland of rats during and after lactation, *Arch. Histol. Jpn.*, 40, 421, 1977.
26. Mather, I. H. and Keenan, T. W., Function of endomembranes and the cell surface in the secretion of organic milk constituents, in *Biochemistry of Lactation*, Mepham, T. B., Ed., Elsevier, Amsterdam, 1983, 231.
27. Foster, R. C., A putative role for cell-cell epithelial contacts in lactose secretion, *Jpn. J. Physiol.*, 29, 651, 1979.
28. Greenberg, G. and Hay, E. D., Epithelia suspended in collagen gels lose polarity and express characteristics of migrating mesenchymal cells, *J. Cell Biol.*, 95, 333, 1982.
29. Emerman, J. T. and Pitelka, D. R., Maintenance and induction of morphological differentiation in dissociated mammary epithelium on floating collagen membranes, *In Vitro*, 13, 316, 1977.
30. Emerman, J. T., Enami, J., Pitelka, D. R., and Nandi, S., Hormonal effects on intracellular and secreted casein in cultures of mouse mammary epithelial cells on floating collagen membranes, *Proc. Natl. Acad. Sci. U.S.A.*, 74, 4466, 1977.
31. Enami, J. and Nandi, S., Secretion of casein in cultures of mouse mammary epithelial cells on floating collagen gels, *J. Dairy Sci.*, 61, 729, 1978.
32. Emerman, J. T., Burwen, S. J., and Pitelka, D. R., Substrate properties influencing ultrastructural differentiation of mammary epithelial cells in culture, *Tissue Cell*, 11, 109, 1979.
33. Enami, J., Yang, J., and Nandi, S., Simultaneous production of casein and mammary tumor virus in mouse mammary epithelial cells grown on floating collagen gels, *Cancer Lett.*, 6, 99, 1979.

34. Yang, J., Guzman, R., Richards, J., Jentoft, V., DeVault, M. R., Wellings, S. R., and Nandi, S., Primary culture of human mammary epithelial cells embedded in collagen gels, *J. Natl. Cancer Inst.*, 65, 337, 1980.
35. Burwen, S. J. and Pitelka, D., Secretory function of lactating mouse mammary epithelial cells cultured on collagen gels, *Exp. Cell Res.*, 126, 249, 1980.
36. Haeuptle, M.-T., Suard, Y. L. M., Bogenmann, E., Reggio, H., Racine, L., and Kraehenbuhl, J.-P., Effect of cell shape change on the function and differentiation of rabbit mammary cells in culture, *J. Cell Biol.*, 96, 1425, 1983.
37. Tonelli, Q. and Sorof, S., Induction of biochemical differentiation in three-dimensional collagen cultures of mammary epithelial cells from virgin mice, *Differentiation*, 22, 195, 1982.
38. Ray, D. B., Horst, I. A., Jansen, R. W., and Kowal, J., Normal mammary cells in long term culture. II. Prolactin, corticosterone, insulin, and triiodothyronine effects of α-lactalbumin production, *Endocrinology*, 108, 584, 1981.
39. Ollivier-Bousquet, M., Early effects of prolactin on lactating rabbit mammary gland. Ultrastructural changes and stimulation of casein secretion, *Cell Tissue Res.*, 187, 25, 1978.
40. Ray, D. B. Horst, I. A., Jansen, R. W., and Kowal, J., Normal mammary cells in long term culture. I. Development of hormone-dependent functional monolayer cultures and assay of α-lactalbumin production, *Endocrinology*, 108, 573, 1981.
41. DeLouis, C. and Denamur, R., Induction of lactose synthesis by prolactin in rabbit mammary gland explants, *J. Endocrinol.*, 52, 311, 1972.
42. Goodman, G. T., Akers, R. M., Friderici, K. H., and Tucker, H. A., Hormonal regulation of α-lactalbumin secretion from bovine mammary tissue cultured *in vitro*, *Endocrinology*, 112, 1324, 1983.
43. Katiyar, V. N., Enami, J., and Nandi, S., Effect of polypeptide hormones on stimulation of casein secretion by mouse mammary epithelial cells grown on floating collagen gels, *In Vitro*, 14, 771, 1978.
44. Markoff, E. and Talamantes, F., The lactogenic response of mouse mammary explants to mouse prolactin and growth hormone, *Endocrinol. Res. Commun.*, 7, 269, 1980.
45. Djiane, J., Houdebine, L.-M., and Kelly, P., Correlation between prolactin-receptor interaction, down-regulation of receptors, and stimulation of casein and deoxyribonucleic acid biosynthesis in rabbit mammary gland explants, *Endocrinology*, 110, 791, 1982.
46. Bolander, J. F., Nicholas, K. R., Wyk, J. J., and Topper, Y. J., Insulin is essential for accumulation of casein mRNA in mouse mammary epithelial cells, *Proc. Natl. Acad. Sci. U.S.A.*, 78, 5682, 1981.
47. Carrington, C. A., Hosick, H. L., Forsyth, I. A., and Dils, R. R., Novel multialveolar epithelial structures from rabbit mammary gland that synthesize milk specific fatty acids in response to prolactin, *In Vitro*, 17, 363, 1981.
48. Shiu, R. P. C. and Paterson, J. A., Alteration of cell shape, adhesion and lipid accumulation in human breast cancer cells (T-47D) by human prolactin and growth hormone, *Cancer Res.*, 44, 1178, 1984.
49. Bussmann, L. E. and Deis, R. P., α-Glutamyltransferase activity in mammary gland of pregnant rats and its regulation by ovarian hormones, prolactin and placental lactogen, *Biochem. J.*, 223, 275, 1984.
50. Srinivasan, N., Govindarajulu, P., and Vanithakumari, G., Influence of sex steroids and prolactin rat mammary gland, *Ind. J. Physiol. Pharmacol.*, 24, 322, 1980.
51. Rillema, J. A., Wing, L.-Y. C., and Cameron, C. M., Effect of various concentrations of prolactin and growth hormone on the magnitude of stimulation of RNA synthesis, casein synthesis and ornithine decarboxylase activity in mouse mammary gland explants, *Horm. Res.*, 15, 133, 1981.
52. Falconer, I. R., Langley, J. V., and Vacek, A. T., Effect of prolactin on ^{86}RB+ uptake, potassium content and [G-^{3}H]quabain binding of lactating rabbit mammary tissue, *J. Physiol.*, 334, 1, 1983.
53. Bisbee, C. A., Prolactin effects on ion transport across cultured mouse mammary epithelium, *Am. J. Physiol.*, 240, *(Cell Physiol. 9)*, C110, 1981.
54. Bisbee, C. A., Transepithelial electrophysiology of cultured mouse mammary epithelium: sensitivity to prolactins, *Am. J. Physiol.*, 241 *(Endocrinol. Metab. 4)*, E410, 1981.
55. Errick, J. E. and Kano-Sueoka, T., Growth responsiveness to prolactin and its loss in normal rat mammary cells in culture, *Mol. Cell. Endocrinol.*, 32, 255, 1983.
56. Bolander, F. F. and Topper, Y. J., Loss of differentiative potential of the mammary gland in ovariectomized mice: identification of a biochemical lesion, *Endocrinology*, 108, 1649, 1981.
57. Topper, Y. J. and Freeman, C. S., Multiple hormone interactions in the development biology of the mammary gland, *Physiol. Rev.*, 60, 1049, 1980.
58. Cowie, A. T., Forsyth, I. A., and Hart, I. C., Eds., *Hormonal Control of Lactation*, Springer-Verlag, New York, 1980.
59. Ensor, D. M., *Comparative Endocrinology of Prolactin*, Chapman and Hall, London, 1978.
60. Jeanrenaud, B., Le Marchand, Y., and Patzelt, C., Role of microtubules in hepatic secretory processes, in *Membrane Alterations As Basis of Liver Injury*, Pepper, M., Bianchi, L., and Reutter, U., Eds., University Park Press, Baltimore, Md., 1977, 247.

61. Guerin, M. A. and Loizzi, R. F., Tubulin content and assembly states in guinea pig mammary gland during pregnancy, lactation, and weaning, *Proc. Soc. Exp. Biol. Med.*, 165, 50, 1980.
62. Dustin, P., *Microtubules*, Springer-Verlag, New York, 1978.
63. Bhattacharyya, B. and Wolff, J., Polymerization of membrane tubulin, *Nature (London)*, 264, 576, 1976.
64. Zisapel, N., Levi, M., and Gozes, I., Tubulin: an integral protein of mammalian synaptic vesicle membranes, *J. Neurochem.*, 34, 26, 1980.
65. Zor, U., Role of cytoskeletal organization of the regulation of adenylate cyclase-cyclic adenosine monophosphate by hormones, *Endocrinol. Rev.*, 4, 1, 1983.
66. Rall, T. W., Sutherland, E. W., and Wasilait, W. O., The relationship of epinephrine and glucogen to liver phosphorylase. III. Reactivation of liver phosphorylase in slices and in extracts, *J. Biol. Chem.*, 218, 483, 1956.
67. Peters, R., Hormones and the cytoskeleton, *Nature (London)*, 177, 426, 1956.
68. Spiegel, A. M. and Downs, R. W., Guanine nucleotides: key regulators of hormone receptor-adenylate cyclase interaction, *Endocrinol. Rev.*, 2, 275, 1981.
69. Rasenick, M. M., Stein, P. J., and Bitensky, M. W., The regulatory subunit of adenylate cyclase interacts with cytoskeletal components, *Nature (London)*, 294, 560, 1981.
70. Kennedy, M. S. and Insel, P. A., Inhibitors of microtubule assembly enhance beta-adrenergic and prostaglandin E1-stimulated cyclic AMP accumulation in S49 lymphoma cells, *Mol. Pharmacol.*, 16, 215, 1979.
71. Patton, S., Stemberger, B. H., and Knudson, C. M., The suppression of milk fat globulin secretion by colchicine: an effect coupled to inhibition of exocytosis, *Biochem. Biophys. Acta*, 499, 404, 1977.
72. Sokka, T. K. and Patton, S., In vivo effects of colchicine on milk fat globule membrane, *Biochem. Biophys. Acta*, 731, 1, 1983.
73. Busson-Mabillot, S., Chambaut-Guerin, A. M., Ovtracht, L., and Muller, P., Microtubules and protein secretion in rat lacrimal glands: localization of short-term effects of colchicine on the secretory process, *J. Cell Biol.*, 95, 105, 1982.
74. Kumar, N., Blumenthal, R., Henkart, M., Weinstein, J. N., and Klausner, R. D., Aggregation and calcium-induced fusion of phosphatidylcholine vesicle-tubulin complexes, *J. Biol. Chem.*, 257, 15137, 1982.
75. Sandborn, E., Koen, P. F., McNabb, J. D., and Moore, G., Cytoplasmic microtubules in mammalian cells, *J. Ultrastruct. Res.*, 11, 123, 1964.
76. Nickerson, S. C. and Keenan, T. W., Distribution and orientation of microtubules in milk secreting epithelial cells of rat mammary gland, *Cell Tissue Res.*, 202, 303, 1979.
77. Nickerson, S. C., Akers, R. M., and Weinland, B. T., Cytoplasmic organization and quantitation of microtubules in bovine mammary epithelial cells during lactation and involution, *Cell Tissue Res.*, 223, 421, 1982.
78. Soifer, D. D., The biology of cytoplasmic microtubules, *Ann. N.Y. Acad. Sci.*, 253, 1, 1975.
79. Sloboda, R. D., Dentler, W. L., Bloodgood, R. A., Telzer, B. R., Granett, S., and Rosenbaum, J. L., Microtubule-associate proteins (MAPs) and the assembly of microtubules *in vitro*, *Cell Motil.*, 3, 1171, 1976.
80. Tash, J. S., Means, A. R., Brinkley, B. R., Dedman, J. R., and Cox, S. M., Cyclic nucleotide and Ca^{2+}-regulation of microtubule initiation and elongation, in *Microtubules and Microtubule Inhibitors 1980*, DeBrabander, M. and DeMay, J., Eds., Elsevier/North-Holland, Amsterdam, 1980, 281.
81. Weisenberg, R., The role of nucleotides in microtubule assembly, *Ann. N.Y. Acad. Sci.*, 253, 573, 1975.
82. Weingarten, M. D., Lockwood, A. H., Hwo, S.-Y., and Kirscher, M. W., A protein factor essential for microtubule assembly, *Proc. Natl. Acad. Sci. U.S.A.*, 72, 1858, 1975.
83. Schliwa, M., Euteneuer, U., Bulinski, J. C., and Izant, J. G., Calcium lability of cytoplasmic microtubules and its modulation by microtubule-associated proteins, *Proc. Natl. Acad. Sci. U.S.A.*, 78, 1037, 1981.
84. Pickett-Heaps, J. D., Aspects of spindle evolution, *Ann. N.Y. Acad. Sci.*, 253, 352, 1975.
85. Ben-Ze'ev, A., Farmer, S. R., and Penman, S., Mechanisms of regulating tubulin synthesis in cultured mammalian cells, *Cell*, 17, 319, 1979.
86. Margolis, R. L. and Wilson, L., Opposite end assembly and disassembly of microtubules at steady state, *in vitro*, *Cell*, 13, 1, 1978.
87. Fuller, G. M., Ellison, J., McGill, M., and Brinkley, B. R., The involvement of mitochondria and calcium in regulating microtubule assembly *in vitro*, *J. Cell Biol.*, 67, 126a, 1975.
88. Baumrucker, C. R. and Keenan, T. W., Membranes of mammary gland. X. Adenosine triphosphate dependent calcium accumulation by Golgi apparatus rich fractions from bovine mammary gland, *Exp. Cell Res.*, 90, 253, 1975.
89. West, D. W., Energy-dependent calcium sequestration activity in a Golgi apparatus fraction derived from lactating rat mammary glands, *Biochem. Biophys. Acta*, 673, 374, 1981.

90. Virk, S. S., Kirk, C. J., and Shears, S. B., Ca^{+2} transport and Ca^{+2}-dependent ATP hydrolysis by Golgi vesicles from lactating rat mammary glands, *Biochem. J.*, 226, 741, 1985.
91. Powell, J. T. and Brew, K., Metal ion activation of galactosyltransferase, *J. Biol. Chem.*, 251, 3645, 1976.
92. Kuhn, N. and White, A., The role of nucleoside diphosphatase in a uridine nucleotide cycle associated with lactose synthesis in rat mammary gland Golgi apparatus, *Biochem. J.*, 168, 423, 1977.
93. Marcum, J. M., Dedman, J. R., Brinkley, B. R., and Means, A. R., Control of microtubule assembly-disassembly by calcium-dependent regulator protein, *Proc. Natl. Acad. Sci. U.S.A.*, 75, 3771, 1978.
94. Kumagai, H. and Nishida, E., The interactions between calcium-dependent regulator protein (calmodulin) and microtubule proteins. Further studies on the mechanism of microtubule assembly inhibition by calmodulin, *Biomed. Res.*, 1, 223, 1980.
95. Anderson, B. M., Osborn, M., and Weber, K., Specific visualization of the distribution of the calcium dependent regulatory protein of cyclic nucleotide phosphodiesterase (modulator protein) in tissue culture cells by immunofluorescence microscopy: mitosis and intercellular bridge, *Eur. J. Cell Biol.*, 17, 354, 1978.
96. Welsch, M. J., Dedman, R., and Brinkley, B. R., Calcium-dependent regulator protein: localization in mitotic apparatus of eukaryotic cells, *Proc. Natl. Acad. Sci. U.S.A.*, 75, 1867, 1978.
97. Welsch, M. J., Dedman, J. R., Brinkley, B. R., and Means, A. R., Tubulin and calmodulin. Effects of microtubule and microfilament inhibitors on localization in the mitotic apparatus, *J. Cell Biol.*, 81, 624, 1979.
98. Smith, J. J., Park, C. S., and Keenan, T. W., Calcium and calcium ionophore A23187 alter protein synthesis and secretion by acini from rat mammary gland, *Int. J. Biochem.*, 14, 573, 1982.
99. Cantin, M. D., Ed., *Cell Biology of the Secretory Process,* Karger, New York, 1984.
100. Means, A. R. and Dedman, J. R., Calmodulin in endocrine cells and its multiple role in hormone action, *Mol. Cell. Endocrinol.*, 19, 215, 1980.
101. Means, A., Tash, J. S., and Chafouleas, J. G., Physiological implications of the presence, distribution, and regulation of calmodulin in eukaryotic cells, *Physiol. Rev.*, 62, 1, 1982.
102. Schubart, U. K., Erlichman, J., and Fleischer, N., Insulin release and protein phosphorylation: possible role of calmodulin, *Fed. Proc.*, 41, 2278, 1982.
103. DeLorenzo, R. J., The calmodulin hypothesis of neurotransmission, *Cell Calcium,* 2, 365, 1981.
104. Means, A. R., Tash, J. S., Chafouleas, J. G., Lagace, L., and Guerriero, V., Regulation of the cytoskeleton by Ca^{2+} calmodulin and cAMP. II. Endocytosis and cytoskeleton, *N.Y. Acad. Sci.*, 69, 383, 1982.
105. Watanabe, K. and West, W., Calmodulin, activated cyclic nucleotide phosphodiesterase, microtubules, and vinca alkaloids, *Fed. Proc.*, 41, 2292, 1982.
106. Riss, T. L. and Baumrucker, C. R., Calmodulin purification and quantitation from bovine mammary tissue, *J. Dairy Sci.*, 65, 1722, 1982.
107. Pizarro, M., Puente, J., and Sapag-Hagar, M., Calmodulin and cyclic nucleotide-phosphodiesterase activities in rat mammary gland during the lactogenic cyle, *FEBS Lett.*, 136, 127, 1981.
108. Hsie, A. W. and Puck, T. T., Morphological transformation of Chinese hamster cells by dibutyryl adenosine cyclic 3′:5′-monophosphate and testosterone, *Proc. Natl. Acad. Sci. U.S.A.*, 68, 358, 1971.
109. Johnson, G. S., Friedman, R. M., and Pastan, I., Restoration of several morphological characteristics of normal fibroblasts in sarcoma cells treated with adenosine-3′:5′-cyclic monophosphate and its derivatives, *Proc. Natl. Acad. Sci. U.S.A.*, 68, 425, 1971.
110. Prasad, K. N. and Hsie, A. W., Morphological differentiation of mouse neuroblastoma cells induced *in vitro* by dibutyryl adenosine 3′:5′-cyclic monophosphate, *Nature (London) New Biol.*, 233, 141, 1971.
111. DiPasquale, A. M., McGuire, J., Moellmann, G., and Wasserman, S., Microtubule assembly in cultivated Greene melanoma cells is simulated by dibutyrl adenosine 3′:5′-cyclic monophosphate or cholera toxin, *J. Cell Biol.*, 71, 735, 1976.
112. Puck, T. T., Waldren, C. A., and Hsie, A. W., Membrane dynamics in the action of dibutyrl adenosine 3′:5′-cyclic monophosphate and testosterone on mammalian cells, *Proc. Natl. Acad. Sci. U.S.A.*, 69, 1943, 1972.
113. Weissman, G., Goldstein, I., Hoffstein, S., and Tsung, P. K., Reciprocal effects of cAMP and cGMP on microtubule-dependent release of lysosomal enzymes, *Ann. N.Y. Acad. Sci.*, 253, 750, 1975.
114. Zurier, R. B., Hoffstein, S., and Weissman, G., Mechanisms of lysosomal enzyme release from human leukocytes. I. Effect of cyclic nucleotide and colchicine, *J. Cell Biol.* 58, 27, 1973.
115. Dustin, P., Secretion, exo- and endocytosis, in *Microtubules,* Springer-Verlag, New York, 1978, 284.
116. Garland, D. L., cAMP inhibits the *in vitro* assembly of microtubules, *Arch. Biochem. Biophys.*, 198, 335, 1979.
117. Rappaport, L., Leterrier, J. F., Virion, A., and Nunez, J., Phosphorylation of microtubule-associated proteins, *Eur. J. Biochem.*, 62, 539, 1976.

118. Jameson, L., Frey, T., Zeeberg, B., Dalldorf, F., and Caplow, M., Inhibition of microtubule assembly by phosphorylation of microtubule-associated proteins, *Biochemistry*, 19, 2472, 1980.
119. Brinkley, B. R., Pepper, D. A., Cox, S. M., Fistel, S., Brenner, K. S. L., Wible, L. J., and Pardue, R. L., Characteristics of centriole- and kinetochore-associated microtubule assembly in mammalian cells, in *Microtubules and Microtubule Inhibitors 1980*, DeBrabander, M. and DeMay, J., Eds., Elsevier/North-Holland, Amsterdam, 1980, 281.
120. Tash, J. S., Lagace, L., Lynch, D. R., Cox, S. M., Brinkley, B. R., and Means, A. R., Role of cAMP-dependent protein phosphorylation in microtubule assembly and function, in *Protein Phosphorylation*, Rosen, O. M. and Krebs, E. G., Eds., Cold Spring Harbor Laboratory, Cold Spring Harbor, N.Y., 1981, 1171.
121. Dallas, P. and Solomon, F., Cytoplasmic microtubule-associated proteins: phosphorylation at novel sites is correlated with their incorporation into assembled microtubules, *Cell*, 30, 407, 1982.
122. Soifer, D., Mack, K., and Chambers, D. A., Photoaffinity labelling of a microtubule-associated cyclic AMP-binding protein, *Arch. Biochem. Biophys.*, 219, 388, 1982.
123. Louis, S. L. and Baldwin, R. L., Changes in the cyclic 3′:5′-adenosine monophosphate system of rat mammary gland during the lactation cycle, *J. Dairy Sci.*, 58, 861, 1975.
124. Sapag-Hagar, M. and Greenbaum, A. L., Changes in the activities of adenyl cyclase and cAMP phosphodiesterase and of the level of 3′:5′-cyclic adenosine monophosphate in rat mammary gland during pregnancy and lactation, *Biochem. Biophys. Res. Commun.*, 53, 982, 1983.
125. Sapag-Hagar, M. and Greenbaum, A. L., The role of cyclic nucleotides in the development and function of rat mammary tissue, *FEBS Lett.*, 46, 180, 1974.
126. Rillema, J. A., Cyclic nucleotides, adenylate cyclase, and cyclic AMP phosphodiesterase in mammary glands from pregnant and lactating mice, *Proc. Soc. Exp. Biol. Med.*, 151, 748, 1976.
127. Loizzi, R. F., Cyclic AMP changes in guinea pig mammary gland and milk, *Am. J. Physiol., (Endocrinol. Metab. 8)*, 245, E549, 1983.
128. Sapag-Hagar, M., Greenbaum, A. L., Lewis, D. J., and Hallowes, R. C., The effects of dibutyryl cAMP on enzymatic and metabolic changes in explants of rat mammary tissue, *Biochem. Biophys. Res. Commun.*, 59, 261, 1974.
129. Rillema, J. A., Possible interaction of cyclic nucleotides with the prolactin stimulation of casein synthesis in mouse mammary gland explants, *Biochem. Biophys. Acta*, 432, 348, 1976.
130. Loizzi, R. F., Cyclic AMP inhibition of mammary gland lactose synthesis: specificity and potentiation by 1-methyl-3-isobutylxanthine, *Horm. Metab. Res.*, 10, 415, 1978.
131. Loizzi, R. F., de Pont, J. J., and Bonting, S. L., Inhibition by cyclic AMP of lactose production in lactating guinea pig mammary gland slices, *Biochem. Biophys. Acta*, 392, 20, 1975.
132. Wilde, C. J. and Kuhn, N. J., Lactose synthesis and the utilization of glucose by rat mammary acini, *Eur. J. Biochem.*, 13, 311, 1981.
133. Perry, J. W. and Oka, T., Cyclic AMP as a negative regulator of hormonally induced lactogenesis in mouse mammary gland organ culture, *Proc. Natl. Acad. Sci. U.S.A.*, 77, 2093, 1980.
134. Park, C. S., Smith, J. J., Eigel, W. N., and Keenan, T. W., Selected hormonal effects on protein secretion and amino acid uptake by acini from bovine mammary gland, *Int. J. Biochem.*, 10, 889, 1979.
135. Lacy, P. E., Howell, S. L., Young, D. A., and Fink, C. J., New hypothesis of insulin secretion, *Nature (London)*, 219, 1177, 1968.
136. Hall, P. F., The role of the cytoskeleton in endocrine function, in *Cellular Regulation of Secretion and Release*, Academic Press, New York, 1982, 195.
137. Lacy, P. E., Finke, E. H., and Codilla, R. C., Cinemicrographic studies on beta granule movement in monolayer culture of islet cells, *Lab. Invest.*, 33, 570, 1975.
138. Kanazawa, Y., Kawazu, S., Ikeuchi, M., and Kosaka, K., The relationship of intracytoplasmic movement of beta granules to insulin release in monolayer-cultured pancreatic beta-cells, *Diabetes*, 29, 953, 1980.
139. Montague, W., Howell, S. L., and Green, I. C., Insulin release and the microtubular system of the islets of Langerhans; effects of insulin secretagogues on microtubule subunit pool size, *Horm. Metab. Res.*, 8, 166, 1976.
140. Pipeleers, D. G., Pipeleers-Marichal, M. A., Sherline, P., and Kipnis, D. M., A sensitive method for measuring polymerized and depolymerized forms of tubulin in tissues, *J. Cell Biol.*, 74, 341, 1977.
141. Pipeleers, D. G., Pipeleers-Marichal, M. A., and Kipnis, D. M., Microtubule assembly and the intracellular transport of secretory granules in pancreatic islets, *Science*, 191, 88, 1975.
142. Shah, J. H., Stevens, B., and Sorenson, B. J., Dissociation of the effects of vincristine on stimulated insulin release and the pancreatic beta-cell microtubular structures in the intact rat, *Diabetes*, 30, 539, 1981.
143. Boyd, A. E., III, Bolton, W. E., and Brinkley, B. R., Microtubules and beta cell function: effect of colchicine on microtubules and insulin secretion *in vitro* by mouse beta cells, *J. Cell Biol.*, 92, 425, 1982.

144. Suprenant, K. A. and Dentler, W. L., Association between endocrine pancreatic secretory granules and *in vitro*-assembled microtubules is dependent upon microtubule-associated proteins, *J. Cell Biol.*, 93, 164, 1982.
145. Labrie, F., Gauthier, M., Pelletier, G., Borgeat, P., LeMay, A., and Gouge, J.-J., Role of microtubules in basal and stimulated release of growth hormone and prolactin in rat adenohypophysis *in vitro*, *Endocrinology*, 93, 903, 1973.
146. Antakly, T., Pelletier, G., Zeytinoglu, F., and Labrie, F., Effects of colchicine on the morphology and prolactin secretion of rat anterior pituitary cells in monolayer culture, *Am. J. Anat.*, 156, 353, 1979.
147. Pelletier, G. and Bornstein, M. B., Effect of colchicine on rat anterior pituitary gland in tissue culture, *Exp. Cell Res.*, 70, 221, 1972.
148. Sherline, P., Lee, Y-C., and Jacobs, L. S., Binding of microtubules to pituitary secretory granules and secretory granule membranes, *J. Cell Biol.*, 72, 380, 1977.
149. Redman, C. M., Banerjee, D., Howell, K., and Palade, G. E., Colchicine inhibition of plasma protein release from rat hepatocytes, *J. Cell Biol.*, 66, 42, 1975.
150. Prentki, M., Gabbiani, G., Chaponnier, C., and Jeanrenaud, B., Microtubules and actin microfilaments in lipoprotein secretion by isolated rat hepatocytes, in *Lipoproteinolism and Endocrine Regulation*, Hessel, L. W. and Krans, H. M. J., Eds., Elsevier/North-Holland, Amsterdam, 1979, 239.
151. Nickerson, S. C., Smith, J. J., and Keenan, T. W., Role of microtubules in milk secretion — action of colchicine on microtubules and exocytosis of secretory vesicles in rat mammary epithelial cells, *Cell Tissue Res.*, 207, 361, 1980.
152. Nickerson, S. C., Smith, J. J., and Keenan, T. W., Ultrastructural and biochemical response of rat mammary epithelial cells to vinblastine sulfate, *Eur. J. Cell Biol.*, 23, 115, 1980.
153. Smith, J. J., Nickerson, S. C., and Keenan, T. W., Metabolic energy and cytoskeletal requirements for synthesis and secretion by acini from rat mammary gland. I. Ultrastructural and biochemical aspects of synthesis and release of milk proteins, *Int. J. Biochem.*, 14, 87, 1982.
154. Akers, R. M. and Nickerson, S. C., Effect of prepartum blockade of microtubule formation on milk production and biochemical differentiation of the mammary epithelium in Holstein heifers, *Eur. J. Biochem.*, 15, 771, 1983.
155. Smith, J. J., Nickerson, S. C., and Keenan, T. W., Metabolic energy and cytoskeletal requirements for synthesis and secretion by acini from rat mammary gland. II. Intracellular transport and secretion of protein and lactose, *Int. J. Biochem.*, 14, 99, 1982.
156. Guerin, M. A. and Loizzi, R. F., Inhibition of mammary gland lactose secretion by colchicine and vincristine, *Am. J. Physiol. (Cell Physiol.)*, 234, C177, 1978.
157. Malaisse-Lagae, F., Amherdt, M., Ravazzola, M., Sener, A., and Hutton, J. C., Role of microtubules in the synthesis, conversion, and release of (pro)insulin, *J. Clin. Invest.*, 63, 1284, 1979.
158. Kemper, B., Habener, J. F., Rich, A., and Potts, J. T., Microtubules and the intracellular conversion of proparathyroid hormone to parathyroid hormone, *Endocrinology*, 96, 903, 1975.
159. Chu, L. L. H., MacGregor, R. R., and Cohn, D. V., Energy-dependent intracellular translocation of proparathormone, *J. Cell Biol.*, 72, 1, 1977.
160. Taylor, A., Maffly, R., Wilson, L., and Reaven, E., Evidence for involvement of microtubules in the action of vasopressin, *Ann. N.Y. Acad. Sci.*, 253, 723, 1975.
161. Dousa, T. P. and Valtin, H., Cellular actions of vasopressin in the mammalian kidney, *Kidney Int.*, 10, 46, 1976.
162. Ollivier-Bousquet, M. and Denamur, M. R. D., Inhibition par la colchicine de la secretion des proteines du lait, *C.R. Acad. Sci. Paris*, 276, 2183, 1973.
163. Patton, S., Reversible suppression of lactation by colchicine, *FEBS Lett.*, 48, 85, 1974.
164. Knudson, C. M., Stemberger, B. H., and Patton, S., Effects of colchicine on ultrastructure of the lactating mammary cell: membrane involvement and stress on the Golgi apparatus, *Cell Tissue Res.*, 195, 169, 1978.
165. Loizzi, R. F., dePont, J. J., and Bonting, S. L., Inhibition by cyclic AMP of lactose production in lactating guinea pig slices, *Biochem. Biophys. Acta*, 393, 20, 1975.
166. Amato, P. A. and Loizzi, R. F., The effects of cytochalasin B on glucose transport and lactose synthesis in lactating mammary gland slices, *Eur. J. Cell Biol.*, 20, 150, 1979.
167. Sherline, P., Bodwin, C. K., and Kipnis, D. M., A new colchicine binding assay to tubulin, *Anal. Biochem.*, 62, 400, 1974.
168. Rapaport, E., Berkley, P. D., and Bucher, N. L. R., Charcoal adsorption assay for measurement of colchicine binding and tubulin content of crude tissue extracts, *Anal. Biochem.*, 69, 92, 1975.
169. Sherline, P. and Mundy, G. R., Role of the tubulin-microtubule system in lymphocyte activation, *J. Cell Biol.*, 74, 371, 1977.
170. Filner, P. and Behnke, O., Stabilization and isolation of brain microtubules with glycerol and dimethylsulfoxide (DMSO), *J. Cell Biol.*, 59, 99a, 1973.

171. Rubin, R. W. and Weiss, G. D., Direct biochemical measurements of microtubule assembly and disassembly in Chinese hamster ovary cells, *J. Cell Biol.*, 64, 42, 1975.
172. Pipeleers, D. G., Pipeleers-Marichal, M. A. P., and Kipnis, D. M., Physiological regulation of total tubulin and polymerized tubulin in tissues, *J. Cell Biol.*, 74, 351, 1977.
173. Beertsen, W., Heersche, J. N. M., and Aubin, J. E., Free and polymerized tubulin in cultured bone cells and Chinese hamster ovary cells: the influence of cold and hormones, *J. Cell Biol.*, 95, 387, 1982.
174. Loizzi, R. F., Ovariectomy-induced tubulin polymerization in pregnant rat mammary gland, *Proc. Soc. Exp. Biol. Med.*, 173, 252, 1983.
175. Loizzi, R. F., Immunofluorescent localization of tubulin in mammary gland from lactating and pregnant guinea pig, *J. Cell Biol.*, 87, 245a, 1980.
176. Weber, K., Bibring, Th., and Osborn, M., Specific visualization of tubulin-containing structures in tissue culture cells by immunofluorescence. Cytoplasmic microtubules, vinblastine-induced paracrystals, and mitotic figures, *Exp. Cell Res.*, 95, 111, 1975.
177. Asch, B. B., Medina, D., and Brinkley, B. R., Microtubules and actin-containing filaments of normal, preneoplastic, and neoplastic mouse mammary epithelial cells, *Cancer Res.*, 39, 893, 1979.
178. Pencek, P. F. and Loizzi, R. F., Morphological characterization and tubulin status of several populations of alveolar cells isolated from lactating guinea pig mammary gland, *J. Cell Biol.*, 91, 325a, 1981.
179. Kraehenbuhl, J. P., Dispersed mammary gland epithelial cells. I. Isolation and separation procedures, *J. Cell. Biol.*, 72, 390, 1977.
180. Amato, P. A. and Loizzi, R. F., The identification and localization of actin and actin-like filaments in lactating guinea pig mammary gland alveolar cells, *Cell Motil.*, 1, 329, 1981.
181. Turner-Pencek, P. F. and Loizzi, R. F., A morphometric study of isolated cell populations from pregnant and lactating guinea pig mammary gland, *J. Cell Biol.*, 95, 39a, 1982.
182. Loizzi, R. F. and Pencek, P. F., Ultrastructure of cell populations isolated from pregnant and lactating mammary gland, *J. Cell Biol.*, 95, 55a, 1982.
183. Loizzi, R. F., Tubulin polymerization in rat mammary gland cells, *J. Cell Biol.*, 99, 42a, 1984.
184. Tokunaka, S., Friedman, T. M., Toyama, Y., Pacifici, M., and Holtzer, H., Taxol induces microtubule-rough endoplasmic reticulum complexes and microtubule-bundles in cultured chondroblasts, *Differentiation*, 24, 39, 1983.
185. Kuhn, N. J., Progesterone withdrawal as the lactogenic trigger in the rat, *J. Endocrinol.*, 44, 39, 1969.
186. Nicholas, K. R. and Hartmann, P. E., Progressive changes in plasma progesterone, prolactin and corticosteroid levels during late pregnancy and the initiation of lactose synthesis in the rat, *Aust. J. Biol. Sci.*, 34, 445, 1981.
187. Nicholas, K. R. and Hartman, P. E., Progesterone control of the initiation of lactose synthesis in the rat, *Aust. J. Biol. Sci.*, 34, 435, 1981.
188. Deis, R. P. and DeLouis, C., Lactogenesis induced by ovariectomy in pregnant rats and its regulation by oestrogen and progesterone, *J. Steroid Biochem.*, 18, 687, 1983.
189. Vermouth, N. T. and Deis, R. P., Prolactin release and lactogenesis after ovariectomy in pregnant rats: effect of ovarian hormones, *J. Endocrinol.*, 63, 13, 1974.
190. Amenomori, Y., Chen, C. L., and Meites, J., Serum prolactin levels in rats during different reproductive states, *Endocrinology*, 86, 506, 1970.
191. Kuhn, N. J., Carrick, D. T., and Wilde, C. J., Lactose synthesis: the possibilities of regulation, *J. Dairy Sci.*, 63, 328, 1980.
192. Simpson, A. A., Simpson, M. H. W., and Kulkarni, P. N., Effect of perphenazine during late pregnancy on prolactin production and lactogenesis in the rat, *J. Endocrinol.*, 57, 431, 1973.
193. Loizzi, R. F., Progesterone withdrawal stimulates mammary gland tubulin polymerization in pregnant rats, *Endocrinology*, 116, 2543, 1985.
194. Josefsberg, Z., Posner, B. I., Patel, B., and Bergeron, J.J. M., The uptake of prolactin into female rat liver. Concentration of intact hormone in the Golgi apparatus, *J. Biol. Chem.*, 254, 209, 1979.
195. Kelly, P. A., Shiu, R. P. C., Robertson, M. C., and Friesen, H. G., Characterization of rat chorionic mammotropin, *Endocrinology*, 97, 621, 1975.
196. Glaser, L. A., Kelly, P. A., and Gibori, G., Differential action and secretion of rat placental lactogens, *Endocrinology*, 115, 969, 1984.

Chapter 10

PROTEIN PHOSPHORYLATION OF PROLACTIN TARGET TISSUE: MAMMARY GLAND

Charles L. Brooks

TABLE OF CONTENTS

I. Introduction 180

II. Methods of Study 180

III. Classification of Protein Kinases 181
- A. Cyclic Nucleotide-Dependent Protein Kinases 181
- B. Calcium-Dependent Protein Kinases 182
- C. Calcium- and Cyclic Nucleotide-Independent Protein Kinases 183

IV. Specific Protein Kinases in Mammary Tissues 184
- A. Protein Kinases Involved in Fatty Acid Synthesis 184
  - 1. Pyruvate Dehydrogenase 184
  - 2. ATP Citrate Lyase 184
  - 3. Acetyl-CoA Carboxylase 185
- B. Mammary Casein Kinases 187
- C. Nuclear Protein Kinases 190
- D. Other Protein Kinase Activities in Mammary Gland 191

References 192

I. INTRODUCTION

The goal of this article is to review protein phosphorylation in one prolactin target tissue, the mammary gland. In relating the molecular actions of prolactin to protein phosphorylations one will realize that few, if any, prolactin-regulated protein kinases or phosphatases have yet been documented. Perhaps researchers have put great efforts into this area and have been fruitless, or perhaps work needs to be conducted to elucidate such reactions. I believe the latter to be the reality, and hope the current review enlightens the reader as to the strategies and methods used to investigate phosphorylation reactions.

The phosphorylation of proteins occurs as an enzymatically mediated posttranslational modification. Phosphate is covalently attached to the side group of either serine, threonine, or tyrosine. Phosphorylation of proteins may be permanent, such as caseins,[1,2] or may be transient, as the phosphorylation and dephosphorylation reactions of substrate enzymes such as glycogen synthase.[3] Phosphorylation is catalyzed by a group of enzymes termed "protein kinases", while the dephosphorylations are mediated by phosphatases. Under laboratory conditions, kinases have been run backwards to dephosphorylate substrates.[4]

Protein kinases have been classified by a number of criteria, including endogenous or exogenous substrate specificity (i.e., phosphorylase kinase or casein kinase I and II), regulator molecules (i.e., cAMP-dependent protein kinases), or phosphorylation site specificities (i.e., tyrosine kinases). Clearly, there is problem in relating a newly discovered protein kinase to those already well characterized. Clarification will occur when investigators thoroughly characterize these enzymes in terms of physical properties, cellular locale, physiological substrate, regulatory molecules, and site specificity. A good example of where this information has been obtained and provided clarity is that of the enzymes of glycogen metabolism.[5]

In many systems, the easiest classification of kinase enzymes is by the mechanisms that regulate these kinases. A few of the well-characterized and easily testable kinase regulatory mechanisms are (1) cyclic adenosine monophosphate regulation, (2) cyclic guanosine monophosphate regulation, (3) calcium regulation (4) calcium-dependent calmodulin regulation, and (5) calcium-activated phospholipid-dependent regulation.

In addition, there are protein kinases that do not seem to be regulated by cyclic nucleotides, calcium, calmodulin, or phospholipids. These latter activities are often controlled by the concentrations of substrate end products and the specific K_ms of the kinase for these molecules. The regulation of protein kinases by the binding of these relatively small ligands is acute, whereas the regulation of the cellular concentration of the protein kinases themselves results in a long-term control of kinase activity. Although a number of enzymes important to mammary secretion are modulated by both long-term and acute regulation,[6,7] the modulation of the amounts and activities of the protein kinases identified in mammary gland have not been studied.

II. METHODS OF STUDY

Protein phosphorylation has been successfully studied in several different experimental systems. These include in vivo studies, whole cell homogenates or subcellular organelles, and isolated protein systems. Each methodology has distinct advantages and disadvantages, but all are required for full characterization of a protein phosphorylation.

Study of protein phosphorylation in vivo requires the injection of radioactive orthophosphate into prepared animals or its introduction into incubations of tissue slices, explants, or various cell preparations. After sufficient time, the cells of interest are manipulated and subsequently homogenized in media that will stop the kinases as well as the phosphatases. The phosphorylated proteins are then examined after electrophoresis, specific immunoprecipitation, cell fractionation, or isolation of a specific substrate. The advantages of the whole cell approach are listed by Garrison[8] and include: (1) the use of kinases and substrates in their physiological state, (2) Hormonal stimulation studies can be performed to catalog all responsive phosphorylations, and (3) Hormonal stimulation studies can be performed despite the incomplete understanding of the reactions mediating the receptor-binding event to the phosphorylation reaction. Several disadvantages of in vivo studies are the kinases, phosphatases, and their controllers are not easily identified by these methods and the identities of the substrates are not always apparent. Finally, in whole animal studies, the applied stimulus may not be that which directly alters protein phosphorylation in the cell type of interest. Studies using isolated cell systems require documentation of the physiological state of the cell preparations. This includes such tests as the continued uptake and metabolism of glucose, equilibration of cellular ^{32}P, and confirmation that the stimulus does not alter the specific activity or amount of labeled nucleotide pool in a cell. On the whole, in vivo studies are necessary to demonstrate the physiological relevance of the process which can then be better characterized by other methodologies.

Protein phosphorylation studies can be performed on cellular homogenates or isolated subcellular fractions. These studies often will allow identification of the ion requirements of a specific phosphorylation reaction and identification of regulators such as calmodulin, cyclic nucleotides, and diacylglycerol, because buffer composition can be manipulated; however, it is a poor system to unambiguously define the reaction. In addition, with the disruption of cell morphology, nonphysiological phosphorylations may occur and physiological phosphorylations may be lost. In homogenates, organelle preparations, and partially pure preparations, the concentrations of endogenous cellular regulators and inhibitors are radically changed, and thus their effects may not be identified. Finally, care must be exercised to add physiologically relevant concentrations of exogenous reaction ingredients.

Ultimately, the use of isolated proteins in totally defined, reconstituted systems will allow the elucidation of the specific characteristics of the phosphorylation reaction. These include determinations of substrate, regulator inhibitor, and ion affinities. These data must then be extrapolated back to knowledge of whole cell concentrations of such substances to determine their physiological relevance.

III. CLASSIFICATION OF PROTEIN KINASES

Protein kinases have been classified by a number of properties, including their substrate specificities,[9] biochemical characteristics,[10] or regulatory mechanisms. Perhaps the simplest method of classification is by the mechanism of regulation followed by identification of their physiological substrate. Ultimately, classification of protein kinases by the physical properties of the enzyme and substrates, such as amino acid sequence, domain, and subunit composition, should allow a more rational classification of protein kinases relating them to their evolutionary histories.

A. Cyclic Nucleotide-Dependent Protein Kinases

The cyclic nucleotide-dependent protein kinases have received the most intensive study. A large number of cyclic nucleotide-dependent protein kinases have been purified and characterized.[11] Cyclic adenosine monophosphate and cyclic guanosine monophosphate have been identified as the regulators of these enzymes.

Cyclic AMP-dependent protein kinases are composed of two catalytic subunits (C) and two regulatory subunits (R). The inactive tetramer (R_2C_2) is activated by cyclic AMP. Two moles of cyclic AMP are bound per mole of regulatory subunit, allowing dissociation of the regulatory and catalytic subunits.[11] The regulatory subunit and holoenzyme are quite acidic, while the catalytic subunit is a basic protein.[12] In addition, a heat-stable inhibitor protein of cyclic AMP-dependent protein kinase is found in many tissues. The inhibitor protein binds to the free catalytic subunit and blocks activity.[13,14] Similar enzymes are present in mammary tissue, as well as a large number of other tissues.[15] A number of substrate proteins for cyclic AMP-dependent protein kinase have been identified.[11,16]

Cyclic guanosine-dependent protein kinase was identified by Kuo and Greengard in 1970.[17] The activity of cyclic GMP-dependent protein kinase was low compared to the cyclic AMP-dependent enzyme, and the physiological substrates of these enzymes were less clearly defined. Several of the best-studied systems include the cyclic GMP-dependent lipase kinase[18] and the phosphorylation of several nuclear high-mobility group proteins.[19,20]

B. Calcium-Dependent Protein Kinases

What were once thought to be a group of calcium-regulated protein kinases are recognized as two distinct types of protein kinases. The first of these are the calcium- and calmodulin-dependent protein kinases whose activities are controlled by combination of the catalytic subunit with a calcium-calmodulin complex.[21] The second category of calcium-requiring enzymes bind calcium directly, are dependent on phospholipid, and are regulated by diacylglycerol. This latter enzyme is termed C-kinase.[22] The control of both these enzymes obviously depends on the concentration of calcium and the affinity of calcium and ligand. Recent articles have reviewed the control of calcium concentrations in secretory cells[23] and mammary epithelial cells.[24]

Calmodulin is a highly conserved, ubiquitous protein of the family that includes parvalbumin, the vitamin D-dependent calcium-binding protein, troponin C, and the S-100 proteins.[25] Calmodulin contains four high-affinity binding sites for calcium (10^{-5} M affinity). Occupation of these sites promotes a conformational change in the protein structure.[26] Calcium-loaded calmodulin binds to several protein kinases, including myosin light-chain kinase,[27,28] a glycogen synthase kinase,[29] tyrosine hydroxylase kinase,[30] synapsin kinase,[31] and tubulin kinase.[32] Calmodulin is the δ subunit of phosphorylase kinase.[33] Mammary tissue has recently been shown to contain calmodulin,[34] and several calcium- and calmodulin-dependent protein kinases have been observed in mammary tissue,[2,35] The calcium- and calmodulin-dependent enzymes, unlike the cyclic AMP-dependent kinases, have different subunit and holoenzyme structures with varying substrate specificities, although several of these enzymes are proving to be similar or perhaps identical.[36]

The calcium- and phospholipid-dependent protein kinase was first noted in brain tissues.[37] The simultaneous addition of anionic phospholipids and calcium increased the kinase activity.[38] Finally, the addition of micromolar concentrations of diacylglycerol dramatically increases the activity of the kinase by decreasing the K_m of calcium far below the resting cellular concentrations.[39] Protein kinase C is present in abundance in many tissues and is a single subunit with a molecular weight by SDS-gel electrophoresis of 82 kdalton.[40,41] The production of diacylglycerol and phosphoinositol by the inositol cycle and the relationship of those compounds to arachadonic acid production and increased intracellular calcium transport has recently been reviewed.[22,42-46] The nature of the physiological substrates for C-kinase is unknown with the exception of glycogen synthase[47,48] and, as yet, this protein kinase has not been identified in mammary tissue. Phospholipase C, however, is present in mammary tissues and causes prolactin-like responses.[49] Phospholipase C is known to mediate the production of diacylglycerol by the inositol cycle.[45]

C. Calcium- and Cyclic Nucleotide-Independent Protein Kinases

A final class of protein kinases are regulated by neither cyclic nucleotides nor calcium. Enzymes in this class can be found in a wide variety of tissues. Chief among this group are casein kinase I and II,[50] mammary casein kinase,[51] pyruvate dehydrogenase kinase,[52] glycogen synthase kinase,[53] and initiation factor IF-2 kinase.[53]

Casein kinase I and II are found in a wide variety of mammalian and avian tissues as well as several plant species.[50] They were initially separated by gradient elution from DEAE-cellulose.[54]

Casein kinase I is found in several subcellular fractions with the molecular weight of this monomer being 37 kdaltons.[55] This enzyme requires magnesium, is substrate specific for ATP, and is found autophosphorylated with 1 to 4 mol of phosphate per mole of enzyme. The phosphorylation site specificity requires a glutamic acid at the N-2 position (two residues toward the amino end),[56,57] as well as an acidic or neutral residue at N-3. When using mixed casein as a substrate, complete *apriori* dephosphorylation of the substrate reduces the rate of casein kinase I by 16-fold. The K_m for these acidic substrates is approximately 1 mg/mℓ.

Casein kinase II has an $\alpha\alpha'\beta_2$ or $\alpha_2\beta_2$ subunit structure, the α subunit being 35 to 44 kdaltons, while the β subunit being 24 to 26 kdaltons.[54] The β subunit is likely the catalytic site,[58] and either ATP (4 to 15 $\mu M = K_m$) or GTP (7 to 40 $\mu M = K_m$) act as phosphate donors. As with casein kinase I, magnesium is required for activity. Threonine is the major phosphorylation site in native casein,[56] while in partially dephosphorylated preparations, serine becomes the preferred site.[57,59] The K_m for the various caseins is approximately 1.1 mg/mℓ.[56] Casein kinase II, whether phosphorylating either serine or threonine, requires either a glutamic acid or phosphoserine in the N + 1 position and either glutamic acid or aspartic acid in the N + 2 position.[54] Casein kinase II isolated from rabbit erythrocytes is inhibited by 2,3-diphosphoglycerate[60] or heparin.[61] These and several other acidic compounds seem to compete with the casein substrate. Polyamines activate this enzyme[62]

The endogenous substrates for casein kinase I and II are similar, although the specific sites of phosphorylation are unique. Substrates include ribosomal proteins, initiation factors, spectrin, RNA polymerase, high-mobility group protein 17, and glycogen synthase. These are reviewed by Hathaway and Traugh.[50]

Pyruvate dehydrogenase is the first enzyme in the pathway leading eventually to lipid synthesis. Pyruvate dehydrogenase, a large complex located on the inner mitochondral wall,[63] produced acetyl-CoA. This enzyme was fist noted to be controlled by a phosphorylation-dephosphorylation cycle in 1969 by Linn et al.[64] The mammary gland seems to regulate pyruvate dehydrogenase by phosphorylation and dephosphorylation cycles controlled by insulin and, perhaps, modified by prolactin. The mammary kinase and phosphatase have been neither isolated nor characterized.

Tyrosine-specific protein kinases have been recently discovered.[65] Only several tenths of 1% of cellular protein phosphorylations are on tyrosines. These protein kinases are associated with both viral- and cellular-transforming genes. Tyrosine-specific kinases are also related to protein receptors involved in the control of cellular growth such as the insulin and epidermal growth factor receptor.[66-68] Despite their association with cellular growth, tyrosine-specific kinases are present in terminally differentiated cells.[69] The substrates for these kinases are not characterized other than the autophosphorylation of growth factor receptors. Several tyrosyl-protein phosphatases have been recognized.[70]

IV. SPECIFIC PROTEIN KINASES IN MAMMARY TISSUES

A. Protein Kinases Involved in Fatty Acid Synthesis

Fatty acid synthesis occurs in the cytosolic fatty acid synthase complex. The substrates for fatty acid synthesis are malonyl-CoA, produced by acetyl-CoA carboxylase from acetyl-CoA. Acetyl-CoA is produced in the mitochondria by pyruvate dehydrogenase and is transported from the mitochondria to the cytosol by a transport system that couples the acetyl group to oxaloacetate to form citrate, which may leave the mitochondria. Once citrate enters the cytoplasm, it is again converted to acetyl-CoA by citrate lyase.

In the mammary gland, enzymes such as acetyl-CoA carboxylase and pyruvate dehydrogenase have been suggested as the rate-limiting steps in fatty acid biosynthesis. The rates of these, and several other enzymes, have been shown to be regulated by phosphorylation-dephosphorylation cycles.

1. Pyruvate Dehydrogenase

The structure and control of pyruvate dehydrogenase has been well studied in liver, muscle, and adipose tissue.[71] An associated protein kinase can be isolated from the enzyme complex, which phosphorylates the α subunit of the pyruvate decarboxylase components.[72] The activity of this kinase is controlled by insulin[73] in vivo and by a yet to be characterized soluble insulin-stimulated second messenger in vitro.[74] A phosphatase has also been described.[71,75] The enzymatic activity of pyruvate dehydrogenase is inactivated by phosphorylation.

In mammary tissue of rats, the total and active fractions of pyruvate dehydrogenase are regulated during pregnancy, lactation, and weaning.[76] This activity possesses many of the sensitivities of pyruvate dehydrogenase examined in other tissues, and therefore was considered to be a similar enzyme complex to those reported in other tissues.[71] The effect of insulin and prolactin manipulation on mammary pyruvate dehydrogenase activity showed that a 24-hr treatment by CB-154 significantly reduced the active pyruvate dehydrogenase pool, while not influencing the total enzyme. Insulin treatment 2 hr prior to the removal of tissues for assay counteracted the effects of CB-154 and increased both active and total pyruvate dehydrogenase activity.[77] Prolactin (2 mg) injected at the time of CB-154 administration failed to alter the active pyruvate dehydrogenase pool or influence the total enzyme. No data on serum hormone concentrations, pyruvate dehydrogenase kinase activity, or pyruvate dehydrogenase-associated phosphate were reported. Short-term reduction of insulin by streptozotocin dramatically reduced the active pyruvate dehydrogenase without affecting the total enzyme. Starvation decreases the active pyruvate dehydrogenase by 75%, while injection of insulin and glucose 30 min prior to sacrifice restored the activity.[78] Using the total and active fractions of pyruvate dehydrogenase in extracts of mitochondria as an assay end point, the kinase is shown to have increased activity in preparations from starved rats, while the phosphatase activity declined. As in other tissues,[71] the site of kinase phosphorylation is on the α subunit of the pyruvate decarboxylase component of the pyruvate dehydrogenase complex.[79]

The data relating prolactin and the regulation of pyruvate dehydrogenase by a phosphorylation/dephosphorylation cycle does not prove such a mechanism. Although decreases in prolactin influence the active pyruvate dehydrogenase pool, this may be accomplished by regulation of the insulin responsiveness of the mammary epithelial cells.[80]

2. ATP Citrate Lyase

Acetyl-CoA is synthesized in the mitochondria of cells and is a major substrate for fatty acid synthesis in various tissues of nonruminant species. This substrate is transported from the mitochondria by conversion to citrate, passage from the mitochondria, and transfer of an acetyl group to CoA to form cytosolic acetyl-CoA.[81] The cytoplasmic enzyme that performs the last reaction is ATP-citrate lyase. Hepatic ATP-citrate lyase is regulated by insulin and glucagon.[82]

Purified rat liver enzyme contains an acid-stable phosphate group.[83] This phosphorylation occurs in vivo and is regulated by insulin and glucagon.[84,85] Therefore, the short-term activity of ATP-citrate lyase may be regulated by a phosphorylation/dephosphorylation cycle.

The lyase enzyme was similar in molecular weight to the rat liver enzyme (116 kdaltons). ATP-citrate lyase has been purified to homogeneity from rat mammary gland and is phosphorylated by the exogenous catalytic subunit of cyclic AMP-dependent protein kinase (from rabbit skeletal muscle).[86] Incubation of the in vitro ^{32}P-labeled ATP-citrate lyase with purified phosphatases removed all ^{32}P, but failed to remove significant amounts of endogenous phosphate. The in vitro phosphorylation of ATP-citrate lyase by the catalytic subunit of cyclic AMP-dependent protein kinase had no effect on the V_{max} or substrate K_m of the lyase enzyme.[87] Incubation of isolated enzyme with γ-labeled[^{32}P]-ATP showed a slow labeling of the enzyme, presumably by the endogenous protein kinase.

The data indicate that the catalytic subunit of the cyclic AMP-dependent protein kinase is probably not an endogenous kinase regulating ATP-citrate lyase activity. The endogenous kinase which copurifies with ATP-citrate lyase has yet to be shown to regulate the lyase activity. No data regarding hormonal control of ATP-citrate lyase in the mammary gland is available, although this activity is modulated during lactation.[7]

3. *Acetyl-CoA Carboxylase*

Acetyl-CoA carboxylase catalyzes the conversion of acetyl-CoA to the main carbon source substrate for fatty acid synthesis, specifically, malonyl-CoA. This reaction is believed to be the rate-limiting step in fatty acid synthesis[88] and controlled by phosphorylation/dephosphorylation reactions.[89,90] Purified acetyl-CoA carboxylase is a biotin- and phosphate-containing enzyme with a reported molecular weight between 240 and 270 kdaltons,[91] is dependent on citrate for activity, and is inhibited by low concentrations of palmityl-CoA.[92]

Liver preparations of acetyl-CoA carboxylase are phosphorylated and inactivated by incubation with magnesium ions and [$\gamma^{32}P$]ATP.[93,94] The acetyl-CoA carboxylase is also phosphorylated by exogenous catalytic subunit of cyclic AMP-dependent protein kinase.[90] The phosphoenzyme may be activated by treatment with exogenous phosphatase.[95] Acetyl-CoA carboxylase becomes phosphate labeled in vivo when fat cells are incubated with radioactive orthophosphate.[96]

Hormones that are known to raise intracellular concentrations of cyclic AMP, such as glucagon in liver or adrenaline in adipocytes, have been shown to increase acetyl-CoA carboxylase phosphorylation in vivo.[89,97,98]

Partially purified preparations of acetyl-CoA carboxylase from bovine mammary gland lost activity when incubated with ATP prior to assay. The enzyme was phosphorylated in vitro. The phosphorylation and inactivation were shown to be tightly correlated.[99] Acetyl-CoA carboxylase, when purified from lactating rat mammary gland, was phosphorylated in the presence of ATP and magnesium at a slow rate. The addition of cyclic AMP to this mixture increased the phosphorylation of acetyl-CoA carboxylase, while the addition of cyclic AMP-dependent protein kinase inhibitor reduced phosphorylation to basal levels.[100] Therefore, at least cyclic AMP-dependent protein kinase and one or more cyclic AMP-independent protein kinases are closely associated with acetyl-CoA carboxylase. In homogenates of lactating rat mammary gland, addition of cyclic AMP failed to alter the activity of acetyl-CoA carboxylase.[101] The disparity of cyclic AMP mediated reduction of the carboxylase in homogenates and purified enzyme preparations may be due to the lack of inhibitors of proteolytic enzymes in the homogenate or due to the presence of vigorous endogenous phosphatase activities. The literature does not resolve this point, as the physical characteristics of the carboxylase or the presence of phosphatase activities were not determined.[101]

Acetyl-CoA carboxylase was phosphorylated at a physiologically relevant rate by the addition of exogenous catalytic subunit of cyclic AMP-dependent protein kinase. Whether this kinase could regulate the carboxylase activity was addressed in a subsequent article[102] using acetyl-CoA carboxylase purified from rat mammary gland and the catalytic subunit of cyclic AMP-dependent protein kinase and its inhibitor protein isolated from rabbit skeletal muscle. In the presence of cyclic AMP-dependent protein kinase inhibitor protein, ^{32}P was slowly incorporated in acetyl-CoA carboxylase accompanied by a slight decrease in activity. The incubation of acetyl-CoA carboxylase with the catalytic subunit of cyclic AMP-dependent protein kinase quickly phosphorylated the carboxylase, corresponding to a dramatic reduction in carboxylase activity. The stochiometry of phosphorylation suggests that one phosphorylation per carboxylase greatly reduces its activity, but that an additional phosphate can be added with little or no change in activity. Protein phosphatase-1 (from rabbit skeletal muscle) removed phosphate from the carboxylase with a corresponding increase in activity.

Tryptic peptide analysis of acetyl-CoA carboxylase suggests that control of the activity of this enzyme occurs by phosphorylation of a site residing on a fragment of pI approximately 6. Phosphorylation at this site reduces the V_{max} and increases the dissociation constant for citrate, each by twofold. Comparison of in vitro sites of phosphorylation of mammary acetyl-CoA carboxylase with those of acetyl-CoA carboxylase phosphorylated in whole fat cell preparations under the influence of adrenalin reveal several tryptic phosphopeptides common to both preparations. These have isoelectric points of approximately 7.1, 4.3, 3.6, 3.4, and 3.1.[103]

Mild trypsin treatment of the enzyme cleaves at 25-kdalton phosphopeptide. Cleavage of this phosphopeptide from the holoenzyme increases the decarboxylase activity, but not to the same degree achieved by treatment with rabbit skeletal muscle protein phosphatase I.[104]

The phosphate content of acetyl-CoA carboxylase has been reported to be as high as 6.2 moles of phosphate per mole of enzyme subunit.[105] This data suggests there may be several phosphorylation sites other than the major pI = 6 tryptic peptide that can regulate the carboxylase activity. This is substantiated by the observation that, although protein phosphatase-1 can remove phosphates from acetyl-CoA carboxylase and increase the specific activity of purified preparations, the subsequent phosphorylations catalyzed by cyclic AMP-dependent protein kinase and the cyclic AMP-independent protein kinase cannot reduce the carboxylase activity to those of nonphosphatase-treated enzymes.

Recently, three cyclic AMP-independent acetyl-CoA carboxylase kinases have been separated from cytosol of lactating rat mammary gland.[106] These three kinases are not effected by cyclic AMP or the inhibitor of cyclic AMP-dependent protein kinase.

The first of these kinases (acetyl-CoA carboxylase kinase A) has an apparent molecular weight of 76 kdaltons. This activity is believed to be identical to kinase-2, earlier identified when it copurified with acetyl-CoA carboxylase.[100] The molecular weight of this kinase is estimated by gel permeation chromatography as being 76 kdaltons. Kinase A phosphorylates in a similar tryptic peptide to both endogenous kinase-2 and the catalytic subunit of cyclic AMP-dependent protein kinase. In a manner again similar to the cyclic AMP-dependent protein kinase, phosphorylation by kinase A can reduce the activity of acetyl-CoA carboxylase in a phosphatase-reversible manner by lowering the reaction rate and increasing the K_a for citrate. This enzyme places 1.5 mol of phosphate per mole of enzyme. It remains to be proven that cyclic AMP-dependent protein kinase and kinase A phosphorylate identical sites, but both seem to phosphorylate a carboxylase domain which results in similar effects. Kinase A appears to have substantially different substrate specificities than either kinases B_1, B_2, or cyclic AMP-dependent protein kinase.

Acetyl-CoA carboxylase kinase B_1 has a molecular weight of approximately 186 kdaltons. The kinase phosphorylates and inactivates acetyl-CoA carboxylase by reduction of the velocity of the reaction, but not by altered carboxylase affinity for citrate. Kinase B_1 phosphorylated the carboxylase on a tryptic peptide not phosphorylated by either kinase B_2 or cyclic AMP-dependent protein kinase. Kinase B_1 autophosphorylated a 26 kdalton subunit when incubated in the presence of magnesium and ATP. Phosvitin and casein were the most active substrates for kinase B_1. Kinase B_1 resembles the previously described casein kinase II in terms of mass, subunit composition, substrate preference, elution from phosphocellulose, and heparin sensitivities.

Acetyl-CoA carboxylase kinase B_2 phosphorylates its substrate and reduces its carboxylase activity in a reversible manner. Phosphorylation neither reduces the reaction velocity under specified conditions nor alters the carboxylase affinity for citrate. Kinase B_2 phosphorylates the carboxylase on a tryptic fragment that neither kinase B_1 nor cyclic AMP-dependent protein kinase can phosphorylate. Kinase B_2 prefers phosvitin and casein as substrates, and phosphorylates phosphorylase kinase as actively as kinase B_1. Kinase B_2 autophosphorylates a 36-kdalton peptide chain. Kinase B_2 is thought to be casein kinase I by similarity of mass, substrate specificity, subunit composition, sensitivity to heparin, and elution from phosphocellulose.

Total (immunoprecipitable) acetyl-CoA carboxylase as well as the activity of this enzyme increases with the initiation of lactation in rat mammary glands.[107] The effects of hormone administration or ablation on the total or active acetyl-CoA carboxylase are not well studied. Two technical problems appear to be encountered when both the total and active fraction of this enzyme are to be quantitated. The first is assessment of proteolysis during enzyme preparation. The second difficulty rests in interpreting the acetyl-CoA carboxylase assay results. Investigators have used the carboxylase activities in the absence or presence of citrate as a measure of the biologically active enzyme, but in these assays and enzyme preparations they have not controlled the activities of any kinases or phosphatases; thus, assay results may not reflect the in vivo acetyl-CoA carboxylase activity. The activity under these conditions will be a valid measure of the active enzyme. Dephosphorylation of the carboxylase by endogenous and/or exogenous phosphatase activities and subsequent assay would provide a measure of total carboxylase activity. Immunoassay may provide an alternative measure of total carboxylase activity.

Despite these reservations, data do suggest that starvation, insulin treatment, or prolactin withdrawal[108,109] can, under certain situations, alter acetyl-CoA carboxylase activities. There is, at present, no data correlating these changes with the in vivo state of acetyl-CoA carboxylase phosphorylation.

B. Mammary Casein Kinases

Study of mammary casein kinases is facilitated by a detailed knowledge of the structural chemistry of the substrate molecules. Bovine caseins are defined as a family of phosphoproteins precipitated at pH 4.6 at 20°C; their nomenclature and description has been reviewed.[110] The sequence data for each bovine casein has been determined along with their associated polysaccharides and sites of phosphorylation.[111-116] The phosphorylation of caseins appears to be permanent in that no mammary phosphatases specific for caseins have been characterized. Phosphorylation of several of the kinases are critical for micelle formation in the presence of calcium,[117] but the phosphorylation of κ casein appear not to be involved in micelle formation.[118,180] The functional significance of κ casein phosphorylation remains to be determined. Perhaps its phosphorylation is involved in the processes regulating export of micellar material from the mammary epithelial cells.[119] The structure and relationship of the various caseins in the milk micelle have not been well defined.[120-122]

Milk from other species all seem to contain similar members of the casein family with differences in relative amounts and characteristics of the predominant proteins.[123] All milks appear to contain κ casein.[124] The individual caseins have been isolated and characterized from rat milk.[125-128] A κ-like casein is also present in this species.[124-128] Caseins from mouse, guinea pig, and other species have also been isolated,[129-133] but less information as to their relationship to the bovine caseins is available. Several of the casein genes from rats have been sequenced[134-136] and a few casein proteins from other species have also been sequenced;[137,138] it appears that the amino acid sequences have diverged, except for the sequences surrounding the sites of phosphorylation. This observation emphasizes the important role that casein phosphorylation likely plays in milk micelle formation, secretion, and uptake in the neonate.

Human milk contains several caseins, of which the major species have been sequenced,[139,140] identifying the sites of phosphorylation. Human casein differs from bovine caseins in that this protein is phosphorylated to a varying degree with between zero and five phosphate groups attached to seryl and threonyl residues. Also, κ casein from human milk has been isolated and partially sequenced.[141]

Identification of mammary casein kinases need to fulfill several criteria. Substrate preference for dephosphorylated caseins of the same species should be demonstrated; further, the ester linkage of phosphate at sites predicted by sequence data needs to be demonstrated. Second, the casein kinase must be associated with the proper organelle; the Golgi apparatus presently seems most appropriate. Third, the casein kinase activity should, under physiological conditions, account for the phosphorylation present in the milk product of the species under investigation.

Preparation of substrate casein is easily accomplished by purification followed by dephosphorylation with spleen phosphoprotein phosphatase,[117,142] potato acid phosphatase,[143] or bovine intestinal alkaline phosphatase.[2] The product of these procedures needs to be analyzed for phosphorus content and effects of proteolysis by SDS or urea-containing gel electrophoresis. I find that, without proteolytic enzyme inhibitors present in the reaction, one often finds multiple bands on gel analysis and a poor substrate in kinase reactions.

The first report of a mammary casein kinase was presented in 1958 by Sundararajan and colleagues.[144] Using an acetone fractionation of rabbit mammary homogenates, these workers demonstrated phosphorylation of exogenous bovine casein with an enzymatic preference for the partially dephosphorylated substrate. This work was completed prior to elucidation of sequence data for the substrate or information concerning the location of phosphorylation. This laprine preparation demonstrated an absolute requirement for magnesium.

After phosphorylation of caseins was established as a post-translational modification,[145] a casein kinase activity was shown to exist in association with the Golgi apparatus of lactating rat mammary tissue.[146,147] The location of this kinase activity seems appropriate in that micelle formation begins to occur in the Golgi apparatus.[148] In whole membranes, the phosphorylation of dephosphorylated α_{s1} casein was preferred to native α_{s1} by four- to fivefold. Cyclic AMP (a high 5-mM concentration) showed no effect on this kinase activity. Further study revealed a Triton-X100® solubilized preparation of Golgi membranes could also preferentially phosphorylate previously dephosphorylated β and κ casein as well. Using dephosphorylated α_{s1} casein as substrate, a 7.6 pH maxima was established and the apparent K_m for ATP and casein was 80 and

12 μM, respectively. In this solubilized Golgi preparation, magnesium, calcium, manganese, or cobalt ions promoted increased phosphorylation. The inhibitor protein for cyclic AMP-dependent protein kinase was without effect. The precise sites of phosphorylation were not reported.

Quickly, a second series of investigations of phosphorylations in bovine mammary gland[149,150] were reported that duplicated many of the original observations in a second species and significantly expanded this area of research. A partially purified casein kinase activity was salt extracted (0.5 M NaCl) from a crude membrane fraction. The extract was fractionated on DEAE-cellulose and three kinase activities were eluted, two of which preferred whole casein to histone as substrate. Dephosphorylated α_{s1} casein was a better substrate for one of these kinases, but a dephosphorylated CNBr casein fragment was the most effective substrate. This substrate produced phosphoseryl residues and was thought to correspond to the phosphorylated residue cluster in positions 64 through 68. Dephosphorylated β casein was also phosphorylated, but mostly on the threonyl residues. In further work, casein kinases were solubilized from bovine membranes enriched for Golgi apparatus with sodium deoxycholate and fractionated by ammonium sulfate. This preparation required calcium or magnesium for activity and preferred dephosphorylated casein substrate. By specific cleavage analysis of α_{s1}-casein, this second preparation faithfully phosphorylated in the presence of calcium, residues 46, 48, 75, and 115 (all serine residing outside the phosphoseryl residue cluster located within residues 64 and 68). The four seryl residues between residues 63 and 69 were not phosphorylated. In contrast, in the presence of magnesium, the cluster Ser–Ile–Ser–Ser–Ser (residues 64 through 68) were phosphorylated by this preparation. Dephosphorylated κ casein was not phosphorylated in this preparation. These results suggest that more than one (possibly three) casein kinases exist.

West and Clegg[151] have studied casein phosphorylation in the rat using isolated Golgi vesicles which utilize endogenous rat caseins. They show that proteins which comigrate in SDS-containing polyacrylamide gels with rat caseins are phosphorylated in preparations of rat Golgi vesicles. In addition, a transient 70-kdalton phosphorylation appears very early after the addition of [$\gamma^{32}P$] ATP (5 sec) and then disappears with a half-time of approximately 20 sec. The investigators suggest this phosphorylation may be a phosphorylated intermediate of the calcium-dependent ATPase known to be located in the Golgi membrane.[152] Further work with this system suggests that the rat casein kinase requires intravesicular ATP and Ca^{2+}, while magnesium is required extravesicularly.[153,154]

The importance of the local amino acid sequence serving as a primary specifier of the phosphorylation site is suggested by the work of Szymanski and Farrell[155] who showed substantial phosphorylation of a tryptic peptide of human β casein or other small peptides containing several residues on either side of the site of phosphorylation. The sequence specificity of casein kinases, originally proposed by Mercier and colleagues,[156] has been championed by Bingham.[53,157] The current idea is that the recognition site for the phosphorylation of a threonyl or seryl residue in casein must require a glutamic acid residue or phosphoseryl residue in the N + 2 position. This recognition site is different than that of casein kinase I or II and cyclic AMP-dependent protein kinase (see appropriate sections). The concept that one mammary casein kinase phosphorylates all caseins by recognition of a specific site is furthered by data obtained from a kinase preparation from Triton-X100-®-solubilized bovine mammary Golgi membranes. Here it is argued that the interaction of the casein kinase with each phosphorylation site has a specific K_m determined by the enzyme and site and modified by divalent ions and adjacent residues. Thus, each site will be phosphorylated at a unique rate.[158]

The existence of multiple mammary casein kinases is a second possibility. This concept is supported by the two different specificities of preparations from bovine mammary gland.[149,150] Further work supporting the multiple enzyme hypothesis is found in substantially purified casein kinase from guinea pig mammary Golgi membranes.[159] Guinea pig casein C is normally phosphorylated at both seryl and threonyl residues in vivo, but the isolated kinase activity only phosphorylates seryl residues in vitro. Finally, work in my lab[2,160] has demonstrated a calcium- and calmodulin-dependent κ casein kinase in both rat and bovine membranes. This activity is associated most highly with the Golgi apparatus and catalyzes the phosphorylation of dephosphorylated κ casein preferentially. It requires magnesium for basal activity and is stimulated by the combination of calcium and calmodulin. In preliminary studies, the enzyme fails to retain activity during Triton-X100® solubilization, while an activity requiring magnesium for the specific phosphorylation of dephosphorylated α_{s1} casein is retained. The two casein kinase activities have been partially separated by DEAE-cellulose chromatography.[160]

Clearly, the purification and characterization of mammary casein kinases needs to be completed in order to resolve the situation. The use of synthetic peptides as substrates during this work will define the recognition sites for each enzyme, as it has been in other kinase systems.[161-163] The use of synthetic peptides as substrates for the various casein kinases will likely work as cleavage fragments of human β casein, and other short peptides are substrates for casein kinases prepared from bovine membranes.[155]

Little is known about the control of expression of the mammary casein kinases. The rat calcium- and calmodulin-dependent κ casein kinase remains high during lactation and falls to low activities within 24 hr of weaning a 20-day-old litter of pups.[2] No other work describes either the expression of these enzymes or the control of their activities by endocrine or other factors.

C. Nuclear Protein Kinases

Cyclic AMP inhibits the growth of several types of estrogen-dependent mammary tumors,[164,165] including the DMBA-induced rat mammary carcinoma. The phosphorylation of nuclear proteins change when rats bearing this tumor are treated by ovariectomy or dibutyryl cyclic AMP.[166,167] Growing tumors predominately phosphorylate a small basic protein which appears coincident with histones 2a, 2b, and 3. Treatment with dibutyryl cyclic AMP or ovariectomy causes regression of most tumors and increases phosphorylation of a protein with an apparent molecular weight of 76,000 daltons. This phosphorylation of a 76-kdalton protein in regressing tumors has been confirmed, and the effect of ovariectomy in these tumors increased the concentration of cyclic AMP and prostaglandin E_2.[168] The predominant kinase activity extracted from nuclei of growing tumors prefers α casein over mixed histones as an exogenous substrate, and is not effected by the inhibitor protein for the catalytic subunit of the cyclic AMP-dependent protein kinase. In contrast, the predominant kinase activity extracted from nuclei of regressing tumors prefers mixed histones as exogenous substrate, and its activity is reduced by the addition of the inhibitor protein to the catalytic subunit of the cyclic AMP-dependent protein kinase. Data suggest the cyclic AMP-dependent protein kinase is increased in the nuclei of regressing DMBA-induced tumors. Studies suggest this increase is due to the translocation of a cytoplasmic holoenzyme-cyclic AMP ternary complex into the nucleus.[169] This theory appears to be in contrast to the data showing that binding of cyclic AMP to the regulatory subunit of this kinase promotes dissociation of the regulatory and catalytic subunits. The role that reduced circulating estrogens and increased cyclic AMP agonists play in this process remains obscure.

In mouse mammary gland explant cultures, insulin stimulates the incorporation of [^{32}P] into both histone (including H-I) and nonhistone nuclear proteins. The addition of prolactin to cultures incubated with insulin and hydrocortisone also increased the phosphorylation of these two groups of nuclear proteins.[170] The type of kinase responsible for these phosphorylations, or whether the endocrine treatments influences the total or active fraction of the enzyme, remains to be explored.

Hohmann[171] further examined the H-I histone phosphorylation in cultured mammary glands from pregnant Balb/C mice and in D1 mammary tumors. Histone H-I is preferentially phosphorylated in mammary tissue at the amino-terminal portion, while the D1 tumor, while displaying similar H-I subtypes, is preferentially phosphorylated at the carboxy-terminal portion. The phosphorylation of this histone seems to be increased by hormonal treatment. The identity of the kinases responsible for these various phosphorylations and their control remains to be explored.

D. Other Protein Kinase Activities in Mammary Gland

In 1971, Majumder and Turkington[172] described the presence of two cytosolic protein kinases in rat mammary gland. One was sensitive to low concentrations of cyclic AMP. Enriched preparation of both kinases preferred histones over caseins as substrates. In this light, perhaps the cyclic AMP-independent activity was the free catalytic subunit of the cyclic AMP-dependent protein kinase. Similar kinase activities were increased seven- to ninefold during pregnancy in mice[173] and, in cultures, prolactin stimulated the accumulation of these activities in the presence of insulin. The induction of these kinase activities was rapid, reaching maximal values by 4 hr. These data were the earliest and one of the few reports of prolactin induction of protein kinases in mammary tissue.

Using the technique of [^{32}P]-labeled cultures with subsequent separation of cellular organelles, these same workers observed phosphorylation of plasma membrane proteins and ribosomal proteins when separated by urea-containing acrylamide gel electrophoresis. Insulin and prolactin were observed to increase the phosphorylation of a large number of plasma membrane proteins and several ribosomal proteins. These phosphorylations occurred subsequent to an induction of both the regulatory and catalytic subunit of the cyclic AMP-dependent protein kinase in the cytosol.[174]

Asaga et al.[175] reported the presence of casein kinase in the cytosol of the rat mammary gland. This activity was not well characterized, but showed a preference of partially dephosphorylated casein as substrate. Histone was a poor substrate. The activity increased two- to threefold during pregnancy and lactation when compared to tissue from virgin rats.

A tyrosine-specific vinculin kinase has been reported in the RMCD rat mammary cell line.[176] Vinculin phosphorylation is believed to be associated with activity of the cellular src gene product (pp60$^{c\text{-}src}$)[177] and related to cell division by controlling cytoskeletal structure at adhesion plaques. The tyrosine phosphorylation of this cytoskeletal element did not vary with the cell cycle.

Finally, mammary myoepithelial cells contained a myosin light-chain kinase that responded to oxytocin.[178] Myosin light-chain kinase is known to be one of the calcium- and calmodulin-dependent protein kinases in other cellular systems.[27,28] Recently, in whole cell experiments, this phosphorylation has been shown to be regulated by oxytocin via a calcium- and calmodulin-dependent mechanism.[179]

REFERENCES

1. Bingham, E. W., Modification of casein by phosphatases and protein kinases, *J. Agric. Food Chem.*, 24, 1094, 1976.
2. Brooks, C. L. and Landt, M., Calcium-ion and calmodulin-dependent kappa casein kinase in rat mammary acini, *Biochem. J.*, 224, 195, 1984.
3. Roach, P. J. Glycogen synthase and glycogen synthase kinase, *Curr. Top. Cell. Regul.*, 20, 45, 1981.
4. Flockhart, D. A., Removal of phosphate from proteins by the reverse reaction, *Methods Enzymol.*, 99, 14, 1983.
5. Cohen, P., Aitken, A., Damuni, Z., Hemmings, B. A., Ingebritsen, T. S., Parker, P. J., Picton, C., Resink, T., Stewart, A. A., Tonks, N. K., and Woodgett, J., Protein phosphorylation and the neural and hormonal control of enzyme activity, in *Posttranslational Covalent Modifications of Proteins*, Johnson, B. C., Ed., Academic Press, New York, 1983, 19.
6. Korsrud, G. O. and Baldwin, R. L., Hormonal regulation of rat mammary gland enzyme activities and metabolite patterns, *Can. J. Biochem.*, 50, 377, 1972.
7. Baldwin, R. L. and Milligan, L. P., Enzymatic changes associated with the initiation and maintenance of lactation in the rat, *J. Biol. Chem.*, 241, 2058, 1966.
8. Garrison, J. C., Measurement of hormone-stimulated phosphorylation in intact cells, *Methods Enzymol.*, 99, 20, 1983.
9. Rabinowitz, M., Protein kinases, in *The Enzymes*, Vol. 6, 2nd Ed., Boyer P. D., Lardy, H., and Myrbäck, K., Eds., Academic Press, New York, 1960, 119.
10. Traugh, J. A., Ashby, C. D., and Walsh, D. A., Criteria for the classification of protein kinases, *Methods Enzymol.*, 38, 290. 1974.
11. Krebs, E. G. and Beavo, J. A., Phosphorylation-dephosphorylation of enzymes, *Annu. Rev. Biochem.*, 48, 923, 1979.
12. Chen, L.-J. and Walsh, D. A., Multiple forms of hepatic adenosine 3′,5′-monophosphate-dependent protein kinase, *Biochemistry*, 10, 3614, 1971.
13. Posner, J. B., Stern, R., and Krebs, E. G., Effects of electrical stimulation and epinephrine on muscle phosphorylase, phosphorylase b kinase and adenosine 3′,5′-phosphate, *J. Biol. Chem.*, 240, 982, 1965.
14. Soderling, T. R., Hickenbottom, J. P., Reimann, E. M., Hunkeler, F. L., Walsh, D. A., and Krebs, E. G., Inactivation of glycogen synthase and activation of phosphorylase kinase by muscle adenosine 3′,5′-monophosphate-dependent protein kinases, *J. Biol. Chem.*, 245, 6317, 1970.
15. Greengard, P. and Kuo, J. F., in *Role of Cyclic AMP in Cell Function*, Greengard, P. and Costa, E., Eds., Raven Press, New York, 1970, 287.
16. Glass, D. B. and Krebs, E. G., Protein phosphorylation catalyzed by cyclic-AMP-dependent and cyclic-GMP-dependent protein kinase, *Annu. Rev. Pharmacol. Toxicol.*, 20, 363, 1980.
17. Kuo, J. F. and Greengard, P., Cyclic nucleotide-dependent protein kinases: isolation and partial purification of a protein kinase activated by guanosine 3′,5′-monophosphate, *J. Biol. Chem.*, 245, 2493, 1970.
18. Khoo, J. C., Sperry, P. J., Gill, G. N., and Steinberg, D., Activation of hormone-sensitive lipase and phosphorylase kinase by purified cyclic-GMP-dependent protein kinase, *Proc. Natl. Acad. Sci. U.S.A.*, 74, 4843, 1970.
19. Linnala-Kankunnan, A. and Mäenpää, P. H., Phosphorylation of high mobility group protein HMG-14 by a cyclic GMP-dependent protein kinase from avian liver nucleoli, *Biochem. Biophys. Acta*, 654, 287, 1977.
20. Walton, G. M., Speiss, J., and Gill, G. N., Phosphorylation of high mobility group 14 protein by cyclic nucleotide-dependent protein kinases, *J. Biol. Chem.*, 257, 4661, 1982.
21. Wolff, D. J. and Brostrom, C. O., Properties and functions of the calcium-dependent regulator protein, *Adv. Cyclic Nucleotide Res.*, 11, 27, 1979.
22. Kaibuchi, K., Kikkawa, U., Takai, Y., and Nishizuda, Y., Calcium and phospholipid turnover in signal transduction, in *Enzyme Regulation by Reversible Phosphorylation — Further Advances*, Cohen, P., Ed., Elsevier, Amsterdam, 1984, 81.
23. Rubin, R. P., *Calcium and Cellular Secretion*, Plenum Press, New York, 1982.
24. Neville, M. C. and Watters, C. C., Secretion of calcium into milk: review, *J. Dairy Sci.*, 66, 371, 1983.
25. Van Eldik, L. J., Zendegui, J. G., Marshak, D. R., and Watterson, D. M., Calcium-binding proteins and the molecular basis of calcium action, *Int. Rev. Cytol.*, 77, 1, 1982.
26. Klee, C. B. and Newton, D. L., Calmodulin-modulated protein phosphorylation, in *Posttranslational Covalent Modifications of Proteins*, Johnson, B. C., Ed., Academic Press, New York, 1983, 61.
27. Hathaway, D. R. and Adelstein, R. S., Human platelet myosin light chain kinase requires the calcium binding protein calmodulin for activation, *Proc. Natl. Acad. Sci. U.S.A.*, 76, 1653, 1979.

28. Walsh, M. P., Vallet, B., Autric, F., and Demaille, J. G., Purification and characterization of bovine cardiac calmodulin-dependent myosin light chain kinase, *J. Biol. Chem.*, 254, 12136, 1979.
29. Ahmad, Z., DePaoli-Roach, A., and Roach, P. J., Purification and characterization of a rabbit liver calmodulin-dependent protein kinase able to phosphorylate glycogen synthase, *J. Biol. Chem.*, 257, 8348, 1982.
30. Yamaguchi, T. and Fujisawa, H., Tyrosine 3-monooxygenase is phosphorylated by Ca^{++}-calmodulin-dependent protein kinase, followed by activation by activator protein, *Biochem. Biophys. Res. Commun.*, 100, 807, 1981.
31. Kennedy, M. B., McGuinness, T., and Greengard, P., A calcium/calmodulin-dependent protein kinase from mammalian brain that phosphorylates synapsin I: partial purification and characterization, *J. Neurosci.*, 3, 818, 1983.
32. Goldring, J. R., Gonzalez, B., and DeLorenzo, R. J., Isolation of brain Ca^{++}-calmodulin tubulin kinase containing calmodulin binding proteins, *Biochem. Biophys. Res. Commun.*, 108, 421, 1982.
33. Cohen, P., Burchell, A., Foulkes, J. G., and Cohen, P. T. W., Identification of the Ca^{++}-dependent modulator protein as the fourth subunit of rabbit skeletal muscle phosphorylase kinase, *FEBS Lett.*, 92, 287, 1978.
34. Riss, T. L., Bechtel, P. J., and Baumrucker, C. R., Calmodulin content of rat mammary tissue and isolated cells during pregnancy and lactation, *Biochem. J.*, 219, 927, 1984.
35. Brooks, C. L. and Landt, M., Calmodulin-dependent protein kinase in acini from lactating rat mammary tissue: subcellular locale, characterization and solubulization, *Arch. Biochem. Biophys.*, 240, 663, 1985.
36. McGuinnes, T. L., Lai, Y., Greengard, P., Woodgett, J., and Cohen, P., A multifunctional calmodulin-dependent protein kinase: similarities between skeletal muscle glycogen synthase kinase and a brain synapsin I kinase, *FEBS Lett.*, 163, 329, 1983.
37. Inove, M., Kishimoto, A., Takai, Y., and Nishizuka, T., Studies on a cyclic-nucleotide-independent protein kinase and its properties in mammalian tissues, *J. Biol. Chem.*, 252, 7610, 1977.
38. Takai, Y., Kishimoto, A., Iwasa, K.. Kawahara, Y., Mori, T., and Nishizuka, Y., Calcium-dependent activation of a multifunctional protein kinase by membrane phospholipids, *J. Biol. Chem.*, 254, 3692, 1979.
39. Takai, Y., Kishimoto, A., Kikkawa, U., Mori, T., and Nishizuka, Y., Unsaturated diacylglycerol as a possible messenger for the activation of calcium-activated, phospholipid-dependent protein kinase system, *Biochem. Biophys. Res. Commun.*, 91, 1218, 1979.
40. Wise, B. C., Raynor, R. L., and Kuo, J. F., Phospholipid-sensitive Ca^{++}-dependent protein kinase from heart, *J. Biol. Chem.*, 257, 8481, 1982.
41. Kikkawa, U., Takai, Y., Minakuchi, R., Inohara, S., and Nishizuka, Y., Calcium-activated, phospholipid-dependent protein kinase from rat brain, *J. Biol. Chem.*, 257, 13341, 1982.
42. Majerus, P. W., Neufeld, E. J., and Willson, D. B., Production of phosphoinositide-derived messengers, *Cell*, 37, 701, 1984.
43. Farese, R. V., Phospholipids as intermediates in hormone action, *Mol. Cell. Endocrinol.*, 35, 1, 1984.
44. Nishizuka, Y., Protein kinases in signal transduction, *Trends Biochem. Sci.*, 9, 163, 1984.
45. Berridge, M. J., and Irvine, R. F.,Inositol triphosphate, a novel second messenger in cellular signal transduction, *Nature (London)*, 312, 315, 1984.
46. Bakers, P. F., Multiple control for secretion, *Nature (London)*, 310, 629, 1984.
47. Roach, P. J. and Goldman, M., Modification of glycogen synthase activity in isolated rat hepatocytes by tumor-promoting phorbol esters; evidence for differential regulation of glycogen synthase and phosphorylase, *Proc. Natl. Acad. Sci. U.S.A.*, 80, 7170, 1983.
48. Ahmad, Z., Lee, F.-T., DePaoli-Roach, A., and Roach, P. J., Phosphorylation of glycogen synthase by the Ca^{++}- and phospholipid-activated protein kinase (Protein kinase C), *J. Biol. Chem.*, 259, 8743, 1984.
49. Rillema, J. A., Prolactin-like actions of phospholipase C on RNA and casein synthesis in mouse mammary gland explants, *Horm. Metab. Res.*, 16, 532, 1984.
50. Hathaway, G. M. and Traugh, J. A., Casein kinases — multipotential protein kinases, *Curr. Top. Cell. Regul.*, 21, 101, 1982.
51. Bingham, E. W. and Farrell, H. M., Phosphorylation of casein by the lactating mammary gland: a review, *J. Dairy Sci.*, 60, 1199, 1977.
52. Linn, T. C., Petit, F. H., and Reed, L. J., α Keto acid dehydrogenase complexes. X. Regulation of the activity of the pyruvate dehydrogenase complex from beef kidney mitochondria by phosphorylation and dephosphorylation, *Proc. Natl. Acad. Sci. U.S.A.*, 62, 234, 1969.
53. Trachsel, H., Ranu, R. S., and London, I. M., Regulation of protein synthesis in a rabbit reticulocyte lysate: purification and characterization of heme-reversible translation inhibitor, *Proc. Natl. Acad. Sci. U.S.A.*, 75, 3654, 1978.
54. Hathaway, G. M. and Traugh, J. A., Cyclic nucleotide-independent protein kinases from rabbit reticulocytes; purification of casein kinases, *J. Biol. Chem.*, 254, 762, 1979.

55. Tao, M., Conway, R., and Cheta, S., Purification and characterization of a membrane-bound protein kinase from human erythrocytes, *J. Biol. Chem.*, 255, 2563, 1980.
56. Tuazon, P. T., Bingham, E. W., and Traugh, J. A., Cyclic nucleotide-independent protein kinases from rabbit reticulocytes, *Eur. J. Biochem.*, 94, 497, 1979.
57. Pinna, L. A., Meggio, F., and Donella-Deana, A., Structure of the sites of substrate proteins undergoing phosphorylation by protein kinases with special reference to liver casein kinases, in *Protein Phosphorylation and Bio-Regulation*, Thomas, G., Podesta, E., and Gordon, J., Eds., Karger, Basel, 1980, 8.
58. Hathaway, G. M., Zoller, M. J., and Traugh, J. A., Identification of the catalytic subunit of casein kinase II by affinity labeling with 5′-p-fluorosulfonylbenzoyl adenosine, *J. Biol. Chem.*, 256, 11442, 1981.
59. Pinna, L. A., Donella-Deana, A., and Meggio, F., Structural features determining the site specificity of a rat liver cAMP-independent protein kinase, *Biochem. Biophys. Res. Commun.*, 87, 114, 1979.
60. Hosey, M. M. and Tao, M., Protein kinases of rabbit and human erythrocyte membranes, *Biochem. Biophys. Acta*, 482, 348, 1977.
61. Hathaway, G. M., Lubben, T. H., and Traugh, J. A., Inhibition of casein kinase II by heparin, *J. Biol. Chem.*, 255, 8038, 1980.
62. Matsumura, S. and Takeda, M., Phosphoprotein kinases from rat liver cytosol, *Biochem. Biophys. Acta*, 289, 237, 1972.
63. Wieland, O. H., The mammalian pyruvate dehydrogenase complex: structure and regulation, *Rev. Physiol. Biochem. Pharmacol.*, 96, 123, 1983.
64. Linn, T. C., Petit, F. H., and Reed, L. J., −Keto acid dehydrogenase complexes. X. Regulation of the activity of the pyruvate dehydrogenase complex from beef kidney mitochondria by phosphorylation and dephosphorylation, *Proc. Natl. Acad. Sci. U.S.A.*, 62, 234, 1969.
65. Cooper, J. A. and Hunter, T., Regulation of cell specific transformation by tyrosine-specific protein kinases: the search for important cellular substrate proteins, *Curr. Top. Microbiol. Immunol.*, 107, 125, 1983.
66. Cohen, S., Carpenter, G., and King, L., Jr., Epidermal growth factor-receptor-protein kinase interactions, *J. Biol. Chem.*, 255, 4834, 1980.
67. Kasuga, M., Karlsson, F. A., and Kahn, C. R., Insulin stimulates the phosphorylation of the 95,000-dalton subunit of its own receptor, *Science*, 215, 185, 1982.
68. Rubin, J. B., Shia, M. A., and Pilch, P. F., Stimulation of tyrosine-specific phosphorylation *in vitro* by insulin-like growth factor-I, *Nature (London)*, 305, 438, 1983.
69. Tuy, F. P. D., Henry, J., Rosenfeld, C., and Kahn, A., High tyrosine kinase activity in normal nonproliferating cells, *Nature (London)*, 305, 435, 1983.
70. Foulkes, J. G., Phosphotyrosyl-protein phosphatases, *Curr. Top. Microbiol. Immunol.*, 107, 163, 1983.
71. Denton, R. M., Randell, P. J., Bridges, B. J., Cooper, R. H., Kerbey, A. L., Pask, H. T., Severson, D. L., Stansbie, D., and Whitehouse, S., Regulation of mammalian pyruvate dehydrogenase, *Mol. Cell. Biochem.*, 9, 27, 1975.
72. Linn, T. C., Pelley, J. W., Pettit, F. H., Hucho, F., Randall, D. D., and Reed, L. J., α Keto acid dehydrogenase complexes, XV. Purification and properties of the component enzymes of the pyruvate dehyrogenase complexes from the bovine kidney and heart, *Arch. Biochem. Biophys.*, 148, 327, 1972.
73. Severson, D. L., Denton, R. M., Pask, H. T., and Randle, P. J., Calcium and magnesium ions as effectors of adipose-tissue pyruvate dehydrogenase phosphate phosphatase, *Biochem. J.*, 140, 225, 1974.
74. Seals, J. R. and Jarett, L., Activation of pyruvate dehydrogenase by direct addition of insulin to an isolated plasma membrane/mitochondria mixture: evidence for generation of insulin's second messenger in a subcellular system, *Proc. Natl. Acad. Sci. U.S.A.*, 77, 77, 1980.
75. Siess, E. A. and Wieland, O., Purification and characterization of pyruvate-dehydrogenase phosphatase from pig-heart muscle, *Eur. J. Biochem.*, 26, 96, 1972.
76. Coore, H. G. and Field, B., Properties of pyruvate dehydrogenase on rat mammary tissue and its changes during pregnancy, lactation and weaning, *Biochem. J.*, 142, 87, 1974.
77. Field, B. and Coore, H. G., Control of rat mammary gland pyruvate dehydrogenase by insulin and prolactin, *Biochem. J.*, 156, 333, 1976.
78. Baxter, M. A. and Coore, H. G., The mode of regulation of pyruvate dehydrogenase of lactating rat mammary gland, *Biochem. J.*, 174, 553, 1978.
79. Baxter, M. A. and Coore, H. G., Specificity of lactating rat mammary gland pyruvate dehydrogenase phosphate phosphatase for the phosphorylated residues of the α-subunit, *Biochem. Soc. Trans.*, 7, 949, 1979.
80. O'Keefe, E. and Cuatrecases, P., Insulin receptors in murine mammary cells: comparison in pregnant and nonpregnant animals, *Biochem. Biophys. Acta*, 343, 64, 1974.

81. Srere, P. A., Citrate enzymes: their structures, mechanisms and biological functions, *Curr. Top. Cell. Regul.*, 5, 229, 1972.
82. Nepokroeff, C. M., Lakshmanan, M. R., Ness, G. L., Muesing, R. A., Kleinsek, D. A., and Porter, J. W., Coordinate control of rat liver lipogenic enzymes by insulin, *Arch. Biochem. Biophys.*, 162, 340, 1974.
83. Linn, T. C. and Srere, P. A., Identification of ATP-citrate lyase as a phosphoprotein, *J. Biol. Chem.*, 254, 1691, 1979.
84. Alexander, M. C., Kowaloff, E. M., Witters, L. A., Dennihy, D. T., and Avruch, J., Purification of a hepatic 123,000-dalton hormone-stimulated ^{32}P-peptide and its identification as ATP-citrate lyase, *J. Biol. Chem.*, 254, 8052, 1979.
85. Janski, A. M., Srere, P. A., Cornell, N. W., and Veech, R. L., Phosphorylation of ATP citrate lyase in response to glucagon, *J. Biol. Chem.*, 254, 9365, 1979.
86. Guy, P. S., Cohen, P., and Hardie, D. G., Rat mammary gland ATP-citrate lyase is phosphorylated by cyclic-AMP-dependent protein kinase, *FEBS Lett.*, 109, 205, 1980.
87. Guy, P. S., Cohen, P., and Hardie, D. G., Purification and physiochemical properties of ATP citrate (pro-S3) lyase from lactating rat mammary gland and studies of its reversible phosphorylation, *Eur. J. Biochem.*, 114, 399, 1981.
88. Howanitz, P. J. and Levy, H. R., Acetyl-CoA carboxylase and citrate cleavage enzyme in the rat mammary gland, *Biochem. Biophys. Acta*, 106, 430, 1965.
89. Witters, L. A., Kowaloff, E. M., and Avruch, J., Glucagon regulation of protein phosphorylation: identification of acyl-CoA carboxylase as a substrate, *J. Biol. Chem.*, 254, 245, 1979.
90. Tipper, J. P. and Witter, L. A., In vitro phosphorylation and inactivation of rat liver acetyl-CoA carboxylase purified by avidin affinity chromatography, *Biochem. Biophys. Acta*, 715, 162, 1982.
91. Goodson, J., Pope, T. S., and Allred, J. B., Molecular weights of subunits of acyl CoA carboxylase in rat liver cytoplasm, *Biochem. Biophys. Res. Commun.*, 122, 694, 1984.
92. Block, K. and Vance, D., Control mechanisms in the synthesis of saturated fatty acids, *Annu. Rev. Biochem.*, 46, 263, 1977.
93. Carlson, C. A. and Kim, K., Regulation of hepatic acyl coenzyme A carboxylase by phosphorylation and dephosphorylation, *Arch. Biochem. Biophys.*, 164, 478, 1974.
94. Carlson, C. A. and Kim, K., Differential effects of metabolites on the active and inactive forms of hepatic acetyl CoA carboxylase, *Arch. Biochem. Biophys.*, 164, 490, 1974.
95. Lee, K. and Kim, K., Regulation of rat liver acetyl coenzyme A carboxylase: evidence for the interconversion between active and inactive forms of enzyme by phosphorylation and dephosphorylation, *J. Biol. Chem.*, 252, 1748, 1977.
96. Brownsey, R. W., Hughes, W. A., Denton, R. M., and Mayer, R. J., Demonstration of the phosphorylation of acetyl-coenzyme A carboxylase within intact rat epididymal fat-cells, *Biochem. J.*, 168, 441, 1977.
97. Lee, K.-H. and Kim, K.-H., Stimulation by epinephrine of in vivo phosphorylation and inactivation of acetyl coenzyme A carboxylase of rat epididymal adipose tissue, *J. Biol. Chem.*, 254, 1450, 1979.
98. Brownsey, R. W., Hughes, W. A., and Denton, R. M., Adrenaline and the regulation of acetyl-coenzyme A carboxylase in rat epididymal adipose tissue: inactivation of the enzyme is associated with phosphorylation and can be reversed on dephosphorylation, *Biochem. J.*, 184, 23, 1979.
99. Smith, A. and Williamson, I. P., Partial purification and some properties of acetyl-coenzyme A carboxylase from bovine mammary tissue, *Trans. Biochem. Soc.*, 5, 737, 1977.
100. Hardie, D. G. and Cohen, P., The regulation of fatty acid biosynthesis: simple procedure for the purification of acetyl-CoA carboxylase from lactating rat mammary gland, and its phosphorylation by endogenous cyclic-AMP-dependent and -independent protein kinase activities, *FEBS Lett.*, 91, 1, 1978.
101. McNeille, E. M., Clegg, R. A., and Zammit, V. A., Regulation of acetyl-CoA carboxylase in rat mammary gland: effects of incubation with Ca^{++}, Mg^{++} and ATP on enzyme activity in tissue extracts, *Biochem. J.*, 200, 639, 1981.
102. Hardie, D. G. and Guy, P. S., Reversible phosphorylation and inactivation of acetyl-CoA carboxylase from lactating rat mammary gland by cyclic AMP-dependent protein kinase, *Eur. J. Biochem.*, 110, 167, 1980.
103. Brownsey, R. W. and Hardie, D. G., Regulation of acetyl-CoA carboxylase: identity of sites phosphorylated in intact cells treated with adrenaline and *in vitro* by cyclic AMP-dependent protein kinase, *FEBS Lett.*, 120, 67, 1980.
104. Guy, P. S. and Hardie, D. G., Regulation of mammalian acetyl-CoA carboxylase: limited proteolysis mimics dephosphorylation, *FEBS Lett.*, 132, 67, 1981.
105. Hardie, D. G. and Cohen, P., Dephosphorylation and activation of acetyl-CoA carboxylase from lactating rabbit mammary gland, *FEBS Lett.*, 103, 333, 1979.

106. Munday, M. R. and Hardie, D. G., Isolation of three cyclic-AMP-independent acetyl-CoA carboxylase kinases from lactating rat mammary gland and characterization of their effects on enzyme activity, *Eur. J. Biochem.*, 141, 617, 1984.
107. Mackall, J. C. and Lane, M. D., Changes in mammary-gland acetyl coenzyme A carboxylase associated with lactogenic differentiation, *Biochem. J.*, 162, 635, 1977.
108. McNeille, E. M. and Zammit, V. A., Regulation of acetyl-CoA carboxylase in rat mammary gland: effects of starvation and of insulin and prolactin deficiency on the fraction of the enzyme in the active form *in vivo*, *Biochem. J.*, 204, 273, 1982.
109. Munday, M. R. and Williamson, D. H., Effects of starvation, insulin or prolactin deficiency on the activity of acetyl-CoA carboxylase in mammary gland and liver of lactating rats, *FEBS Lett.*, 138, 285, 1982.
110. Whitney, R. M., Brunner, J. R., Ebner, K. E., Farrell, H. M., Josephson, R. V., Morr, C. V., and Swaisgood, H. E., Nomenclature of the proteins of cow's milk, fourth revision, *J. Dairy Sci.*, 59, 795, 1976.
111. Fiat, A. M., Alais, C., and Jolles, P., The amino-acid and carbohydrate sequences of a short glycopeptide isolated from bovine k-casein, *Eur. J. Biochem.*, 27, 408, 1972.
112. Mercier, J.-C., Grosclaude, F., and Ribadeau-Dumas, B., Structure primaire de la caséine α_{s1}-bovine, *Eur. J. Biochem.*, 23, 41, 1971.
113. Brignon, G., Ribadeau-Dumas, B., Mercier, J.-C., Pelissier, J. P., and Das, B. C., Complete amino acid sequence of bovine α_{s2}-casein, *FEBS Lett.*, 76, 274, 1977.
114. Mercier, J.-C., Brignon, G., and Ribadeau-Dumas, B., Structure primaire de la caséine kB bovine, *Eur. J. Biochem.*, 35, 222, 1973.
115. Ribadeau-Dumas, B., Brignon, G., Grosclaude, F., and Mercier, J.-C., Structure primaire de la caséine β bovine, *Eur. J. Biochem.*, 25, 505, 1972.
116. Groves, M. R., Gordon, W. G., Greenberg, R., Peterson, R. F., and Jenness, R., Sequencing β-casein C: isolation of a large fragment after cleavage of thiotrifluoroacetylated β-casein C, *J. Dairy Sci.*, 58, 301, 1975.
117. Bingham, E. W., Farrell, H. M., and Carroll, R. J., Properties of dephosphorylated α_s casein. Precipitation by calcium ions and micelle formation, *Biochemistry*, 11, 2450, 1972.
118. Pepper, L. and Thompson, M. P., Dephosphorylation of α_s and k caseins and its effect on micelle stability in the k-α_s casein system, *J. Dairy Sci.*, 46, 764, 1963.
119. Lee, EY.H.-P., Lee, W. H., Kaetzel, C. S., Perry, G., and Bissell, M. J., Interaction of mouse mammary epithelial cells with collagen substrate: regulation of casein gene expression and secretion, *Proc. Natl. Acad. Sci. U.S.A.*, 82, 1419, 1985.
120. Hill, R. J. and Wake, R. G., Amphiphile nature of k-casein as the basis for its micelle stabilizing property, *Nature (London)*, 221, 635, 1969.
121. Farrell, H. M., Models for casein micells formation, *J. Dairy Sci.*, 56, 1195, 1973.
122. Horisberger, M. and Vauthey, M., Localization of k-casein on thin sections of casein micelles by the gold method, *Histochemistry*, 80, 9, 1984.
123. Jenness, R., Comparative aspects of milk proteins, *J. Dairy Res.*, 46, 197, 1979.
124. Kotts, C. and Jenness, R., Isolation of k-casein-like proteins from milks of various species, *J. Dairy Sci.*, 59, 816, 1976.
125. Woodward, D. R., Rat casein: isolation and characterization of two major fractions, *Comp. Biochem. Physiol.*, 57B, 365, 1977.
126. Hirose, M., Kato, T., Omori, K., Maki, M., Yoshikawa, M., Sasaki, R., and Chiba, H., Purification and properties of a major casein component of rat milk, *Biochem. Biophys. Acta*, 667, 309, 1981.
127. Hirose, M., Kato, T., Omari, K., Takeuchi, M., Yoshikawa, M., Sasaki, R., and Chiba, H., Purification and characterization of four components of rat caseins, *Biochem. Biophys. Acta*, 671, 139, 1981.
128. Visser, S., Jenness, R., and Mullin, R. J., Isolation and characterization of caseins from rat milk, *J. Dairy Sci.*, 64, 559, 1981.
129. Green, M. R. and Pastewka, J. V., Molecular weights of three mouse milk caseins by sodium dodecyl sulfate polyacrylamide gel electrophoresis and k-like characteristics of a fourth casein, *J. Dairy Sci.*, 59, 1738, 1976.
130. Dayal, R., Hurlimann, J., Suard, Y. M. L., and Kraehenbuhl, J.-P., Chemical and immunochemical characterization of caseins and the major whey proteins of rabbit milk, *Biochem. J.*, 201, 71, 1982.
131. Vissar, S., Jenness, R., and Mullin, R. J., Isolation and characterization of β and γ-caseins from horse milk, *Biochem. J.*, 203, 131, 1982.
132. Yamauchi, K., Azuma, N., and Shimizu, M., Comparison of casein of cynomolgus monkey (*Macaca fascicularis*) with human casein, *Comp. Biochem. Physiol.*, 75B, 287, 1973.
133. Craig, R. K., McIlreavy, D., and Hall, R. L., Separation and partial characterization of guinea pig milk, *Biochem. J.*, 173, 633, 1978.

134. Blackburn, D. E., Hobbs, A. A., and Rosen, J. M., Rat β casein cDNA: sequence analysis and evolutionary comparisons, *Nucl. Acids Res.*, 10, 2295, 1982.
135. Hobbs, A. A. and Rosen, J. M., Sequence of rat α and β-casein mRNAs: evolutionary comparison of the calcium-dependent rat casein multigene family, *Nucl. Acids Res.*, 10, 8079, 1982.
136. Nakhasi, H. L., Grantham, F. H., and Gullino, P. M., Expression of k-casein in normal and neoplastic rat mammary gland is under the control of prolactin, *J. Biol. Chem.*, 259, 14894, 1984.
137. Gaye, P. and Gautron, J.-P., Amino terminal sequences of the precursors of ovine caseins, *Biochem. Biophys. Res. Commun.*, 79, 093, 1977.
138. Mercier, J.-C. and Chobert, J.-M., Comparative study of the amino acid sequences of the caseinomacropeptides from seven species, *FEBS Lett.*, 72, 208, 1976.
139. Groves, M. L. and Gordon, W. G., The major component of human casein: a protein phosphorylated at different levels, *Arch. Biochem. Biophys.*, 140, 47, 1970.
140. Greenberg, R., Groves, M. L., and Dower, H. J., Human β casein amino acid sequence and identification of phosphorylation sites, *J. Biol. Chem.*, 259, 5132, 1984.
141. Chobert, J.-M., Mercier, J.-C., Bahy, C., and Haze, G., Structure primaire du caseinomacropeptide des caseins k porcine et humaine, *FEBS Lett.*, 72, 173, 1976.
142. West, D. W. and Dalgleish, D. G., A kinetic analysis of the dephosphorylation by bovine spleen phosphoprotein phosphatase (EC 3.1.3.16) of a phosphopeptide derived from β casein, *Biochem. Biophys. Acta*, 438, 169, 1976.
143. Bingham, E. W., Farrell, H. M., and Dahl, K. J., Removal of phosphate groups from casein with potato acid phosphatase, *Biochem. Biophys. Acta*, 429, 448, 1976.
144. Sundararajan, T. A., Sampath Kumar, K. S. V., and Sarma, P. S., Some properties of protein phosphokinase from rabbit mammary gland, *Biochem. Biophys. Acta*, 29, 449, 1958.
145. Turkington, R. W. and Topper, Y. J., Casein biosynthesis: evidence for phosphorylation of precursor proteins, *Biochem. Biophys. Acta*, 127, 366, 1966.
146. Bingham, E. W., Farrell, H. M., and Basch, J. J., Phosphorylation of casein; role of the Golgi apparatus, *J. Biol. Chem.*, 247, 8193, 1972.
147. Bingham, E. W. and Farrell, H. M., Casein kinase from the Golgi apparatus of lactating mammary gland, *J. Biol. Chem.*, 249, 3647, 1974.
148. Berry, K. E., Hood, L. F., and Patton, S., Formation of casein micelles in Golgi vesicles of mammary tissue, *J. Dairy Sci.*, 54, 911, 1971.
149. Chew, L. F. and MacKinlay, A. G., Histone and casein kinases of lactating bovine mammary gland, *Biochem. Biophys. Acta*, 359, 73, 1974.
150. MacKinley, A. G., West, D. W., and Mason, W., Specific casein phosphorylation by a casein kinase from lactating bovine mammary gland, *Eur. J. Biochem.*, 76, 233, 1977.
151. West, D. W. and Clegg, R. A., Transient formation of a phosphoprotein during autophosphorylation of rat mammary gland Golgi vesicles, *Biochem. Biophys. Acta*, 690, 290, 1982.
152. Baumrucker, C. R., Calcium transport in lactation, in *Lactation: A Comprehensive Treatis*, Vol. 4, Larson, B., Ed., Academic Press, New York, 1978, 463.
153. West, D. W. and Clegg, R. A., Casein kinase activity in rat mammary gland Golgi vesicles: phosphorylation of endogenous caseins, *Eur. J. Biochem.*, 137, 215, 1983.
154. West, D. W. and Clegg, R. A., Casein kinase activity in rat mammary gland Golgi vesicles. Demonstration of latency and requirement for a transmembrane ATP carrier, *Biochem. J.*, 219, 181, 1984.
155. Szymanski, E. S. and Farrell, H. M., Isolation and solubilization of casein kinase from Golgi apparatus of bovine mammary gland and phosphorylation of peptides, *Biochem. Biophys. Acta*, 702, 163, 1982.
156. Mercier, J.-C., Grosclaude, F., and Ribadeau-Dumas, B., Primary structure of bovine caseins: a review, *Milchwissenschaft*, 27, 402, 1972.
157. Bingham, E. W., Role of mammary casein kinases in the phosphorylation of milk proteins, *J. Dairy Sci.*, 46, 181, 1979.
158. Bingham, E. W. and Groves, M. L., Properties of casein kinase from lactating bovine mammary gland, *J. Biol. Chem.*, 254, 4510, 1979.
159. Pascall, J. C., Boulton, A. P., and Craig, R. K., Characterization of a membrane-bound serine-specific casein kinase isolated from lactating guinea-pig mammary gland, *Eur. J. Biochem.*, 119, 91, 1981.
160. Brooks, C. L., unpublished data.
161. Kemp, B. E., Bylund, D. B., Huang, T.-S., and Krebs, E. G., Substrate specificity of the cyclic AMP-dependent protein kinases, *Proc. Natl. Acad. Sci. U.S.A.*, 72, 3448, 1975.
162. Kemp, B. E., Benjamini, E., and Krebs, E. G., Synthetic hexapeptide substrates and inhibitors of 3′:5′-cyclic AMP-dependent protein kinase, *Proc. Natl. Acad. Sci. U.S.A.*, 73, 1038, 1976.
163. Yeaman, S. J., Cohen, P., Watson, D. C., and Dixon, G. H., The substrate specificity of adenosine 3′:5′-cyclic monophosphate-dependent protein kinase of rabbit skeletal muscle, *Biochem. J.*, 162, 411, 1977.

164. Pastan, I. H., Johnson, G. S., and Anderson, W. B., Role of cyclic nucleotides in growth control, *Annu. Rev. Biochem.*, 44, 491, 1975.
165. Cho-Chung, Y. S. and Gullino, P., In vivo inhibition of growth of two hormone-dependent mammary tumors by dibutryl cyclic AMP, *Science*, 183, 87, 1974.
166. Cho-Chung, Y. S. and Redler, B. H., Dibutyrl cyclic AMP mimics ovariectomy: nuclear protein phosphorylation in mammary tumor regression, *Science*, 197, 272, 1977.
167. Cho-Chung, Y. S., Redler, B. H., and Lewallen, R. P., Nuclear protein phosphorylation and hormone-dependent mammary tumor regression following dibutyryl cyclic adenosine 3′:5′-monophosphate treatment or ovariectomy, *Cancer Res.*, 38, 3405, 1978.
168. Foecking, M. K., Abou-Issa, H., Webb, T. E., and Minton, J. P., Concurrent changes in growth-related biochemical parameters during regression of hormone-dependent rat mammary tumors, *J. Natl. Cancer Inst.*, 71, 773, 1983.
169. Cho-Chung, Y. S., Archibald, D., and Clair, T., Cyclic AMP receptor triggers nuclear protein phosphorylation in a hormone-dependent mammary tumor cell-free system, *Science*, 205, 1390, 1979.
170. Turkington, R. W. and Riddle, M., Hormone-dependent phosphorylation of nuclear proteins during mammary gland differentiation *in vitro*, *J. Biol. Chem.*, 244, 6040, 1969.
171. Hohmann, P. G., Phosphorylation of HI histones in normal and transformed mouse cells, *Cancer Biochem. Biophys.*, 7, 45, 1983.
172. Majumder, G. C. and Turkington, R. W., Adenosine 3′,5′-monophosphate-dependent and -independent protein phosphokinase isoenzymes from mammary gland, *J. Biol. Chem.*, 246, 2650, 1971.
173. Majumder, G. C. and Turkington, R. W., Hormone regulation of protein kinases and adenosine 3′,5′-monophosphate-binding protein in developing mammary gland, *J. Biol. Chem.*, 246, 5545, 1971.
174. Majumder, G. C. and Turkington, R. W., Hormone-dependent phosphorylation of ribosomal and plasma membrane proteins in mouse mammary gland *in vitro*, *J. Biol. Chem.*, 247, 7207, 1972.
175. Asaga, H., Muto, K., Homareda, H., and Komine, S., Casein kinase in cytosol of rat and mouse mammary epithelial cells isolated from histone kinase by $MgCl_2$ treatment and influence of pregnancy and lactation on the enzyme activity, *Endocrinol. Jpn.*, 28, 265, 1981.
176. Rosok, M. J. and Rohrschneider, L. R., Increased phosphorylation of vinculin on tyrosine does not occur during the release of stress fibers before mitosis in normal cells, *Mol. Cell. Biol.*, 3, 475, 1981.
177. Sefton, B. M., Hunter, T., Ball, E. H., and Singer, S. J., Vincullin: a cytoskeletal target of the transforming protein of Rous sarcoma virus, *Cell*, 24, 165, 1981.
178. Olins, G. M. and Bremel, R. D., Phosphorylation of myosin in mammary myoepithelial cells in response to oxytoxin, *Endocrinology*, 110, 1933, 1982.
179. Olins, G. M. and Bremel, R. D., Oxytocin-stimulated myosin phosphorylation in mammary myoepithelial cells: roles of calcium ions and cyclic nucleotides, *Endocrinology*, 114, 1617, 1984.

Chapter 11

PROLACTIN, GROWTH FACTORS, AND CELL GROWTH*

Charles S. Nicoll, Thomas R. Anderson, Nora J. Hebert, and Sharon M. Russell

TABLE OF CONTENTS

I. Introduction 200

II. Experiments and Results 201
- A. Relationship Between the Dry Weight Measurement and the DNA Content of the Crop-Sac Musosa 201
- B. Effects of Systemically Administered Pituitary Hormones and Extracts on the Local Response of Crop-Sac to Prolactin Injection 201
- C. Analysis of Serum from Pigeons for Prolactin-Synergizing Activity 202
- D. On the Nature of the Circulating Synlactin Activity 202
- E. Relative Potencies of Insulin-Like Growth Factors as Prolactin Synergists 204
- F. Secretion of Prolactin-Synergizing Activity by the Liver In Vitro 204
- G. Direct Effects of PRL on Hepatic Secretion of Synlactin In Vivo 205
- H. Mitogenic Effects of PRL, IGFs, and Hepatic Factors on the Mammary Gland 205
- I. Effects of Other Growth Factors 206

III. Discussion 208

References 210

* This work was supported by NSF grants PCM 8203583 and DCB 8510774.

I. INTRODUCTION

Prolactin (PRL) is the most versatile of the hormones of the vertebrate pituitary gland by virtue of the fact that it has a greater variety of physiological actions among the vertebrates than all other pituitary hormones combined.[1,2] These effects of the hormone have been grouped into seven categories (Table 1).

It is generally assumed that the growth-promoting effects of PRL on its target organs (Table 2) involve a direct action on the cells that constitute them. Indeed, convincing evidence that PRL directly stimulates mammary growth in vivo was obtained by Lyons and colleagues soon after the hormone was purified.[3,4] It is also well established that prolactin has a direct growth-promoting effect on the crop-sac of pigeons and doves.[5,6] However, despite these positive in vivo results, it has been difficult to show a direct mitogenic effect of PRL on mammary epithelium in vitro;[7] only a few studies have reported a stimulatory effect of PRL on mammary cell division in culture.[8,9] In addition, we have made numerous unsuccessful attempts over the past 25 years to demonstrate an in vitro mitogenic effect of PRL on the pigeon crop-sac.[45] Although PRL stimulates growth of the amphibian tail fin in vivo[1,2,10] and it antagonizes thyroxine-induced regression of this structure, PRL has only the latter effect in vitro.[11] Thus, PRL again fails to stimulate growth by direct action on the cells or tissue in vitro. These results suggest that a factor(s) that is present in vivo is missing from the in vitro experimental situations that have been studied.

Additional evidence which suggests that other factors may be involved in the growth-promoting actions of PRL on its target organs comes from comparisons of dose-response relationships of the hormone on the growth of the pigeon crop-sac mucosa. The crop-sac of birds is an expanded portion of the esophagus which is used to store food. In pigeons and doves the mucosal lining of this organ proliferates in response to PRL. Clumps of dead epithelial cells are shed into the crop lumen and this "milk" is used to feed the young. This response to PRL is used as a bioassay for the hormone, which can be injected either subcutaneously or intramuscularly at a site distant from the organ to induce mucosal proliferation systemically,[14] or the hormone can be injected into the skin overlying the crop to promote localized growth of the mucosal epithelium in the local microassay.[5,6] When PRL is injected subcutaneously at a site distant from the crop-sac, a much steeper dose-response slope is obtained than when the hormone is injected directly over the organ by the local microassay procedure.[6,14-16] If the systemically administered PRL were acting in the same way as the locally applied hormone, the slopes of the two dose-response relationships should be essentially the same, even though substantially more hormone is required to be effective systemically because it is diluted in the circulatory system. There are several possible explanations for this difference in slope. The systemically administered hormone may be modified in the circulation to become more active, or it may have a longer half-life than it does locally. Alternatively, it may stimulate the production of another factor that acts synergistically with the PRL on the crop-sac mucosal epithelial cells. Another possibility is that the higher doses of the PRL that are required for a systemic response may contain sufficient amounts of other pituitary hormones that act synergistically with the PRL (either directly or indirectly) to augment the response of the crop-sac to PRL. The work of Bates et al.[17] gives credence to this latter possibility. They showed that in hypophysectomized pigeons, the response of the crop-sac to systemically injected PRL was greatly reduced compared to that of intact birds. The responsiveness of the tissue to PRL was restored by treating the birds with GH, thyroxine, and a corticosteroid.

Table 1
GENERAL CATEGORIES OF THE ACTIONS OF PROLACTIN

Regulation of water and electrolyte balance
Control of growth and/or development
Metabolic effects
Reproductive actions
Effects on integumentary (ectodermal) structures
Synergistic and antagonistic interactions with steroid hormones
Synergistic and antagonistic interactions with thyroid hormones

Table 2
SOME TARGET ORGANS THAT GROW IN RESPONSE TO PROLACTIN IN DIFFERENT VERTEBRATE GROUPS

Class	Organs
Mammals	Mammary gland, prostate, seminal vesicles
Birds	Crop-sac of pigeons and doves
Amphibians	Tail fin of larval anurans and of urodeles; gills, nuptial and cloacal pads
Fish	Seminal vesicles in some species, kidney

II. EXPERIMENTS AND RESULTS

A. Relationship Between the Dry Weight Measurement and the DNA Content of the Crop-Sac Mucosa

In the local crop-sac assay procedure for PRL, the dry weight of a 4-cm-diameter disc of mucosal epithelium is used as an index of growth.[6] In order to verify that this measurement is a reliable index of growth of these cells, correlation and regression analyses were performed on the DNA content and dry weight of mucosal epithelial tissue samples. This analysis showed that the dry weight measurement was highly correlated with the DNA content of the mucosal epithelial cells.[16] Hence, measuring mucosal dry weight is a reliable (although indirect) index of the cell number present in that tissue.

B. Effects of Systemically Administered Pituitary Hormones and Extracts on the Local Response of Crop-Sac to Prolactin Injection

These experiments were conducted to determine whether factors present in pituitary extract or any of the known pituitary hormones, either individually or in combination, could act synergistically with locally injected PRL when they were given systemically. The preparations were injected into the loose skin between one of the legs and the lower abdomen of pigeons according to the dose schedule shown in Table 3. The doses of the purified hormones used were determined from the amount of each that we would expect to be present in the pituitary powder, based on data in the literature.[18] A large "priming" dose was given on the 1st day and this was followed by smaller "maintenance" doses on the subsequent 3 days. During those 3 days, PRL was injected locally twice daily over one side of the crop-sac of each pigeon, and saline injections were given over the contralateral crop. This procedure allowed us to determine whether the systemic treatment alone had any effect on the crop-sac by measuring the mucosal dry weight of the saline-injected side. In addition, we could determine, in the same pigeons,

Table 3
EFFECTS OF SYSTEMIC TREATMENT WITH DIFFERENT PITUITARY HORMONE PREPARATIONS ON THE LOCAL CROP-SAC RESPONSE TO PRL

	Dose of preparation — (μg/kg/day)		
Systemic treatment	Priming dose (day 1)	Maintenance dose (days 2—4)	Local PRL response (mg)
Saline	1 mℓ	1 mℓ	7.2 ± 0.2
Pituitary powder	10,000	1,000	17.2 ± 0.7
Prolactin	400	40	19.3 ± 0.7
Growth hormone	1,000	100	12.8 ± 1.0
ACTH	24	2.4	7.3 ± 0.3
LH, FSH, TSH	50	5	6.6 ± 0.8

Note: The results represent the mucosal dry weight (MDW) of the PRL-treated hemicrop minus the MDW of the contralateral saline-injected hemicrop. Each group contained six to eight pigeons.

whether the systemic treatment affected the response to the locally applied PRL which was injected on the contralateral side.

The various hormone preparations caused little or no stimulation of the mucosa of the saline-injected side at the doses tested.[18] In addition, the pituitary glycoprotein hormones (FSH, LH, and TSH) and ACTH did not affect the response of the contralateral side to the locally administered PRL, either individually or in different combinations (Table 3). However, GH, PRL, and pituitary extract significantly increased the local response to PRL (Table 3). These results suggest that the augmenting effects of the pituitary extract and of the PRL and GH may be the result of stimulating the production of some blood-borne factor or factors. To test this possibility, the following experiment was conducted.

C. Analysis of Serum from Pigeons for Prolactin-Synergizing Activity

Pigeons were injected with saline or ovine prolactin (10 mg/day for 3 days) and then killed by decapitation. The blood was collected and the serum was separated from it and diluted to a concentration of 25% with physiological saline before being tested for crop-sac-stimulating activity. It was also tested to determine whether it could augment the local response to a low dose of PRL. The results of this experiment are shown in Table 4. The serum from neither the saline-treated nor the PRL-injected pigeons had significant direct crop-sac-stimulating activity. In addition, the low dose of PRL had minimal stimulatory activity. When the serum from saline-treated birds was combined with the low dose of PRL, significant augmentation of the local PRL response was observed. However, the serum from PRL-treated birds showed a much greater augmentation of the local response to the hormone. These results indicate that serum from saline-injected pigeons contains a factor or factors that can act synergistically with PRL on the pigeon crop-sac mucosa. Injections of PRL increase the serum concentration of this synergistic activity. Because the serum has no direct stimulatory activity by itself, but it shows PRL-synergizing activity, we named this activity "synlactin".[18,19]

D. On the Nature of the Circulating Synlactin Activity

Prolactin and GH have similar effects on a variety of target organs.[20] Since both hormones increased the responsiveness of the crop-sac to the locally applied PRL (Table 3), we considered some of their common actions that might be involved in this

Table 4
EFFECTS OF SERUM FROM SALINE- OR PRL-INJECTED PIGEONS ON THE LOCAL CROP-SAC RESPONSE TO A DOSE OF 2.0 μg OF PRL

Material injected[a]	Local response MDW (mg)[b]
Saline	13.4 ± 0.7
PRL	14.0 ± 0.8
Control serum	13.5 ± 0.4
PRL serum	15.5 ± 0.9
Control serum and PRL	18.4 ± 1.1
PRL serum and PRL	23.6 ± 1.5

[a] The control serum was from saline-treated pigeons and the PRL serum was from birds given systemic injections of PRL. The sera were diluted to a concentration of 25% with medium 199 prior to injection.

[b] MDW = mucosal dry weight.

Table 5
EFFECTS OF HUMAN SM-C ON THE LOCAL PIGEON CROP-SAC RESPONSE TO PRL

Material injected	Total dose (μg)	Response MDW[a] (mg)
Saline	—	11.1 ± 0.4
PRL	1.0	13.6 ± 0.8
SM-C	8.0	13.7 ± 0.7
PRL and SMC	1.0 + 8.0	23.9 ± 1.9

[a] MDW = mucosal dry weight.

synergism. It has been reported that PRL and GH increase the level of somatomedin (SM) activity in the serum of hypophysectomized rats[21] and hypopituitary dwarf mice,[22] and that both hormones stimulate the secretion of SM from the isolated perfused rat liver.[23,24] Accordingly, it seemed worthwhile to investigate the possible involvement of SMs or insulin-like growth factors (IGFs) as prolactin synergists. At this stage of our investigations, purified preparations of SM were not available, but we could obtain ample quantities of proinsulin, which is a member of the IGF family of molecules.[25]

When bovine proinsulin was injected systemically at a dose of 2 mg on day 1 and 0.2 mg on days 2 to 4, it was as effective at synergizing with locally applied PRL as either systemically injected PRL or GH.[18] These results indicate that the synergist that is present in the circulation of PRL-treated pigeons (Table 4) and which is apparently stimulated by systemic injections of either PRL of GH (Table 3) could be an IGF.

After completing the experiments with proinsulin, we obtained some highly purified human SM-C. It was tested along with the proinsulin for direct synergistic effects in the pigeon crop-sac local microassay. As shown in Table 5, a high dose of SM-C itself had a slight stimulatory effect on the crop-sac epithelium, and a low dose of PRL had a similar degree of stimulatory activity. When the two hormones were combined, they showed a striking degree of synergism. It would require at least a tenfold increase in the dose of prolactin by itself to achieve the same degree of stimulation that was produced by PRL plus SM-C. Similar results were obtained with proinsulin.[16,19]

Table 6
EFFECTS OF MEDIUM IN WHICH SLICES OF LIVER FROM SALINE-, GH- OR PRL-TREATED PIGEONS OR RATS WERE INCUBATED ON THE LOCAL CROP-SAC RESPONSE TO A DOSE OF 1.0 μg PRL

Material injected	Local crop-sac response — MDW[a] (mg)	
	Pigeon experiment	Rat experiment
Saline	13.5 ± 0.5	11.6 ± 0.7
PRL (1.0 μg)	17.8 ± 2.2	14.7 ± 0.6
Control LIM[b] plus PRL	16.1 ± 1.7	14.2 ± 0.7
GH-LIM plus PRL	18.6 ± 1.9	15.8 ± 0.6
PRL-LIM plus PRL	26.7 ± 2.0	17.9 ± 0.9

[a] MDW = mucosal dry weight.
[b] LIM = liver incubation medium.

E. Relative Potencies of Insulin-Like Growth Factors as Prolactin Synergists

Insulin, proinsulin, human SM-C, and relaxin (another member of the IGF family[25]) were tested at two doses for their ability to augment the crop-sac response to local injection of 1.0 μg of ovine PRL. The proinsulin was used as a reference preparation. The relative potencies of these IGFs were as follows: SM-C = 436%, proinsulin = 100%, relaxin = 45%, and insulin = 1%.[16,19] We also tested different preparations of multiplication-stimulating activity (MSA), which is the rat equivalent of SM-A or IGF II,[26] and found that it was inactive as a PRL synergist. These results indicate that the pigeon crop-sac receptor for IGF is unusual because, in other systems, proinsulin is usually much less active than insulin, and relaxin generally has very little activity.[27]

F. Secretion of Prolactin-Synergizing Activity by the Liver In Vitro

The liver is considered to be a major source of IGFs in postnatal mammals,[28] but other organs can produce these factors both pre- and postnatally.[29-31] Accordingly, we investigated whether the liver or kidney of pigeons and female rats that were treated with either GH or PRL could secrete a PRL synergist in vitro. The liver and kidney tissue from these animals was cut into slices about 0.5 mm thick and these were incubated in medium 199 for 3 to 4 hr at a concentration of 50 mg of tissue per milliliter. The tissue was removed, the medium centrifuged, and the supernatant filtered through Whatman No. 1 paper. The medium was then diluted 1:5 or 1:10 with medium 199 and was tested for crop-sac-stimulating activity in the local assay.[6] In addition, the medium samples were injected along with 1.0 μg of ovine PRL to test for PRL-synergizing activity. The medium in which slices of pigeon or rat liver or kidney were incubated had no PRL-like activity when it was injected over the crop-sac alone.[32,33] The kidney incubation medium from none of the treated rats or pigeons augmented the local response of the crop-sac to the direct mitogenic action of PRL (Table 6). Likewise, the liver incubation medium (LIM) from the pigeons and rats that were injected with saline or GH did not contain synlactin activity. In contrast, the LIM from the PRL-treated rats and pigeons showed a striking augmentation of the response to PRL (Table 6).

The ability of liver slices from rats in different physiological states to secrete this PRL-synergizing activity in vitro was also evaluated. For this experiment we used liver from young adult male and female rats, and from pregnant and lactating females. The results in Table 7 show that the liver medium from the male and virgin female rats had

Table 7
EFFECTS OF MEDIUM IN WHICH SLICES OF LIVER FROM RATS IN DIFFERENT PHYSIOLOGICAL STATES HAD BEEN INCUBATED ON THE LOCAL CROP-SAC RESPONSE TO PRL

Material injected	MDW[a] (mg)
Saline	9.1 ± 0.6
1.0 μg PRL	13.0 ± 0.8
PRL plus ♂ LIM[b]	13.5 ± 1.1
PRL plus virgin ♀ LIM	14.1 ± 0.9
PRL plus pregnant ♀ LIM	18.6 ± 1.2
PRL plus lactating ♀ LIM	19.1 ± 1.1

[a] MDW = mucosal dry weight.
[b] LIM = liver incubation medium.

no PRL-synergizing activity. In contrast, the liver from the pregnant and the lactating rats did secrete significant amounts of a PRL synergist.

G. Direct Effects of PRL on Hepatic Secretion of Synlactin In Vivo

The results described in the previous sections led us to investigate whether PRL could modify its direct proliferative action on the pigeon crop-sac by an indirect (i.e., liver-mediated) effect in vivo.[33] Pigeons were equipped with catheters that were inserted into one of their intestinal veins. The catheters were attached to a coil of tubing that was connected to a 7-day osmotic minipump which was left in the abdominal cavity of the birds. The coil of tubing contained segments of solvent or of GH or PRL in the solvent, and these segments were separated by segments of air. By this means, solvent or the hormone-containing solution was infused into the hepatic portal blood supply in pulses of 2-hr duration, with a 4-hr interpulse interval. Thus, the liver of conscious, free-moving pigeons was directly exposed each day for 7 days to four, 2-hr pulses of PRL, GH, or solvent carried to it via the hepatic portal blood. Control pigeons were equipped with similar coils and minipumps, but the catheter was inserted into the external jugular vein so that the infused PRL would not be carried directly to the liver. During the last 3 days of the 7 days of infusion, the pigeons received twice-daily injections of PRL or saline over the contralateral lobes of the crop-sac. The results in Figure 1 show that intrahepatic infusion of PRL, but not GH, caused a striking augmentation of the crop-sac response to the local injections of PRL. Pulsatile infusion of the same dose of PRL into the external jugular vein did not have this effect. Accordingly, it appears to be mediated by the liver.

H. Mitogenic Effects of PRL, IGFs, and Hepatic Factors on the Mammary Gland

Our findings that PRL and IGFs act synergistically to promote growth of the crop-sac, and that the liver secretes synlactin activity, led us to investigate whether somatomedins and/or other liver factors might be involved in regulating the growth of other PRL-responsive target organs. Accordingly, we investigated the mitogenic effects of IGFs, PRL, and liver factors on rat mammary gland explants in vitro.

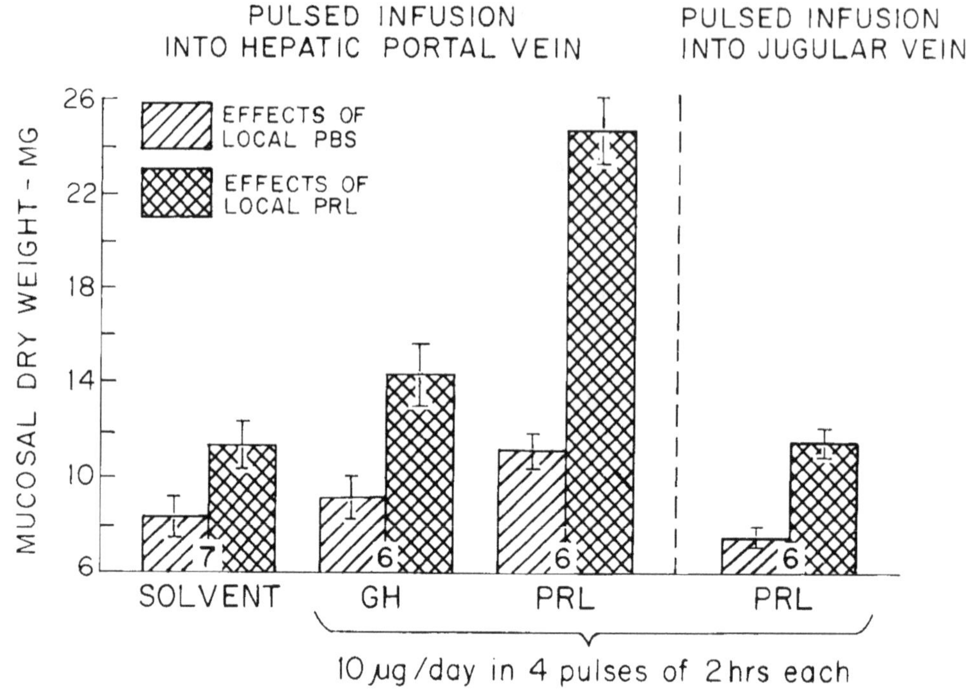

FIGURE 1. Effects of pulsatile infusion of solvent, oGH, or oPRL into the hepatic portal circulation on the local response of the pigeon crop-sac to saline or 1.0 µg of oPRL. The effects of pulsatile infusion of the same dose of oPRL into the external jugular vein are shown by the two columns on the right. The number of pigeons in each group is shown in the bottom of each column, and the vertical lines at the top of each column represent the SEM.

In order to conduct these studies, we developed a new mammary mitogenic assay. Other investigators have found it difficult to show mitogenic effects of PRL on mammary explants from pregnant rats or mice in vitro.[7] We reasoned that this difficulty was due to the fact that the pregnant glands are, in effect, "saturated" with endogenous mitogenic factors, including placental lactogens. Accordingly, rats that were 14 to 16 days pregnant were hysterectomized to eliminate some of these factors. Their mammary glands were removed 16 to 18 hr later and cut into explants, which were incubated in medium 199 for 6 hr. ^{3}H-labeled thymidine was present for the entire incubation period or was added during the last 4 hr of incubation. The mammary explants were then processed to isolate the DNA, which was counted for ^{3}H activity.

The effects of incubating the mammary explants in medium in which liver slices had been previously incubated are shown in Table 8. The medium containing products from the male and virgin female liver slices had no mitogenic activity on the mammary explants. In contrast, the medium from the incubation of liver slices from pregnant rats doubled ^{3}H-thymidine incorporation into DNA, and that from the lactating females increased incorporation by about 50%. These results are of considerable significance because they indicate that hepatic factors may be involved in mammary growth. Since only liver from pregnant and lactating females had this activity, it is unlikely that the mitogen is SM-C, because male and virgin female liver should also secrete this IGF.

I. Effects of Other Growth Factors

The specificity of the PRL-synergizing activity of the IGFs was evaluated by testing other growth factors. These were injected by themselves or with PRL in the local crop-sac assay.[34] The results obtained with four growth factors and two hormones (PRL

Table 8
EFFECTS OF MEDIUM IN WHICH SLICES OF LIVER FROM RATS IN DIFFERENT PHYSIOLOGICAL STATES HAD BEEN INCUBATED ON DNA SYNTHESIS IN RAT MAMMARY EXPLANTS IN VITRO

Liver medium	CPM/mg DFDW[a]
Control 199	610 ± 70
Adult ♂	580 ± 40
Adult ♀	620 ± 80
Pregnant ♀	1220 ± 120
Lactating ♀	860 ± 90

[a] DFDW = defatted dry weight. N = 5 in each group.

Table 9
EFFECTS OF GROWTH FACTORS, PROLACTIN, AND RELAXIN ON THE GROWTH OF THE CROP-SAC MUCOSAL EPITHELIUM IN VIVO

Material injected[a]	N	Mucosal dry weight (mg) 0.8-μg dose	8.0-μg dose	Slope[b]
PRL	5	18.7 ± 1.8	28.9 ± 3.6	10.2 ± 2.0[c]
EGF	6	14.5 ± 0.7	19.6 ± 1.4	5.0 ± 0.9[c]
MSA	5	11.0 ± 1.5	13.8 ± 1.5	2.8 ± 1.7
Relaxin	6	13.3 ± 1.0	13.7 ± 1.4	0.4 ± 0.7
PDGF	4	14.0 ± 2.0	14.8 ± 0.4	0.8 ± 2.3
FGF	5	11.5 ± 1.6	11.4 ± 1.4	−0.1 ± 0.8

[a] EGF = epidermal growth factor; MSA = multiplication-stimulating activity; PDGF = platelet-derived growth factor; FGF = fibroblast growth factor.
[b] The slope is the difference in the response between the high and the low doses as the two doses span one log interval.
[c] $p < 0.01$.

and relaxin) are shown in Table 9. Of these six peptides, only PRL and EGF had crop-sac-stimulating activity by themselves. However, the slope of the dose-response relationship produced by the EGF was only about one half of that produced by PRL. When PDGF, FGF, and MSA were tested for PRL-synergizing activity, they were all found to be inactive.[34] We had previously found that relaxin augments the local crop-sac response to PRL, but MSA was inactive.[16,19]

The effects obtained with EGF plus PRL were particularly interesting. Despite the fact that both of these factors are independently mitogenic on the crop-sac, they showed no interaction when they were tested in combination in the local crop-sac assay (Table 10). Thus, the effects of EGF are distinct from those of the IGFs because it is

Table 10
GROWTH-PROMOTING INTERACTIONS AMONG PRL, EGF, AND PROINSULIN (PI) IN THE PIGEON CROP-SAC IN VIVO

Material injected	MDW[a] (mg)
Saline	12.1 ± 0.7
PRL (1.0 μg)	14.6 ± 0.6
EGF (1.0 μg)	17.1 ± 0.7
PRL + EGF	15.7 ± 0.6
Saline	13.3 ± 0.9
PI (10 μg)	13.6 ± 0.6
PRL (1.0 μg)	15.8 ± 1.0
PRL + PI	24.1 ± 1.5
EGF (10 μg)	17.6 ± 1.4
EGF + PI	17.8 ± 1.1

[a] MDW = mucosal dry weight.

mitogenic on the crop-sac mucosal epithelium by itself, but it does not act synergistically with PRL. It was of interest, therefore, to determine whether EGF would interact with an IGF on the crop-sac epithelium because IGFs have no effect by themselves, except at very high doses, where the responses are very small.[16,19] As can be seen from Table 10, proinsulin did not affect the proliferative effect of EGF; hence, the latter is not simply mimicking the actions of PRL. It remains to be established whether EGF has any physiological role in the growth response of the crop-sac mucosal epithelium to PRL and synlactin.

III. DISCUSSION

Our results indicate that the growth-promoting effects of PRL on the pigeon crop-sac involve both direct and indirect mechanisms. The direct effect may involve sensitization of the mucosal epithelial cells to the mitogenic action of a synergist.[19] The indirect effect involves stimulation of the secretion of a synergist from some other source, which may be the liver.

Several lines of evidence support the suggestion of an indirect effect. When PRL was injected systemically at a dose which had little or no stimulatory effect on the crop-sac, the responsiveness of the organ to the direct mitogenic action of PRL was greatly increased (Table 3). The systemically administered PRL also increased the serum level of a PRL synergist (Table 4). Because systemically injected proinsulin could mimic the augmenting effect of systemic PRL,[18] the possibility arose that the synergistic activity in the serum of the PRL-treated birds could be an IGF. This suggestion was substantiated by the demonstration that several IGFs augmented the crop-sac response to PRL when they were combined with the hormone and injected directly into the organ[16,19] (Table 5). The results with the liver incubation medium indicate that this organ could be the source of the PRL synergist (Tables 6 and 7). However, the most convincing physiological evidence that PRL can indirectly affect the growth of one of its peripheral target organs by an action on the liver was provided by the experiment on intrahepatic infusion of the hormone[33] (Figure 1). Thus, the mechanism of the growth-promoting effects of PRL would appear to have much in common with that of GH.

However, in the case of the latter hormone, an IGF is thought to mediate[28] its growth-promoting effects (hence the term "somatomedin"), but IGFs alone have little or no activity on the crop-sac;[16,19] their mode of action appears to be primarily synergistic. Hence, we have named the PRL-synergizing activity *synlactin.*[19]

Our results with the mammary gland explants also indicate that an IGF may be involved in the growth of this target organ of PRL. Somatomedin-C and PRL did have direct mitogenic effects on mammary explants in vitro.[16] More significantly, medium from incubations of liver from pregnant and lactating rats had mammogenic activity, but the liver medium from male and virgin female rats did not (Table 8). These results with the liver medium in the mammary assay are consistent with our findings with the crop-sac. Only the medium from incubates of liver from pregnant and lactating rats had significant synlactin activity (Table 7). It should be noted here that the synergist that is secreted by the liver is not PRL itself because the liver incubation medium alone was inactive in the crop-sac assay,[16,32,33] and it had no detectable PRL activity in a radioimmunoassay (RIA) for rat PRL.[32] Furthermore, it is unlikely that SM-C is responsible for this synergism because analysis of similar medium samples in an RIA for human SM-C showed that the liver of virgin and pregnant females secretes similar amounts of that IGF.[3]

Our findings with the liver incubation medium samples and with intrahepatic infusion of PRL are significant in relation to previous studies which were of undetermined physiological significance. Ovine and rat PRL were reported to stimulate the secretion of bioactive SM (i.e., sulfation factor activity) by perfused liver of male rats.[23,24] Surprisingly, PRL was 20 times more potent in this regard than was GH.[23] Furthermore, PRL injections restore the serum levels of bioactive SM in hypophysectomized rats,[21] but PRL-treated rats do not grow.[21,35-37] Thus, the SM activity that is secreted by the liver in response to PRL cannot be the same as the SM that is involved in the promotion of somatic growth. Accordingly, the PRL-stimulated SM activity is probably involved in regulating other functions.

Our results presented in this chapter indicate that the PRL-stimulated SM activity in rats and mice[21-24] may be related to the synlactin activity that we have found to be secreted by the liver of rats and pigeons. The role of this factor in these animals may be to stimulate the growth of PRL-responsive organs (i.e., the mammary gland and crop-sac, respectively) and it does so in concert with PRL. In view of the evidence that PRL-dependent SM (synlactin-activity) is secreted by the liver, other data in the literature take on new significance.

It has been reported that the hepatic receptors for PRL increase during pregnancy in rats,[38,39] mice,[40] and rabbits.[41] Our results indicate that these receptors could be concerned with synlactin secretion by the liver in response to PRL. Thus, during pregnancy, when the plasma levels of lactogenic hormones are high (i.e., either from the pituitary and/or from the placenta), the liver is presumably sensitized to PRL, as evidenced by the increase in receptors for the hormone. If these receptors are involved in synlactin secretion, then the condition of pregnancy would favor high circulating levels of the growth factor. The synlactin could then act in concert with PRL and ovarian steroids to promote mammary growth. Anderson et al.[18] have suggested that a similar mechanism may operate to promote growth of the crop-sac of pigeons and doves during egg incubation when plasma PRL levels are rising.[42,43] A schematic representation of the synlactin hypothesis as we presently envision it is shown in Figure 2.

Analysis of the effects of various growth factors on the responsiveness of the pigeon crop-sac mucosa to PRL showed that only some of the IGFs could augment the response. Of these several growth factors, only EGF had significant growth-promoting activity by itself, but its effects were different from those of PRL in several respects, and it did not interact with either PRL or PI. Even though EGF can act as an inde-

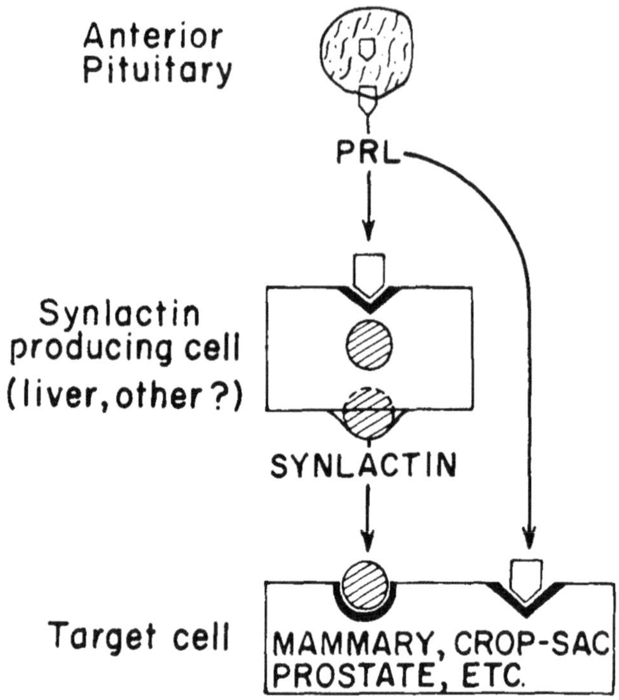

FIGURE 2. Schematic diagram of the synlactin hypothesis adopted from Laron's[44] diagram of the somatomedin hypothesis.

pendent mitogen on the crop-sac, it remains to be determined whether it has any physiological role in the growth of this organ.

REFERENCES

1. Nicoll, C. S., Physiological actions of prolactin, in *Handbook of Physiology, Section 7: Endocrinology*, Vol. 4, Part 2, Knobil, E. and Sawyer, W., Eds., Williams & Wilkins, Baltimore, Md., 1974, 253.
2. Clarke, W. C. and Bern, H. A., Comparative endocrinology of prolactin, *Horm. Protein Pept.*, 8, 105, 1980.
3. Lyons, W. R., Li, C. H., and Johnson, R. E., The hormonal control of mammary growth and lactation, *Rec. Prog. Horm. Res.*, 14, 219, 1958.
4. Lyons, W. R., The direct mammotrophic action of lactogenic hormone, *Proc. Soc. Exp. Biol. Med.*, 51, 308, 1942.
5. Lyons, W. R., Preparation and assay of mammotropic hormone, *Proc. Soc. Exp. Biol. Med.*, 35, 645, 1937.
6. Nicoll, C. S., Bioassay of prolactin. Analysis of the pigeon crop-sac response to local prolactin injection by an objective and quantitative method, *Endocrinology*, 80, 541, 1967.
7. Elias, J. J., The role of prolactin in normal mammary gland growth and function, *Horm. Protein Pept.*, 8, 37, 1980.
8. Mukherjee, A. S., Washburn, L. L., and Banerjee, M. R., Role of insulin as a "permissive" hormone in mammary gland development, *Nature (London)*, 246, 159, 1973.
9. Banerjee, M. R., Wood, B. G., and Kinder, D. L., Whole mammary gland organ culture: selection of appropriate gland, *In Vitro*, 9, 128, 1973.
10. White, B. and Nicoll, C. S., Hormonal control of amphibian metamorphosis, in *Metamorphosis: A Problem in Development*, Gilbert, L. I. and Frieden, E., Eds., Plenum Press, New York, 1981, 363.

11. Derby, A. and Etkin, W., Thyroxine-induced tail resorption *in vitro* as affected by anterior pituitary hormones, *J. Exp. Zool.*, 169, 1, 1968.
12. Beams, H. W. and Meyer, R. K., The formation of pigeon "milk", *Physiol. Zool.*, 4, 486, 1931.
13. Patel, M. D., The physiology of the formation of "pigeons milk", *Physiol. Zool.*, 9, 129, 1936.
14. Nicoll, C. S., Bioassay of prolactin. Analysis of the pigeon crop-sac response to systemic prolactin injection by an improved method of response quantification, *Acta Endocrinol.*, 60, 91, 1969.
15. Bates, R. W., Garrison, M. M., and Cornfield, J., An improved bioassay for prolactin using adult pigeons, *Endocrinology*, 73, 217, 1963.
16. Nicoll, C. S., Anderson, T. R., Hebert, N., and Russell, S. M., The synlactin hypothesis: the growth-promoting action of prolactin on its target organs involves synergism with an insulin-like growth factor, in *Prolactin — Basic and Clinical Correlates*, MacLeod, R. M., Thorner, M. O., and Scapagnini, U., Eds., Liviana Press, Padua, Italy, 1985, 393.
17. Bates, R. W., Miller, R. A., and Garrison, M. M., Evidence in the hypophysectomized pigeon of a synergism among prolactin, growth hormone, thyroxine and prednisone upon weight of the body, digestive tract, kidney and fat stores, *Endocrinology*, 71, 345, 1962.
18. Anderson, T. R., Pitts, D. S., and Nicoll, C. S., Prolactin's mitogenic action on the pigeon crop-sac mucosal epithelium involves direct and indirect mechanisms, *Gen. Comp. Endocrinol.*, 54, 236, 1984.
19. Anderson, T. R., Rodriguez, J., and Nicoll, C. S., The synlactin hypothesis: prolactin's mitogenic action may involve synergism with a somatomedin-like molecule, in *Insulin-like Growth Factors/Somatomedins*, Spencer, E. M., Ed., de Gruyter, Berlin, 1983, 71.
20. Nicoll, C. S., Prolactin and growth hormone: specialists on one hand and mutual mimics on the other, *Perspect. Biol. Med.*, 25, 369, 1982.
21. Bala, R. M., Bohnet, H. G., Carter, J. N., and Friesen, H. G., Effect of ovine prolactin on serum somatomedin bioactivity in hypophysectomized female rats, *Can. J. Pharmacol.*, 56, 984, 1977.
22. Holder, A. T. and Wallis, M., Actions of growth hormone, prolactin and thyroxine on serum somatomedin-like activity and growth in hypopituitary dwarf mice, *J. Endocrinol.*, 74, 223, 1977.
23. Francis, M. J. O. and Hill, D. J., Prolactin-stimulated production of somatomedin by rat liver, *Nature (London)*, 255, 167, 1974.
24. Hill, D. J., Francis, M. J. O., and Milner, R. D. G., Action of rat prolactin on plasma somatomedin levels in the rat and on somatomedin release from perfused rat liver, *J. Endocrinol.*, 75, 137, 1977.
25. Blundell, T. L., Bedarkar, S., Rinderknecht, E., and Humbel, R. E., Insulin-like growth factor: a model for tertiary structure accounting for immunoreactivity and receptor binding, *Proc. Natl. Acad. Sci. U.S.A.*, 75, 180, 1978.
26. Spencer, E. M., Ross, M., and Smith, B., The identity of human insulin-like growth factors I and II with somatomedins C and A and homology with rat IGF I and II, in *Insulin-like Growth Factors/Somatomedins*, Spencer, E. M., Ed., de Gruyter, Berlin, 1983, 81.
27. King, G. L. and Kahn, C. R., Non-parallel evolution of metabolic and growth-promoting functions of insulin, *Nature (London)*, 292, 644, 1981.
28. Daughaday, W. H., Growth hormone and the somatomedins, in *Endocrine Control of Growth*, Daughaday, W. H., Ed., Elsevier, New York, 1981, 1.
29. McConaghy, P. and Dehnel, J., Preliminary studies of "sulfation factor" production by rat kidney, *J. Endocrinol.*, 52, 587, 1972.
30. D'Ercole, A. J., Applewhite, G. T., and Underwood, L. E., Evidence that somatomedin is synthesized by multiple tissues in the fetus, *Devel. Biol.*, 75, 315, 1980.
31. Stracke, H., Schulz, A., Rossol, D., and Schatz, H., Effect of growth hormone on osteoblasts and demonstration of somatomedin-C/IGF-1 in bone organ culture, *Acta Endocrinol.*, 107, 16, 1984.
32. Nicoll, C. S., Hebert, N. J., and Russell, S. M., Lactogenic hormones stimulate the liver to secrete a factor that acts synergistically with prolactin to promote growth of the pigeon crop-sac mucosal epithelium *in vivo*, *Endocrinology*, 116, 1449, 1985.
33. Mick, C. C. W. and Nicoll, C. S., Prolactin directly stimulates the liver *in vivo* to secrete a factor (synlactin) which acts synergistically with the hormone, *Endocrinology*, 116, 2049, 1985.
34. Anderson, T. R., Mayer, G. L., Hebert, N. J., and Nicoll, C. S., Interactions among prolactin and growth factors *in vivo* on the growth, differentiation and morphology of the pigeon crop-sac mucosal epithelium, in preparation.
35. Thorngren, K. G. and Hanssen, L. I., Bioassay of growth hormone and prolactin preparations by determination of longitudinal bone growth with tetracycline, *Acta Endocrinol. (Copenhagen)*, 76, 35, 1974.
36. Russell, S. M., Vodian, M. A., Hughes, J. P., and Nicoll, C. S., Electrophoretic separation of forms of rat growth hormone with different bioassay and radioimmunoassay activities: comparison of intraglandular and secreted forms, *Life Sci.*, 23, 373, 1978.
37. Glasscock, G. F. and Nicoll, C. S., Hormonal control of growth of the infant rat, *Endocrinology*, 109, 176, 1981.

38. Kelly, P. A., Posner, B. I., Tsushima, T., and Friesen, H. G., Studies of insulin, growth hormone and prolactin binding: ontogenesis, effects of sex and pregnancy, *Endocrinology,* 95, 532, 1974.
39. Sasaki, N., Tanaka, Y., Imai, Y., Tsushima, T., and Matsuzaki, F., Different characteristics of solubilized lactogen receptors from livers of pregnant and non-pregnant female rats, *Biochemistry,* 203, 653, 1982.
40. Sasaki, N., Yasuo, I., Tsushima, T., and Matsuzaki, F., Regulation of somatotrophic and lactogenic binding sites in mouse liver membranes, *Acta Endocrinol.,* 101, 574, 1982.
41. Fix, J. A., Leppert, P., and Moore, W. V., Growth hormone and prolactin binding to rabbit liver plasma membranes, *Horm. Metab. Res.,* 13, 508, 1981.
42. Goldsmith, A. R., Edwards, C., Koprucu, M., and Silver, R., Concentrations of prolactin and luteinizing hormone in plasma of doves in relation to incubation and development of the crop-gland, *J. Endocrinol.,* 90, 437, 1981.
43. Cheng, M. F. and Burke, W. H., Serum prolactin levels and crop-sac development in ring doves during a breeding cycle, *Horm. Behav.,* 17, 54, 1983.
44. Laron, Z., Somatomedin, insulin, growth hormone and growth: a review, *Isr. J. Med. Sci.,* 18, 823, 1982.
45. Nicoll, C. S., unpublished data.

INDEX

A

A23187, 114—117
Abetalipoproteinemia, 78
Ablation studies, 169
[^{14}C]-Acetate, 109, 112—113
Acetyl-CoA, 185
Acetyl CoA carboxylase, 54, 185—187
Acinar cells, 166
Acini, 116
Actinomycin D, 54, 143—144
Activation of molecular events, 1—4
S-Adenosylmethionine, 140
S-Adenosylmethionine decarboxylase, 140, 142, 145
Adenylate cyclase, 78—79, 157—158, 169
Adrenal ablation, 128
Adrenal corticoids, 138
Adrenal gland prolactin receptors, 72
Adrenal medulla, 72
α-Adrenergic receptors, 79
Adrenocorticotropic hormone, 73
Age-related changes, 90
Aldosterone, 122
Aliphatic alcohols, 91—93
Alpha-adrenergic receptors, 79
Alveolar cells, 108, 115, 165—166
Alveolar development, 138
Alveolar morphogenesis, 130
Alveolar secretory cells, 166
Alveolar structures, regression of, 122
Alveoli, 108
Amino acid sequence, 189
α-Aminoisobutyric acid uptake, 79
Amphibians
  metamorphosis in, 2, 45, 49—51
  tail fin, 200—201
Androgens, 36, 72—73
Antibodies, 2
Antibodies to receptor, 16
Antidiuretic hormone, 163
Anti-ovine prolactin, 30
Anti-ovine prolactin antibodies, 10
Antitubulin, 166
Antitubulin immunofluorescence, 161
Antizyme, 141
Apical plasma membrane, 108
Arachidonic acid, 58, 80—83, 96—97, 113, 147, 182
[^{3}H]-Arachidonic acid, 113
Arcuate nucleus, 32
Arginase, 140, 142, 144
Arginine, 140
Aspirin, 49
ATP, 163, 187
ATPase, 189
ATP citrate lyase, 184—185
Autoradiography, 6, 57
  identification of receptors, 6, 11
  intracellular prolactin binding, 70
  prolactin localization by, 8

B

B_6 analog, 127
BALB/c female mice, 122—123
Basal 2-deoxy D-glucose uptake, 79
Basal glucose uptake, 79
Beta cells, 162, 163
Beta granule, 162—163
Bilateral adrenalectomy, 128
Bioassay for prolactin, 2, 200
Blood-CSF barrier, 28—30
Bovine growth hormone, 73
Bovine milk, 108
Bovine pituitary glands, 2
Bradykinin, 81—83
Brain, 27—40, 44
  CSF, 28—30
  dopamine, 31—35
  electrical activity, 30—31
  luteinizing hormone-releasing hormone, 35—36
  male sex behavior and, 36
Bromocriptine, 36, 94—95, 98
2-α-Bromoergocryptine (CB-154), 128
Broodiness, 44
Brooding, 52
Bulk transport, 154—155
Butanol, 92—93

C

Ca^{2+}-ATPase, 160, 169
[^{14}C]-acetate, 109, 112—113
Calcium, 107—108, 114—119, 180, 190
  microtubules, 159—161
  phosphorylation, 189
  tubulin assembly and disassembly, 158—159
  uptake, 170
Calcium-dependent protein kinases, 182—183
Calcium-dependent regulator protein, 160
Calcium enzyme, 115
Calcium-independent protein kinases, 183
Calcium influx, 116
Calcium ion, 3, 108, 111, 116, 138
Calcium ionophore, see also A23187, 111, 114, 116—117
Calmodulin, 115, 169, 180, 182, 190
  mammary tissues, 117
  microtubules, 160—161
Casein, 56, 60, 96, 111, 122, 138—139, 156, 187
  genes, 126, 128
  phosphorylation of proteins, 180
  secretion, 164
  synthesis, 111—112, 139, 146, 162, 169

A23187, 116
calcium and prolactin, 115
inhibition of initiation of, 154
stimulation of, 156
β-Casein gene expression, 130
Casein kinase, 187—191
Casein kinase I and II, 180, 183, 187
Casein micelles, 115
Casein mRNA, 3, 130
Castration, 72—73
Caudate-putamen, 35
CB-154, 184
cDNA, 124—125, 130—131
cDNAcsn, 124
Cell differentiation, 55—56, 157, 161
Cell division, see also Mitogenesis, 49
Cell function, 139—140
Cell growth, 139—140, 199—212
Cell membrane-associated phenomena, 78—79
Cell membranes, 108—110
Cell polarity, 164
Cell proliferation, see also Mitogenesis, 52, 55—56, 161
β Cells, 162, 163
Cellular regulation, 109—111
Centrioles, 159
Cerebrospinal fluid (CSF), 28—30, 37
CHAPS, 9—10
Chemotactic peptide receptors, 79
Chicks, 45
Chloride cells, 46—47
Cholesterol metabolism, 98
Choline kinase activity, 147
Chromosomes, 159
Cichild fishes, 47
Circadian rhythm, 43—44, 53
Citrate, 185, 187
Clones, 58
CNS neurons, 30
Cobalt, 189
Colchicine, 154, 158, 161, 163—164, 169
inhibition of tubulin synthesis, 159
milk flow suppression, 164
Colchicine binding assay, 157, 164—165, 167
Collagen gels, 156
Collagen synthesis, 50
Columbid crop, 51—58
Corpora lutea, 71
Corticosteroid-free medium, 123—125, 127
Corticosteroids, 47
Corticosterone, 36
Cortisol, 109, 114, 125—126, 130, 139, 143, 146—147
single injection, 128
C:P ratio, 77—78, 98
Critical tubulin concentration, 159
Crop milk, 2, 43—44, 51, 53
Crop-sac, see also Pigeons, 200—203, 207—208
Crop stimulation, 44
Crop tissue, 44
Cryptic receptors, 19, 96
Cultured mammary cells, 115
Cultured mammary tissues, 111
Cultured mouse mammary gland explants, 108—109, 111
Cyclase activation, 158
Cyclases, 162
Cyclic AMP, 44, 48, 56, 138, 145, 157, 161, 163
inhibition of growth of mammary tumors, 190
guinea pig milk, 162
pregnancy, 162
regulator of lactogenesis, 139
Cyclic AMP-dependent phosphodiesterase, 117
Cyclic AMP-dependent protein kinases, 161—162, 180—182, 185—187, 191
Cyclic AMP-independent protein kinase, 147, 185—186
Cyclic AMP phosphodiesterase, 160—161
Cyclic GMP, 138, 145, 147, 161, 167
conversion of GTP to, 113
Cyclic GMP-dependent phosphodiesterase, 117
Cyclic nucleotide, 3—4, 48, 161—162, 167
Cyclic nucleotide-dependent protein kinases, 181—182
Cyclic nucleotide-independent protein kinases, 183
Cyclic nucleotide-mediated actions of prolactin, 133
Cycloheximide, 32, 144
Cyclooxygenase, 110
Cyclosporine, 145
Cytochalasin B, 164, 169
Cytoplasmic disorganization, 164
Cytoskeletal system, 157, 162, 168—169
Cytosol receptors, 129

D

Dacron raft, 122
DB-cyclic AMP, 167
2-Deoxy D-glucose uptake, 79
Dephosphorylation, 180, 183, 187—188
Depolymerization, 160
Desaturation of receptors, 8
Detergent binding to receptors, 9
Dexamethasone, 72
^{3}H-Dexamethasone, 127
1,2 Diacylglycerides, 111
Diacylglycerol, 110, 182—183
1,2 Diacylglycerol, 110
Diacylglycerol lipase, 110
1,4-Diamino-2-butene, 141—142
Dienoic acid, 80
Diethyl stilbesterol, 88
Differentiation, 55—57, 60, 156, 166
mammary cells, 122
Differentiative processes, 108
α-Difluoromethylornithine (DFHO), 141—142, 146
Diglyceride, 110—111, 113, 115
Dimeric (free) tubulin, 159, 165, 169

7,12-Dimethylbenzanthracene (DMBA)-induced rat mammary tumors, 70—71, 73
1,6-Diphenylhexatriene (DPH), 75—76, 83
Diphosphatidylinositol, 110
Dissociable binding sites, 2
Dissociation, 8, 122—124
DNA, 139
synthesis, 17, 48, 53—54, 111—112
inhibition of initiation of, 154
insulin stimulated, 144
liver incubation of rat mammary explants, 206—207
polyamine accumulation in mammary gland, 143
cDNA, 124—125, 130—131
cDNAscn, 124
Dogs, 139
Dopamine, 31—35
"Dot-blot" hybridization technique, 130
Doves, 2, 45, 51—52, 200—201
Ductal glands, 122

E

EDTA chelation of calcium, 116
EGTA, 115, 159
Electrical activity of hypothalamus following prolactin, 30—31
Emiocytosis, 154—155
Endocrine ablation, 138
Endocrine control of prolactin receptors, 10—15
Endocrine target organ, 122
Endocytosis, 154, 156, 163, 169
Endomembrane flow, 156—157
Enzyme activation, 169
Enzymes, 79, 139
Epidermal growth factor, 53, 138, 139, 207
Erythrocyte membranes, 98
Erythrocytes, 77, 143
Essential fatty acids (EFA), 80—81
Estradiol-17β, 122, 167—168
Estrogens, 71, 73, 123, 168
Ethanol, 91, 93—95
Evolution of prolactin mechanism in nonmammalian vertebrates, 42—45
Exocytosis, 108, 154, 156—158, 162—164
Explant cultures, 162

F

Fat, 43, 54, 108
Fatty acid synthesis, 162, 184—187
Fatty acid synthetase, 54
Feedback system, 31—33, 37
Fetal rat lung, 109
FGF, 207
Fishes, 45
Fluidizing lipids, 76—77
Fluorescence polarization, 74—76, 86, 88
Follicle-stimulating hormone (FSH), 11, 72
Free and polymerized tubulin pools, 165, 167
Free fatty acids, 110—111, 158
Functional differentiation, 122, 131, 138

G

Galactorrhea-amenorrhea, 35
Galactosyltransferase, 155
Gene expression, 54—55, 58, 60, 127
assessment of role of hormones regulating, 122
glucocorticoid and prolactin, synergistic action of, 121—135
mammary gland development in vivo, 130—131
rabbit mammary cells, 4
Gene regulation, 58
Gestation, 86, 88, 129, 131, 138
Gill, 46—50
Glomerulus, 48
Glucagon, 185
Glucocorticoids, 44, 51, 72, 147
cytoplasmic structures, 139
receptors, 126—128
stimulation, 130, 132
synergistic action of prolactin and, 121—135
Glucose-6-phosphate dehydrogenase activity, 147
Glucose stimulation, 163
Glucose uptake, 79, 162
β-Glucuronidase, 50
Glutamate dehydrogenase, 50
Glycerol, 159, 165
Glycogen synthase kinase, 183
Golgi apparatus, 15, 52, 92, 94—95, 154—156, 160, 163, 167, 169
calcium uptake, 170
mammary casein kinase, 188
Golgi complex, see Golgi apparatus
Golgi vesicles, 160, 169—170, 189
Gonadal steroids, 44
Gonadotropins, 35—37, 42, 71
β-Granule, 162—163
Granulosa cells, 71
Growth factor receptors, 56
Growth factors, 199—212
Growth hormone, 42, 48—51, 53—54, 73, 123, 138, 163
Growth-promoting effects of prolactin, 200—201, 208
GTP, 113, 159, 160
Guanylate cyclase, 113
Guanyl nucleotide, 158
Guinea pig, 130, 158, 162, 166, 190

H

Haloperidol, 33, 35
Hamsters, 139
Hepatic factors, mitogenic effects on mammary gland, 205—206

Hepatic membranes, 79—90
Hepatic metabolism, 45
Hepatic portal blood infusion, 205
Hepatic prolactin receptors, 73—74
Hepatic receptors, 209
Hepatic secretion of synlactin in vivo, 205
Hepatocytes, 80
Histone, 189
Histone nuclear proteins, 190
Histone phosphorylation, 191
Hormones, 15, 17—18, 109, 122, 138
Human chorionic gonadotropin, 11, 29, 89
Human erythrocyte, 79
Human growth hormone, 2
Human luteinizing hormone receptors, 79
Human placental lactogen, 29
Human polymorphonuclear leukocytes, 79
Human prolactin, 2
Human SM-C, 203
Hyaluronic acid synthesis, 50
Hybridization analysis of RNA, 124
α-Hydrazino-δ-aminovaleric acid, 146
α-Hydrazine-ornithine, 141—142
Hydrocortisone, 130, 167
1α-Hydroxylase, 45
Hyperphagia, 44
Hyperprolactinemia, 28, 36—37, 90, 98
Hypophysectomy, 72—73
Hypoprolactinemia, 90, 98
Hypothalamus, 30—31

I

Identification of receptors, 6, 8—9, 11
^{125}I-labeled prolactin, 8, 11, 70, 72, 154
Immunocytochemistry (IC), 6, 15—16, 70, 73
Immunofluoresence, 161, 163, 165, 166
Inactivation of prolactin, 3
Index of growth, 201
Indomethacin, 49, 83—85, 87, 90, 145
Induced lactogenesis, 167—168
Inositol, 110
Inositol biphosphate, 111
Inositol diphosphate, 111
Inositol monophosphate, 111
Inositol phosphate, 111
Inositol phosphatides, 111, 113
Inositol triphosphate, 110—111
Insulin, 109, 114, 122—123, 125—126, 130, 138, 142—143, 146—147
  acetyl-CoA carboxylase activities, 187
  ATP citrate lyase activity, 185
  phosphorylation, 190—191
  pyruvate dehydrogenase activity, 184
  secretion, 162
Insulin-like growth factors (IGFs), 53, 55—56, 203—206, 208
Insulin-stimulated 2-deoxy D-glucose uptake, 79
Insulin-stimulated glucose uptake, 79
Integument, see also Skin, 46—47
Internalization of prolactin, 3, 8, 15, 154, 169
Internalization of prolactin-receptor complex, 163
Intestine, 46, 48
Intracellular mediators, 3
Intracellular organelle membranes, 108
Intracellular prolactin binding, 70, 73
Intracellular vesicle, 15
Intracerebroventricular (i.c.v.) injection of prolactin, 32—34
Intrahepatic infusion of prolactin, 205—206, 208—209
Ionic strength, tubulin assembly and disassembly, 159
Iontophoretical application of prolactin, 31
Islet cells, 162
3-Isobutyl-1-methylxanthine, 145
Isolated mammary gland cell populations, 165

K

Kidneys, 46—50, 72
Killifish, 45, 47

L

α-Lactalbumin, 122, 138—139, 155—156
Lactate dehydrogenase, 50
Lactating bovine mammary cells, 109
Lactating mammary cells, 169
Lactating rate, 116, 168
Lactation, 35, 71, 73, 86—88, 98, 108, 138, 142
  alveolar cells, 115
  calmodulin, 117
  induced in pseudopregnant rabbits by bovine pituitary gland extract, 2
  phospholipid changes during, 108—109
  polymerized tubulin, 165
  putrescine concentration, 143
  spermidine concentration, 143
  tubulin polymerization during, 164—166
Lactogenesis, 122—124, 129, 132, 139, 143
Lactogenic activities, 2
Lactogenic hormone mixture, 123, 125
Lactogenic hormones, 2, 128
Lactogenic processes, 116
Lactose, 138, 160, 162, 168—169
Lactose synthetase, 156
Lateral mobility of membrane proteins, 78
[^{3}H]-Leucine incorporation into casein-rich phosphoprotein, 112—113
Leukotrienes, 110, 111
Leuteolytic effects, 2
Leuteotropic effects, 2
Leydig cells, 72
Linoleic acid, 80
Lipase, 111
Lipid, 78—79, 138, 146
Lipogenesis, 45
Lipoprotein globule, 52, 54

Lipoprotein lipase, 54
Liposomes, 158
Lipoxygenase, 110
Liver, 45, 50, 145, 204—205, 207—209
Lobuloalveolar growth, 138
Lobuloalveolar morphogenesis, 122, 126, 131
Low molecular weight peptides, 3—4
Lumicolchicine, 164
Luteal cells, 71
Luteinizing hormone (LH), 72
Luteinizing hormone-releasing hormone, 35—36
Lutropin, 11
Lymphocytes, 145
Lysed cell system, 161
Lysophospholipids, 81, 96—97
Lysosomal degradation, 169
Lysosomes, 50

M

Macrophage, 79
Magnesium, 185, 187, 189—190
Male reproductive organs, 72
Male sex accessory glands, 2
Male sex behavior, 36—37
Malonyl-CoA, 185
Mammary casein kinases, 187—190
Mammary cells, 108, 112—117, 122
Mammary gland, see also Mammary tissues, 2, 42—43, 45, 58, 138—139
  acini, 162, 164
  alveolar cells, 108
  biopsies, 165
  development in vivo, 130—131
  duct, 138
  epithelial cells, 71
  explants, 111, 112
  functional differentiation, 122, 131, 138
  guanylate cyclase, 113
  hormones for lactogenesis, 122
  lactating rats, 116
  lobuloalveolar morphogenesis, 122—124, 126
  prolactin binding, 116
  microtubules, 164—168
  milk, 44
  milk product formation, 108
  mitogenic effects of PRL, IGFs and hepatic factors on, 205—206
  ornithine decarboxylase activity, 117
  phospholipase inhibitors, 112
  phosphorylation, 179—198
  polyamine accumulation in, 141—144
  prolactin receptor, 12—14, 70—71, 73
  slices, 164, 167
  tubulin, 164—165, 167—168
Mammary growth, 200—201
Mammary mitogenic assay, 206
Mammary tissues, see also Mammary gland, 14
  acetyl-CoA carboxylase, 185—187
  ATP citrate lyase, 184—185
  calcium, 115
  calmodulin, 117
  casein kinases, 187—190
  fatty acid synthesis, 184—187
  lactational processes, 109
  metabolic actions of prolactin in, 113
  metabolic processes, 112
  nuclear protein kinases, 190—191
  organ cultures, 138
  ornithine decarboxylase activity in, 114—115
  other protein kinase activities in, 191
  phospholipid metabolism, 108
  prolactin effects in, 108
  protein kinase C, activation of, 115
  protein kinases in, 184—191
  pyruvate dehydrogenase, 184
Mammary tumors, 12—14, 17, 70, 190, 191
Mammogenic activity, 209
Mammogenic hormone medium, 123—125, 127, 130
Manganese, 189
Maternal behavior, 36
Median eminence, 28, 31—33, 35
Membrane lipid microviscosity, 74—78, 98
  correlation between changes in prolactin receptors, prostaglandin synthesis and, 79—96
  factors modulating, 76—78
  fluorescence polarization technique to determine, 74—76
  modification by prolactin, 96—98
  modification of cellular functions mediated by changes in, 78—79
Membranes, see also specific types
  availability, 156
  calcium function, 115
  cycling, 156
  fluidity, 77
  fluidizers, 90—93
  fusion, 164
  inositol phosphatides, 113
  localization of tubulin in, 157
  phospholipids, 108—112
  structure, 108, 112
Membrane tubulin, see Tubulin
Mercurated CTP, 128
Mercurated RNA, 128
Metabolic processes, 108, 112—113
Metamorphosis, 2
Methionine, 140
Methylated phospholipids, 18
1,1′-(Methylethanediylidenedinitrilo)-bis (3-aminoguanidine) (MBAG), 141—142
Methylglyoxal *bis*(guanyl-hydrazone) (MGBG), 141—142, 146—147
α-Methylornithine, 141—142
Methyl xanthines, 162
Microfilaments, 164
Microtubule-altering drugs, 154, 158, 163—164
Microtubule-associated protein kinase, 167
Microtubule-associated proteins (MAPs), 159, 161, 163, 169

Microtubule organizing centers (MTOCs), 159
Microtubules, 153—155, 157—158, 169
  assembly, 158—159
  association with tubulin, 157
  calcium, 160—161
  calmodulin, 160—161
  cyclic nucleotides, 161—162
  depolymerization, 160
  disassembly, 158—159
  formation, 159
  mammary gland, 164—168
  polymerization-depolymerization cycles, 159
  regulation, 159—162
  secretion, 162—163
  steady-state situation, 160
  transport, 162—163
  α tubulin dimers, 158
  β tubulin dimers, 158
  vectorial distribution, 164
Microtubule-stabilizing (MTS) buffer, 165
Microviscosity index, 77
Microviscosity parameter, 92, 94
Milk, 108, 109, 115
  fatty acids, 156
  flow, 164, 169
  production, 143
  protein, 164
  secretion, 165
Milk-protein genes, 122, 131
Mitogenesis, 44, 53, 167, 205—206, 208
Mitosis, 159
Molar ratio of cholesterol to phospholipid, 77—78, 98
Molecular cloning, 58
Molecular hybridization, 124, 132
Molecular processes, 1—4
Molecular weight estimates, prolactin receptors, 9—10
Monophosphatidylinositol, 110
Morphogenesis, 122—124
Morphometric analyses, 166
Motor function, 35
Mouse bone marrow cells, 79
Mouse brain, 79
Mouse mammary gland explants, 114
Mouse mammary tissues, 13, 108—109, 112—116
Mouse 3T3 fibroblasts, 79
mRNA, 55—59, 124—132, 147
mRNAcsn, 124—129
MTW9, 70
Mucous cell, 47
Multiplication-stimulating activity (MSA), 204, 207
Myoepithelial cells, 166, 191
Myosine light-chain kinase, 191

N

NaCl, 46
Na^+-K^+-activated ATPase, 47
Na^+-K^+-ATPase, 48, 79
Na^+ pump, 47
Neoplastic transformation, 122
Neuroleptic drugs, 35
Neurons, 30, 31
Nigroneostriatal, 35
Nocadazole, 159
Noncoordinate expression, 131—132
Nondissociable binding sites, 2
Nonhistone nuclear proteins, 190
Nonmammalian species, 2
Nonmammalian vertebrates
  amphibian metamorphosis, 49—51
  columbid crop, 51—58
  evolution of prolactin mechanisms, 42—45
  models of prolactin action in, 41—67
  physiology of prolactin mechanisms, 42—45
  teleost osmoregulation, 45—49
Novikoff hepatoma cell, 79
Nuclear acceptor site, 126
Nuclear protein kinases, 190—191
Nucleic acids, 139
5′-Nucleotidase, 79
Nurturing behavior, 44

O

ODC, see Ornithine decarboxylase
Opercles, 47, 49
Organ cultures of mammary tissue, 138
Ornithine, 140
Ornithine decarboxylase (ODC), 42, 44, 53, 140, 142
  activity, 111—112, 114
    A23187, effects of, 117
    calcium ion, effect of, 116
    hormonal regulation of, 144—145
    mammary gland, 117
    mammary tissues, 114—115
    mouse mammary tissues, 115
    phospholipase A_2, 112
    phospholipase C, 113
    prolactin deprivation, 143
    stimulation of, 156
    TPA, effect of, 114
Ornithine transcarbamylase, 50
Osmoregulation, 2, 45—49, 58
Osmotic minipump, 205
Ovariectomy, 168
Ovary, 2
  prolactin receptors, 11, 71
  steroid hormone, 129, 138
Ovine prolactin, 2
Oxytocin, 191

P

Parabromphenacyl bromide (BPB), 111—112

Parental behavior, regulation of, 2
Parental care, 44
Parenting, 58
Parturition, 71, 88, 117, 130, 131, 139
PDGF, 207
Permeability, 48
Phagocytosis, 79
Pharyngeal mucosae, 46—47
Phorbol esters, 113—114, 117
Phosphatases, 180
Phosphates, 111
Phosphatidic acid, 111
Phosphatidylcholine, 113
Phosphatidylinositides, 110—111
Phosphatidylinositol (PI), 109
Phosphatidylinositol biphosphate, 111
Phosphatidylinositol monophosphate, 111
Phosphodiesterases, 162
Phosphoinositol, 182
Phospholipase A_1, 110
Phospholipase A_2, 19, 80—83, 96—97, 110, 147
  effects on metabolic processes in mammary cells, 112—113
  hormonal regulation, 145
  inhibition of, 111
  metabolic processes, 108
  relationship to phospholipase C, 115
Phospholipase C, 18, 96, 110—115, 182—183
Phospholipase D, 110
Phospholipases, see also specific types, 108, 111—113, 115, 117
Phospholipid-dependent protein kinases, 182—183
Phospholipids, 107—119, 139, 180
  calcium ion fluxes, 108
  changes during lactation, 108—109
  general role in cellular regulation, 109—111
  metabolism, 108—109
Phosphoprotein synthesis, 109, 111
Phosphorylation, see also Protein kinases, 110—111, 113, 159, 161—162, 179—198
  study methods, 180—181
Phosvitin, 187
Pigeon, see also Columbid crop, 44—45, 51, 53, 57, 200—207
Pituitary-derived prolactin, 29
Placental lactogens, 2, 43, 60, 71, 170
Plasmalemma, 154, 156—158, 167, 169—170
Plasma membrane, 2—3, 79, 92, 94—95, 98, 108, 113
Plasminogen activator, 147
Platelets, 77, 78
$[^{32}P]O_4$, 108—109
Polyamines, 3, 133, 139—148, 183
Polymerization-depolymerization cycles, 159, 165
Polymerization of tubulin into microtubules, 159—162
Polymerized tubulin, 159
Polypeptides, 122, 125—126, 138, 147
Polyunsaturated fatty acid, 80, 110
Polyunsaturated free fatty acid, 110—111
Post-translational modification, 43, 180
Pregnancy, 31, 33, 35, 43, 71, 73, 86—88, 108, 139, 142
  cyclic AMP levels, 162
  hepatic receptors for prolactin, increase of, 209
  mammary duct development, 138
  putrescine concentration, 143
  regulators of lactogenesis, 139
  spermidine concentration, 143
Pregnant mare serum gonadotropin, 88—89
Pregnant mice, 114, 122
Preneoplastic transformation, 122
Primary action of prolactin, 108
Proestrous surge, 31
Progesterone, 71, 122—123, 167
  alveolar development, 138
  antagonistic influence of, 132
  inhibitor of lactogenesis, 129
  regulator of lactogenesis, 139
  replacement, 168
  withdrawal, 168—169
Progesterone to glucocorticoid ratio, 129
Proinsulin, 203
Prolactin
  binding, 74, 116
    capacity, 18—19, 88
    sites, 70—71
  cell, 31
  localization, 6, 8, 15—16
  receptors, see Receptors
  synergists, 202—205, 207—208
  target cells, 2—3
  withdrawal, 187
Prolactin-dissociable "receptors", 2
Prolactin-induced endocytosis, 169
Prolactin-receptor complex, 163
Prolactin-stimulated SM activity, 209
Proliferative actions, male sex accessory glands, 2
Propanol, 92—93
Prostacyclin, 82
Prostaglandin, 3, 48, 57—59, 96—97, 111, 138—139, 145
  synthesis, 79—98, 110
Prostaglandin E synthesis, 85, 87, 89
Prostaglandin E_1, 78, 144
Prostaglandin $F_{2\alpha}$ synthesis, 85, 87, 89
Prostaglandin I_2, 82
Prostate gland, 11, 90—92, 95
Prostatic membranes, 90—96
Prostatic prolactin receptors, 72—73
Prostatic tumorogenesis, 70
Prostatotrophic factor, 73
Protein, 116, 139, 145, 162, 164
Protein kinase, see also Phosphorylation, 113, 161, 180
  acetyl-CoA carboxylase, 185—187
  activity, 157
  casein kinase, 187—190
  classification, 180—183
  fatty acid synthesis, 184—187
  mammary tissues, 184—191

nuclear, 190—191
Protein kinase C, 111, 113—115, 182
Protein phosphorylation, see Phosphorylation
Protochordate, 43
Pseudopregnancy, 31
Pseudopregnant rabbits, 2, 111
Purification of prolactin receptors, 10
Puromycin, 54
Putrescine, 139—141, 143, 144
Pyridoxal-5'-phosphate, 127—128
Pyruvate dehydrogenase, 184
Pyruvate dehydrogenase kinase, 183

Q

Quinacrine, 111—112, 145

R

R323OAC, 70
Rabbits, 111—112
lactogenic response, 139
mammary gland, 4, 116, 129
neutrophils, 79
Radioimmunoassay, 52, 56
Radio-receptor assays, 6, 52
Rat
adipocyte membrane, 79
brain membrane, 79
hepatic plasma membrane, 79
lactogenic response, 139
liver, 10, 15
mammary gland, 12—14, 17, 117, 129, 206—207
pituitary membrane, 79
testicular membrane, 79
Ratio of phospholipid to cholesterol, 84—85, 88—89
Ratio of protein to lipid, 77—78
Receptor genes, 51
Receptor protein, 126
Receptor (prolactin), see also other types, 2, 43—44, 51, 53, 109, 138, 145, 170
adrenal gland, 72
antibodies to, 16
characterization, 9—10
correlation between prostaglandin synthesis, membrane lipid microviscosity and changes in, 79—96
cryptic, 19, 96
desaturation of, 8
detergent binding to, 9
down regulation, 16—18
EDTA chelation of calcium, 116
endocrine control of, 10—15
hepatic, 73—74
heterogeneity, 60
identification, 6, 8—9
intracellular vesicle, 15
isolation, 9—10
kidney, 48, 72
list, in various species and tissues, 7
major target tissues, 71—74
mammary glands, 12—14, 70—71, 73
mammary tumors
mice, 13
rat, 12—14, 17
membrane lipid microviscosity, 74—78
molecular weight estimates, 9—10
ovarian, 11, 71
phospholipase inhibitors, effects of, 111—112
plasma membrane, 2—3
prostatic, 11, 72—73
purification, 10
rat liver, 10, 15
regulation, 5—26
identification of receptors, 6, 8—9
mechanisms of, 15—19
prolactin, 16
seminal vesicle, 72
solubilized, 9—10
tadpole, 42, 50
testes, 11—12, 72
up regulation, 17, 54
Recombinant DNA, 55
Red blood cell function, 78
Regression of alveolar structures, 122
Relaxin, 207
Renal, see Kidney
Replacement therapy, 138
Residual steroid hormone, 126
Reticulocyte lysate, 55—56, 58
Retrograde infusion, 164
Reverse pinocytosis, 108, 155
Reverse transformation, 161
Rhesus monkey corpus luteum membrane, 79
Ribosomes, 124
Rigidizing lipids, 76—77
RNA
[^{3}H]-uridine incorporation into, 113
hybridization analysis of, 124
polyamine interaction with, 139
splicing, 43
synthesis, 15, 54, 96, 111—112, 115, 132, 146
A23187, 116
calcium and prolactin, 115
cyclic GMP, 113
spermidine synthase, 145
stimulation of, 156
translation assays, 124
mRNA, 55—59, 124—132, 147
mRNAcsn, 124—129
Rotational mobility of membrane proteins, 78
Rough endoplasmic reticulum (RER), 52, 139, 155—156, 163—164, 167

S

Second messenger, 45, 138, 160

Second metamorphosis, 43, 49
Secretion, 160—161
  microtubules, 162—163
  prolactin, 156—157
  rates, 116
  regulation of, 153—157, 171—178
Secretory processes, 108
Secretory structures, 122
Secretory vesicles, 155—156, 158, 163—164, 166
Seminal vesicle prolactin receptors, 72
Sepharose beads, 70
Serotonin receptors, 79
Serum-free medium, 122
Serum to CSF ratios, 29—30
SH-agarose affinity column chromatography, 128
Sialic acid, 47
Single-point assays, 8
Skin, see also Integument, 47, 49
Snell dwarf, 73—74
Sodium efflux, 46
Solubilized prolactin receptors, 9—10
Somatomedins, 44, 53, 203, 205, 209
Spermidine, 112—113, 139—142, 145—147
Spermidine synthase, 140, 142, 145
Spermine, 139—142
Spermine synthase, 140
Sphingomyelin to lecithin ratio, 78
Spindle fibers, 159—160
Spur-cell anemia, 78
Starvation, 143
Steady state conditions, 160
Steady-state polarization, 92, 94
Stereotyped nesting, 44
Steroid hormones, 122, 125—127, 129, 138, 167
Stimulus-secretion coupling, 160
Striated neurons, 35
Subcellular fractions, 92, 94
Subcellular sites, milk protein regulation, 130
Suckling, 31
Sucrose, 159, 165
Sulpiride, 28
Surfactant formation, 109
SV-3T3 cells, 161
Synergism, 56
Synlactin, 53, 202—205, 208—210

T

Tadpoles, 42, 44, 49, 50
Tailfin, 50
Target cell responsiveness, 43
Target organs, 200—201
Tau proteins, 159
Taxol, 154, 167
Teleost, 2, 43—49
Testes, 11—12, 72
Tetrapods, 45
Theca cells, 71
3T3 cells, 161
Thyroid hormone, 126
Thyroid status, 72
Thyroid-stimulating hormone, 42
Thyrotropin-releasing hormone, 28
Thyroxine, 44, 49, 51, 72
TIDA neurons, 32—33, 35
TIDA of arcuate nucleus, 31
Tight junctions, 156
Total casein, 124
TPA, 113—115, 117
Transcriptional control, 128
Transcripts, 128
Transferrin receptors, 79
Translation assays, 124
Transplantable mammary tumors in rat, 70
Transplantation, 32, 33
Transport, 79, 160, 162—163
Treadmilling, 160
Trifluoperazine, 117
Triglycerides, 108, 109, 111—112, 115, 116
Triton X-100, 9—10
Tryptic peptide, 186
Tuberoinfundibular neuronal system, 33, 35, 37
Tubulin, 153—155, 157—159, 161, 171—178
  antibody, 165
  assays, 166
  content of mammary gland, 168
  polymerization, 164—169
  synthesis, 159, 165, 167, 169
α Tubulin dimers, 158
β Tubulin dimers, 158
Tubulin/microtubule interactions, 168—171
Tubulin-solubilizing (TS) buffer, 165
Type IIa hyperlipoproteinemia, 78
Tyrosine aminotransferase, 50, 127
Tyrosine hydroxylase, 32
Tyrosine kinases, 180
Tyrosine-specific protein kinases, 183
Tyrosine-specific vinculin kinase, 191

U

Uncoupling of synthesis from secretion, 162
Unlabeled prolactin, 8
Unsaturated fatty acids, 109—110
Uptake, 79, 156
Urea cycle, 140
[^{3}H]-Uridine incorporation into RNA, 112—113
Urinary bladder, 46, 48—49

V

Vasopressin, 46, 163, 170
Ventral prostate, 72
Vertebrate target organs, 42
Vertical displacement of membrane proteins concept, 77—78
Vinblastine, 154, 158—159, 163—164
Vincristine, 154, 158, 164

W

Water, 163
Waymouth's synthetic medium, 122
Whey-acidic proteins (WAP), 122, 130—131
Whey proteins, 130, 132
Whole mammary gland, 122—125, 127—128, 130, 132

X

Xanthophore pigment dispersion, 45

Z

Zona fasciculata, 72
Zona reticularis, 72
Zwittergent, 9—10

www.ingramcontent.com/pod-product-compliance
Lightning Source LLC
LaVergne TN
LVHW081322300926
849161LV00008B/1848
9781315890357